RESEARCH IN SCIENCE EDUCATION –
PAST, PRESENT, AND FUTURE

Research in Science Education – Past, Present, and Future

Edited by

Helga Behrendt
Helmut Dahncke

Faculty of Education,
University of Kiel, Germany

Reinders Duit
Wolfgang Gräber
Michael Komorek
Angela Kross

IPN – Institute for Science Education,
University of Kiel, Germany

and

Priit Reiska

University of Educational Sciences,
Tallinn, Estonia

SPRINGER-SCIENCE+BUSINESS MEDIA, B.V.

A C.I.P. Catalogue record for this book is available from the Library of Congress.

ISBN 978-90-481-5631-3 ISBN 978-0-306-47639-6 (eBook)
DOI 10.1007/978-0-306-47639-6

Printed on acid-free paper

Contents

Preface

This volume includes articles based on papers presented at the Second International Conference of the European Science Education Research Association (E.S.E.R.A.) held in Kiel, August 31 to September 4, 1999. About 300 colleagues, virtually from around the world - with a particular European focus - participated. Some 200 papers were presented. Three pages synopses of these papers were published in Proceedings of the conference (edited by Michael Komorek, Helga Behrendt, Helmut Dahncke, Reinders Duit, Wolfgang Gräber and Angela Kross). They are available from the IPN homepage: http://www.ipn.uni-kiel.de.

The participants were asked to submit contributions to the present volume. It contains the invited plenary lectures and a selection of the submitted contributions based on reviews by an international board and the editors. The volume mirrors main lines of research in science education in Europe and around the world. The invited lectures provide overviews of the growth of science education research from the past to the present, including views of future developments. Major emphasis of empirical research still seems to be students' conceptions and conceptual change. About half of the contributions fall into that category. In addition, most of the remaining contributions deal with various cognitive issues of teaching and learning science. It was surprising for us that the number of studies on affective issues and gender differences was much smaller than expected.

The Second International Conference of the European Science Education Research Association was gratefully supported by the German Science Foundation (Deutsche Forschungsgemeinschaft). These funds allowed us to invite eminent plenary lecturers. We would also like to acknowledge support by Kluwer Academic Publishers and the International Journal of Science Education. Finally, the Institute for Science Education at the University of Kiel and the Faculty of Education of the University of Kiel supported the conference in various ways.

We are also most grateful to a number of colleagues who helped to organize the conference and the production of the present volume. The following list includes colleagues who served as members of the board of reviewers for the papers submitted to the conference and/or as members of the board of reviewers for the contributions of the present volume:

Philip Adey (London, UK)
Silvia Caravita (Rome, Italy)
Pierre Clement (Lyon, France)
Justin Dillon (London, UK)
Richard Duschl (London, UK)
Manfred Euler (Kiel, Germany)
Hans Fischer (Dortmund, Germany)
Helmut Fischler (Berlin, Germany)
Peter Häußler (Kiel, Germany)
Gustav Helldén (Kristianstad, Sweden)
Mercé Izquierdo (Barcelona, Spain)
Ulrich Kattmann (Oldenburg, Germany)
Tom Koballa (Athens, GA, USA)

Peter Labudde (Bern, Switzerland)
Piet Lijnse (Utrecht, The Netherlands)
Brunhilde Marquardt-Mau (Kiel, Germany)
Martine Meheut (Paris, France)
Robin Millar (York, UK)
Hans Niedderer (Bremen, Germany)
Jonathan Osborne (London, UK)
Leif Östman (Uppsala, Sweden)
Dimitris Psillos (Thessaloniki, Greece)
David Pushkin (Hackensack, NY, USA)
Christoph von Rhöneck (Ludwigsburg, Germany)
Helga Stadler (Wien, Austria)
Wieslav Stavinski (Krakow, Poland)
Ruth Stavy (Tel Aviv, Israel)
Elke Sumfleth (Essen, Germany)
David Treagust (Perth, Australia)
Rod Watson (London, UK)

Finally, we would like to gratefully acknowledge the tremendous effort made by Ulrike Hennig in transforming the contributions into a form that may be printed. We also like to mention that Valerie Reed carefully checked - and if necessary improved - the English of the contributions and that John-Philip Josten organized the communication with the contributors and reviewers.

The Editors

Part 1: Views and Visions of Science Education Research

Science Education Researchers and Research in Transition: Issues and Policies[1]

Dimitris Psillos
School of Education, Aristotle University of Thessaloniki, Greece

Abstract

Significance and characteristics of science education research is discussed from the perspectives of three modes for approaching science education: (1) The practical mode; (2) the technological mode; and (3) the scientific mode. The practical mode concerns teachers' experiences in practice. The technological mode draws on policy makers' attempts to improve science education. Finally, the scientific mode denotes the contribution of science education as a research domain in its own right to the further development of science education. It will be argued that it is necessary to link the major concerns of all three modes in order to meet the various difficulties of improving science teaching and learning and proposals will be made for developing relevant ESERA policies.

Researchers and research in science education

The apt title of the second ESERA conference is "Science Education Research: Past, Present, and Future". My interpretation of this title is that the time has come to reflect thoroughly on what we all, as members of a developing research community, are practising and what our special contribution to the quality of science education is. This means that, at the turn of this century and indeed at the threshold of a new millenium, it is time to discuss and debate the special characteristics of Science Education Research as a disciplinary activity and Science Education Researchers' practices and their implications for improving science teaching and learning. On this issue I will offer some exploratory thoughts which I hope will stimulate further debate and lead to the development of appropriate policies for ESERA.

The role of Science Education Research and researchers has to be seen in the context of rapid changes taking place in several European countries (Psillos, 1999). Briefly, the impact of the information society, economic globalisation and scientific and technological knowledge all imply that European countries have entered a transitional phase towards a new form of society beyond current short term forecasts. A learning society is emerging in which people's ability to learn will have an increasingly important effect on the course of their lives (C.E.C., 1995). In such a

[1] Presidential Address

H. Behrendt et al. (eds.), Research in Science Education – Past, Present, and Future, 11–16.

future society scientific knowledge will be an increasing part of the everyday life of European citizen. Appreciating scientific knowledge and knowing how to use scientific information will become as important as knowing what that information is.

In this context, at the European Educational Policy level, developmental programs such as Socrates are being applied, promoting cross-border school networks and widening the concept of the educational institution. Education and training become the object of large-scale research programs such as *Targeted Socio-Economic Research* (*TSER*), which aim at creating a European pool of skilled educational researchers. At the researcher level, in our field, we are witnessing the recent proliferation of international professional associations or groups aiming at promoting the quality of education by research and development. ESERA is one prominent example whose members are mainly discipline-based researchers.

My first remark is that discipline-based science education researchers are not the only ones who actively contribute to the education of the public in science. Science teachers at all levels, natural scientists, policy makers, pedagogues and psychologists are among the communities which contribute in their own specific ways to improvement in the quality of science education of students. As late-comers in the area of Science Education and technically competent in research matters, discipline-based researchers sometimes feel that they are in a better position to know the solutions to problems concerning student needs regarding scientific understanding. At other times, they feel that their voice is not heard outside their research labs and this may result in tensions with the other communities mentioned above. I would like to draw the attention of the Science Education Research community to the fact that in the context of the emerging learning society, it is of utmost importance to discuss and debate our distinctive contribution, practices and relations with the other communities involved in science education (in particular with policy makers and natural scientists apart from science teachers).

Modes for approaching science education

My second remark concerns the various practices applied in Science Education. I will briefly discuss three practices by exploring representative cases. At the classroom level, let us consider, for example, the case of laboratory teaching. Let us suppose that a science teacher faces a management problem during a laboratory chemistry session in upper high school. He/she has to take quick and immediate decisions as to whether he/she will ask his/her students to skip a particular part in a laboratory exercise. In such a situation, the problem this teacher is facing is specific and depends on a number of contextual factors such as the age of the students, their number, their interests and scientific background. It also depends on the material conditions such as the amount of apparatus, the quality of the chemicals and the structure of the worksheets. The combination of all these factors at the level of the educational system, results in a specific situation, even a unique one.

At first, the teacher has to understand the situation in terms of any indications he/she receives from his/her students during the session. Such as their work progress, complaints and off-task behaviour. His/her personal experience of comparable situations in past years is a valuable source upon which he/she draws in order to make quick sense of what is happening and to judge the severity of the problem. The teacher will also draw to some degree on his/her general experience and practical ways that he/she has gradually developed in order to get along with students' misbehaviour. He/she may have to apply such tacit knowledge in a creative and imaginative way in order to find a solution. The approach he/she opts for will depend on his/her hermeneutic capacity but also on his/her repertoire of possible effective teaching strategies. In any case he/she has to provide an immediate and effective solution to the problem he/she is facing otherwise the whole session will possibly become educationally insignificant or even harmful.

In the above situation the teachers' approach to a science education problem is based on a practical mode, which is difficult to generalize or model. Most likely in the above situation as well as in other classroom incidents, disciplinary knowledge and stable patterns developed after years of successful practice tend to guide the teachers' approach rather than any formal educational theory. For example, it is interesting to notice that, in the case of labwork, similar and stable patterns of practice appear over and over again across several European countries.

At the policy level, let us consider the following case. A group of curriculum developers are asked by their local educational authority to develop guidelines and new science units for lower secondary school. As the responsible body for educational administration and development in its geographical rural region, the local authority is sensitive to voters' aspirations concerning environmental problems in the area. The developers are asked by the local authority to develop new units on science topics that will take into account environmental themes and problems aiming at sensitizing students to environmental protection issues. The curriculum developers have to take into account recent data from environmental measurements and choose the appropriate themes in order to adapt the new units to local environmental needs. For example, some of the units may include tasks requiring the students to perform outdoor atmospheric pollution measurements using hand-held devices. The developers may have to draw on several sources of data as well as their experience concerning student motivation and engagement in environment-based scientific tasks. The developers will integrate all these sources of information and develop new units possibly including innovative materials, which potentially involve students not only in hands-on experimentation but also in environment protection activities. Usually, the new curriculum has to be prepared, tried out and delivered within a limited time schedule. It must fulfil the local needs in science education for some time to come.

In the above case, the policy makers' approach to a science education matter is based on a technological mode broadly aiming at the improvement of science education at a regional level. The developers rely on both previous experience and new data. They probably involve design criteria in their project that are loosely related to formal educational theories in the form of principles about curriculum development. Such knowledge has to be applied in a creative yet somewhat tacit way by the developers in order to be able to integrate all the various inputs into an averagely effective policy. Disciplinary knowledge plays an important role in such practice.

Finally, at the research level, let us consider the case of a researcher who is interested in students' understanding of voltage, potential difference and electromotive force in introductory university teaching. The researcher may take advice from published investigations concerning students' understanding in the field of electricity. He/she will then apply his/her technical expertise to design a new piece of research. The researcher will apply established research methods in order to obtain new data. For example, he/she may develop and validate a special questionnaire and then administer it to a random sample of students. The results will be analyzed and discussed in terms of previous studies and formal theories linked to students' domain-specific thinking. The discussion will aim to provide an explanation or a model of the students' underlying thinking and may lead to deeper questions to be taken up in a future study. The study will probably be published and communicated to the community of researchers. Implications on teaching may or may not be drawn from the findings of the study.

In the above case, the researcher's approach to an issue concerning understanding of science by students aims at the long-term development of our interpretations regarding science education phenomena. It is a research activity which is based on analytical thinking, aiming at uncovering tacit knowledge and stable practices of the science community. Internal consistency and coherence are primary values. Results and conclusions, as powerful generalisations, are addressed primarily to a local and /or international community of researchers. They may inform policy and stimulate teachers and policy makers to think about science teaching and learning in new and imaginative ways and can lead to the reconstruction of scientific knowledge for educational purposes. However, they can hardly provide teachers with solutions for their immediate needs within their unique contexts and constraints.

Science education is potentially promoted in all of the situations. However, the practices applied have distinctive characteristics the nature of which could be an object of debate and inquiry in our research community. The practical mode, the technological mode and the scientific mode are followed by the various science educators but differ accordingly to the problems which must be resolved, the types of knowledge which are employed, the methodologies which are applied and the expected outcomes.

Dynamics and relations between the modes

My last remark concerns development and communication problems between the various modes. The practical mode is the oldest and probably the most influential approach to science education. Every day, we are all involved in one way or another in teaching activities which require effective practical solutions. The technological mode has developed considerably since the large-scale curriculum projects which appeared in science in the sixties. It has been established in several countries as a means for designing and applying effective policies. Particularly when innovations are planned. The scientific mode is more recent, probably less influential at the teaching and policy level but is developing and acquiring its own status. While in the past there was only the practical mode, gradually the technological and the scientific mode have developed such that, at present all three modes are followed in science education. It is my belief, moreover, that in the emerging learning society the scientific mode has great potential for growth.

What are the relations between all of the modes currently applied in science education? Are the three practices linked to each other or do different communities follow their own preferred modes? Represented diagrammatically, are the three modes being applied in parallel or are there instances where the lines cross or meet? I suggest that there are serious communication gaps between the various modes.

Different languages and techniques are employed which are difficult to translate from one mode to the other. To put it in more simple terms, it is not self-obvious how to translate research results into practice or develop effective policies on a strong research basis.

How may the communication gaps be reduced. In particular between the scientific and the practical mode This is an open question, that has recently drawn the attention of some researchers. For the research community, I consider that it is important to pay attention to the inherent difficulties in linking the various modes and to take action that facilitates their exploration and discussion in public.

In this context I, will describe an attempt to bring together various communities and enhance valuable interactions among practitioners following all three modes. The *Laboratory in Science Education* (LSE) was a two-year project funded by the European Union between 1996 and 1998. Six countries and seven research groups comprising a total of 36 researchers were involved. Besides senior researchers, several practising teachers were involved in the project who at that time were carrying out their doctoral theses (Serre et al., 1998). The final meeting of the project took place in Thessaloniki, Greece, in May 1998. Communities involved in the final meeting included researchers, teachers and policy makers. The latter group was difficult to select since "policy maker" is a rather loosely defined category, particularly when one has to select from a European pool. Finally, colleagues from Ministries, natural scientists from scientific associations, universities and the European Union were invited to attend the meeting.

One problem that the organising group in Thessaloniki had to face was how to involve the policy makers in an international meeting to discuss research in which they had not participated or which possibly was not linked to their own close interests. The organisers opted to ask the policy makers to be actively involved in workshops, to act as discussants to research papers and to provide their perspective on the implications of the research. In other words, the policy makers were not considered as passive recipients of thoughtful research but as mediators and creators of knowledge whose perspective would throw light on the importance of research. It turned out that this was not an easy task. However, the result was beneficial for all participants, as summarised in a comment from a policy maker to the organisers at the end of the meeting: "It is the first time I started to understand your language." The organisers and other researchers turned the comment around on the policy maker, arguing that they themselves had had to think through and plan thoroughly how to make clear their case and the possible specific implications really linked with their results.

My interpretation of the above situation is that there is a long road ahead towards developing fruitful and mutual interactions between the various modes. I consider, though, that it is an important step forward to enlarge the formal discussion and inquiry within our community on relations with natural scientists and policy makers apart from teachers, (see the contribution by Dahncke, Duit, Gilbert, Östman, Psillos, and Pushkin in this volume).

Recommendations and the role of ESERA

In this presentation, I have argued that the time has come to further develop our understanding of our practices, their strengths and their limitations in the context of a rapidly changing Europe. With regard to ESERA, one step with this aim in mind has been to gain insights on members' practices. In previous years, at the policy level, working groups have been established aiming at monitoring members' interests, educational activities and patterns of terminology for communication in an international context. However, not much has been done with regard to developing links with other communities, especially science education policy makers. Perhaps the task is difficult but any efforts may be rewarding if our vision is science education which takes advantage of all modes in order to keep the "utopia for education alive" (UNESCO 1999).

References

Commission of European Community, C.E.C. (1995). *White paper on education and training.* Brussels.

Psillos, D. (1999). Educational demands and science education research: The role of ESERA. In M. Bandiera, S. Caravita, E. Torraca & M. Vicentini, Eds., *Research in Science Education in Europe* (pp.1-6). Dordrecht, The Netherlands: Kluwer.

Rapport à l'UNESCO de la Commission Internationale sur l'Education pour le 21er siècle (présidée par J.Delors). L'Education: un trésor est caché dedans. (Greek translation) Center for Educational Research, Athens 1999.

Séré, M.-G., Leach, J., Niedderer, H., Paulsen, A., Psillos, D., Tiberghien, A., & Vicentini, M., Eds. (1998). *Final Report: Improving Science Education: Issues and research on innovative empirical and computer-based approaches to labwork in Europe.* EC Targeted Socio-Economic Research Programme, Project PL 95-2005.

Research in Science Education in Europe: Retrospect and Prospect[1]

Edgar W. Jenkins
Centre for Studies in Science and Mathematics Education, University of Leeds, UK

Abstract

This paper offers a personal overview of research in science education in Europe. It is suggested that such research is a relatively new domain, characterised by considerable diversity. The paper explores some of the institutional, conceptual and methodological dimensions of this diversity and identifies a number of issues around which science education in Europe might develop in the early years of the twenty-first century.

Introduction

This paper makes no claim to reviewing research in science education in Europe with a breadth and thoroughness that might satisfy a professional historian. Instead, attention will be focused on a number of aspects of research in science education in Europe that seem to me to be of interest and worthy of comment. The limitations of such an approach are, of course, considerable. There is a risk not only of idiosyncrasy and bias, but of failing to do justice to some of the more subtle differences in the research in science education that has been undertaken in different parts of Europe. This partiality will become particularly evident when looking forward, rather than backward, and when offering some necessarily personal thoughts about research in science education in the early years of the next century. Few tasks are more perilous than predicting the future and there is no shortage of people who have got it spectacularly wrong.

What is presented in this paper, therefore, is in no sense definitive. The aim is to encourage a critical reflection upon the past and perhaps identify at least some of the points of a compass that may help navigate the future of research in science education. There are, of course, many insights into science education research in Europe available from the ESERA conference held in Rome (Bandiera, Caravita, Torraca, & Vicentini, 1999). The present paper acknowledges those insights and offers a number of comments upon science education as a field of research as it has developed in Europe in recent years.

A relatively new field of activity

Perhaps the most obvious comment to make about research in science education in Europe is its relative newness. No European country can claim a long tradition of work in this field, although there is often a much longer history of scholarly study of some other aspects of education, notably in pedagogic, didactics and the history of education. There are also many examples of research undertaken in the first half of the twentieth century that can properly be categorised as science education, but much of this, as far as I have been able to establish, has been the work of individuals

[1] Plenary Address

H. Behrendt et al. (eds.), Research in Science Education – Past, Present, and Future, 17–26.

or quasi-government committees. More significantly, such work does not reflect the existence of communities or groups of research workers within science education that are familiar to us today. In the United Kingdom, for example, it was the later 1960s that saw the establishment of the first professorial appointments in science education, a development that owed much to the attempts to introduce and support large-scale reform of school science curricula. It was the same decade which saw the emergence of new journals devoted to science education and the development of postgraduate courses in a number of UK universities. Elsewhere, such as in Israel, Italy or the Scandinavian countries, the 1970s might be a more appropriate decade in which to locate the origins of science education as a field of research. Small numbers of research workers began to investigate various aspects of science education, their work sometimes being undertaken within Science Education Groups or Centres, some of which were set up with direct government funding and with quite specific research and development priorities. Collaborative work within, and, increasingly, between groups and Centres became more common and the appearance of the *European Journal of Science Education* (now the *International Journal of Science Education*) in 1979 reflected the need for a non-American outlet for the expanding body of research in science education being undertaken in Europe and elsewhere. It should perhaps be noted, that the Editorial Board of the *European Journal* included two members from IPN, Gerhard Schaefer and Karl Frey, the latter serving as its Chairman. In 1974, my then colleague, David Layton, took the bold and imaginative step of publishing the research review journal, *Studies in Science Education*, the 33 volumes of which so far published provide promising material not only for researchers but also for those who wish to chart the fortunes of science education as field of research.

Three comments are perhaps appropriate about the relative newness of research in science education in Europe. Firstly, it suggests a need for some caution in assessing the contribution such research might make to educational policy and/or practice, an issue to which I shall return later. Secondly, it stands in marked contrast to the much longer, and very different, history of such research in the United States of America. The various *Digests of Investigations on the Teaching of Science* cover research in the USA from as early as 1906 to 1957, and they reflect a research tradition that was almost exclusively quantitative and empirical in its methodology and largely behaviourist and positivist in its psychology and philosophy (Curtis, 1926, 1931, 1939; Boenig, 1969; Lawlor, 1970). Thirdly, newness necessarily entails substantial and important diversity. For example in the topics chosen for investigation, in the assumptions made, in the methodology used, in the institutional location within which the research is undertaken and in the research and professional backgrounds of those undertaking the research. All these features are evident in the literature of the last few decades as science education has sought to establish its identity and authority as a field of research.

Diversity in research in science education

It is difficult not to be impressed by the wide range of topics that researchers in science education have chosen to investigate in the last thirty or so years. They have related to teachers, students, textbooks, pedagogy, curriculum, assessment, evaluation and, within each of these fields, the diversity is compounded. For example, few fields, if any, have ignored, gender issues, although far too few have accommodated, or even attempted to accommodate a wider international perspective. Unquestionably, the bulk of the research has also been concerned with science teaching at school, rather than college or university level and with formal education, rather than informal or non-formal learning. Broadly speaking, secondary science education has had more attention from researchers than elementary or primary science education, although this has changed rapidly in some countries in recent years. Also in evidence are perspectives drawn principally from philosophy, psychology and sociology, although a number of other disciplines, including history, anthropology and economics are also represented. Comparative studies are relatively rare, although the work of the Third International Mathematics and Science Study (TIMSS) and the OECD PISA Project have brought an unprecedented level of interest in student achievement in many countries and in the factors that seem to lie behind the differences that cross-national testing appears to establish (see Harlen's chapter in this volume and Shorrocks-Taylor & Jenkins, in press).

In the last couple of decades or so, of course, the science education literature has been dominated by research findings concerned with children's understanding and learning of scientific phenomena and it has become almost impossible to escape any reference to constructivism among the papers published in the research journals.

Phillips has claimed that 'Across the broad fields of educational theory and research, constructivism has become something akin to a secular religion'. Claiming, that 'whatever else it may be', constructivism is a powerful folk tale about the origins of human knowledge, Phillips adds that 'Like all religions, it has many sects - each of which harbors some distrust of its rivals' (Phillips, 1995, 5). There is no doubt that constructivist views of learning represent the most marked psychological influence on science education in recent years and it would be interesting to speculate why this should be so. I will confine myself, however, to pointing out that an influence of this kind is by no means new and to warning that it is the dubious privilege of each new wave of learning theorists to rewrite the history of how science used to be taught to suit their own, current, agenda.

Work drawing upon constructivist perspectives is perhaps the nearest that has emerged to a research paradigm within science education, although it would be unwise to apply this Kuhnian label too readily, and it may be more applicable in the USA, with its dominant behaviourist tradition, than within Europe. Quite simply, there remains far too much about which there is disagreement at a fundamental level. There have, of course, been shifts over time in the focus of attention of those working within this field. Early exploratory work, followed by replication studies, led by the 1980s to something of an emphasis upon how students' ideas about a range of natural phenomena might be changed. Today, there is an interest in how students acquire these ideas, together with a greater understanding that 'alternative' and scientifically incorrect models of a range of phenomena are adequate for many everyday purposes. A view borne out by research in fields as different as the public understanding of science (Layton, Jenkins, Davey, & Macgill, 1993), psychology (Lave, 1988) and the nature of practical (Holzner & Marx, 1979) and professional knowledge (Wilson, Shulman, & Richert, 1987). For some critics (e.g. Matthews, 1998), constructivist perspectives on teaching, learning and the nature of knowledge are linked with philosophical concerns about relativism and the standing and authority of science and scientific knowledge in a world in which both are often seen to be under assault. It is a debate which, in my judgement, continues to be hindered, rather than helped, by intemperate language and fuelled by some of the extravagant and unsupported claims made on behalf of so-called constructivist approaches to teaching and learning.

The diversity in science education research to which I have just referred is not without its problems, although it is important to acknowledge that the field is by no means static. There is a real danger that researchers in science education will increasingly talk past each other, rather than to, each other and I suspect that we have all been at conferences where this has been the case. Beyond this, the emergence of science education as a field of research in its own right marks off researchers in this field from the practitioners, i.e. those who teach science in schools or other institutions, with all that this implies. There may also be questions of standards, and work that needs replication and development has often been published and then ignored, or more worryingly, offered in support of some element of science education policy. At the heart of the matter, however, is the following question: what sort of research domain is science education ? It is to this question that I would now like to turn attention.

Science education: what sort of research domain ?

I don't want to try to answer this question in some abstract way but by reference to the science education literature as I have read it. I am, of course, making no claim to have read all of it, and, here, as always, my judgement is coloured by my own professional and academic background and my inability to read as much of the literature published in the various European languages as I would wish. As with any research domain, science education is characterised by the research issues that it addresses and I have already indicated that these issues are very diverse. It is, however, perhaps just possible to identify two rather different traditions in the research that has been undertaken in Europe within the past thirty or so years. Although this work has not, of course, been uninfluenced by developments elsewhere. For example, in countries such as Australia, Canada, New Zealand and the USA. At the risk of some oversimplification, the two traditions can be described as pedagogic and empirical and it is perhaps worth noting that there are some parallels with research in mathematics education where Bishop (1992) identifies a third tradition which he associates with the scholastic philosopher.

The pedagogic tradition has, at its primary focus, the direct improvement of practice, practice here being understood as the teaching of science. Improved learning is assumed to follow from improved teaching, and the evidence for improved teaching lies in such issues as enhanced student motivation, attendance or level of achievement.

There may be some modest questionnaire or other form of evaluation but there is no rigorous research design, partly because no grand theoretical explanation or underpinning is sought and it is the practitioners themselves, i.e. the teachers who require and offer judgements about improvements in their practice. Such improvements cannot be transferred in some simple way to other classrooms, laboratories or teachers. Ideas, however, can be shared and, if judged appropriate, adapted for use in a different context. The work is close to the classroom or laboratory, and its point of reference is how to make some aspect of science more interesting to, and effective for, students.

The empirical tradition in science education research, always much more evident in the USA than in Europe, has weakened considerably in the last thirty years. It is associated with positivism and seeks the 'objective data' needed to understand and influence an assumed educational reality, close familiarity with which lies at the heart of the pedagogic tradition. While such weakening reflects a growing recognition of the inadequacy of the view that science teaching can be reduced to a science, it almost certainly owes more to the failure of the traditional empirical approach to 'deliver the goods', i.e. to raise the standards of learning required of educational systems. Today, there is an improving, although, still very limited, understanding of the complexity of teaching and learning and the interactions that are involved in teaching science in classrooms and laboratories. It remains a struggle for researchers in science education to enter successfully the practitioners' world and, in my judgement, little general progress has been made in working with practitioners in developing the conceptual tools needed to raise the quality of students' learning. This is not, of course, to ignore important initiatives such as the Project for Enhancing Effective Learning (PEEL) in Australia (Baird & Northfield, 1992) and the Cognitive Acceleration through Science Education (CASE) initiative in the United Kingdom (Adey & Shayer, 1994). In both cases, it is interesting to note that, although the teacher-researcher collaboration underpinning the relevant programmes is an application of ideas derived principally from research in cognitive psychology, the initiatives resonate with wider political shifts in a number of European countries to draw practising teachers more closely into the process of raising standards or even, as in the case of England and Wales, to give them direct responsibility for research, including access to, and control of, research funding.

How distinct are the two traditions to which I have referred? Within the Anglo-American community of researchers in science education, many of the differences can be quite marked, although I would not wish to draw boundaries that are too firm and there is, of course, some common membership of the two communities. The differences are found in the journals in which the research is published, the institutional location of the researcher, and the conferences which he or she attends. Using chemical education as my illustration, there are those who teach chemistry in schools, colleges or universities who publish papers concerned with chemical education. However, they would see themselves as teachers, not researchers, and their chosen journal might be *Education in Chemistry, Journal of Chemical Education* or the *School Science Review.* Typically, they are not members of the British, American or other national Educational Research Associations, and they are unlikely to be found at meetings such as the ESERA conference, preferring instead the European Conference on Research in Chemical Education (ECRICE) which started in Montpellier in 1992 or the International Conferences on Chemical Education held every two years, interchangeably with the ECRICE symposium. The researchers remain close to the academic discipline of chemistry and many, I suspect, would strongly resist any attempt to classify them as social, rather than natural, scientists. I would argue that parallel accounts could be given for physics and biology.

Outside the Anglo-American community, my suspicion is that these distinctions are much less securely grounded. This is partly because of the importance within the wider European tradition of *didactics*, a term that has no precise equivalent in the English language and which leads to sometimes seemingly inseperable problems of communication and translation. To an English audience, any reference to the 'didactics of a discipline' is usually very puzzling, not least because the word didactics is commonly associated with a rather direct and authoritative style of teaching rather than having anything to do with the grammar and syntax of a scholarly discipline and how it can be best taught. One consequence of this difference between English and wider European traditions may be that, outside the United Kingdom, rather more researchers in science education work in, or in close association with, academic science departments and thus remain in closer contact with developments in the parent scientific discipline. Dialogue between educational researchers within, and others outside the didactic tradition (including, most obviously, myself) is long overdue and the IPN has already made a distinguished contribution to the debate (Hopmann & Riquarts, 1995).

As a research domain, therefore, science education is diverse, methodologically, conceptually and institutionally. For some, such diversity will be welcome, and be seen as reflective of intellectual vigour and cultural sensitivity. Others may wish for a better focused research agenda for the science education research community, with a sharper sense of the major problems facing science education in Europe and how research can best contribute to their solution. There are perhaps echoes here of a wider European debate. My only comment here, however, would be that I see no conflict between, on the one hand, developing a clearer sense of a European research agenda in science education and, on the other, fostering the independence and cultural diversity necessary to promote research of a high quality in the many countries and regions represented within ESERA and beyond. There are already many welcome initiatives that promote both collaborative research and a wider understanding of the research traditions in different parts of Europe. I have in mind cross-national research projects, the initiatives supported by the European Union and the European Summer Schools for Ph.D. students in science (see, for example, Lijnse, 1993). There would also, I suspect, be relatively little difficulty in drawing up, at least in broad terms, an agenda for research in science education in Europe. Once again, there are lessons that might be learnt from colleagues in mathematics education (Freudenthal, 1983; Sierpinska & Kilpatrick, 1999), and constructing such an agenda might serve, among much else, the useful purpose of focusing our own minds about the kinds of research field to which we claim professional allegiance and which is reflected in ESERA.

Science and the science education research community

Perhaps the most significant difference between what I have overdrawn as the pedagogic and empirical research traditions within science education lies in the change that has taken place in the past few decades in the relationships between science education and science itself. The change seems to be much more marked in those countries most strongly influenced by work in the Anglo-American tradition.

The underlying issues here can be approached, at least indirectly, by asking whether researchers in science education and professional, practising scientists think about science in the same way. For those science education researchers working within the pedagogic tradition to which I have already referred (and certainly to most of those working in science education in England a generation ago), the question would have made little or no sense. If you want evidence that matters have changed I would refer you to the reaction of Nobel Laureate, Professor Ledermann, to the comments of science education researchers from various parts of the world upon the document produced by the International Council of Scientific Unions (ICSU) on capacity building in science.[2] The document, the responses and Ledermann's reaction are all in a recent volume of *Studies in Science Education*. For Ledermann, scientific knowledge is true, objective and universal. Any suggestion of feminist or multicultural science would be a nonsense. In contrast, for many of the commenting science education researchers, each of these adjectives is open to question. The science education research community, although for the reasons I have already indicated, not all of it, is deeply involved with issues stemming from fundamental questions about the nature of science has been quick to recognise some of the possible implications for science education of some feminist scholarship, of some historians of science and of work in cognitive psychology and the sociology of knowledge. The result has been a burgeoning literature that questions or rejects the view of science held by most of the scientific community and opens up new approaches to pedagogy and to structuring the science curriculum. Regrettably, this literature has not always done justice to the views of those who challenge some of the more radical and post-modern perspectives upon science and science education. As an example, to argue for a greater appreciation of 'other ways of understanding' the natural world is an entirely legitimate undertaking. To argue that science courses should be redesigned to accommodate such 'other views' is quite another matter that raises a different set of issues.

Science education research: an underlying assumption

The *International Handbook of Science Education* was compiled to synthesise and reconceptualise past research and theorising in science education, provide practical implications for improving science education, and suggest desirable ways to advance the field in the future (Fraser & Tobin, 1998, xiii).

Laudable though these aims are, the subsequent chapters in the *Handbook*, despite their diversity, seem to assume that science education as a field of activity is exclusively concerned with practice of teaching and learning, together with supporting activities such as assessment, evaluation and teacher education. Correspondingly, research in science education is about improving practice, whether this relates to promoting greater equity, making more effective use of educational technology or developing more informative instruments for formative, diagnostic or summative evaluation. This is a view of research in science education with a long history and it is one that is strongly influenced by the empirical tradition that has dominated research in science education in the USA throughout the twentieth century. Its contemporary manifestation in the United Kingdom is the search for 'evidence-based practice' but its antecedents lie in an essentially nineteenth century belief in a science of education which presents the solution to educational problems as a matter of gathering objective and empirically-tested evidence.

---

[2] The International Council of Scientific Unions (ICSU) was established in 1931 and is based in Paris. It has a national membership of 95 multidisciplinary bodies (scientific research councils or science academies) and 25 international, single-discipline Scientific Unions. In 1993, it replaced its Committee on the Teaching of Science with a more wide-ranging Committee for Capacity Building in Science (CCBS).

I want to suggest that, despite the central importance of teaching and learning, this is too narrow a view of research in science education and that the assumption upon which it rests is still found in much European science education research towards the end of the twentieth century. The narrowness is of different kinds, stemming, on the one hand, from a neglect of other kinds of research in science education that are not concerned directly, or only very indirectly, with improving practice and, on the other, an over-technical and instrumental approach to teaching and learning. A somewhat more generous view of research in science education is that it is concerned with that which 'critically informs...judgements and decisions/ in order to improve action' (Bassey, 1995). This broader perspective has the merit of accommodating a number of strands of research in science education that are all too easily overlooked by an over-emphasis upon the teaching and learning of science in the classroom or laboratory, e.g., science education and economic development, sociological perspectives on science curriculum reform, and historical and policy related studies. Yet, even in the more generous definition I have just cited, research in science education has to be justified by reference to 'improving action'. Understanding for its own sake is seemingly insufficient.

Relating policy and practice

Some of the research which I have read in preparation for this conference seems to me to make a number of assumptions about the relationships between policy and practice that are unsustainable. Many of these assumptions can be reduced to 'If only teachers would do this...' then the problem being addressed would be solved. Other research reports seem to assume that research should influence policy in some straightforward and direct way and fail to recognise, among much else, that science education policy is shaped by much more than research findings. In broad terms, the literature suggests that the policy-practice interface is seriously under-researched in science education. This is significant since it is an area of outstanding importance to researchers and policy-makers alike, and it is one urgently in need of greater theoretical clarity. The rich potential for comparative studies is perhaps obvious.

Looking forward

Much has been achieved to promote science education as a field of research within Europe. In identifying some issues that might deserve attention in the years ahead, it will be necessary to build upon those foundations. In summary form, I offer the following. They are not, in any sense, personal recommendations:

- The development of a clearer sense of a European science education research agenda and of its articulation with the concerns of national governments and international agencies.
- A greater understanding of the relationships between policy and practice, allied with a greater humility of what research can realistically contribute to both policy and practice.
- A shift of emphasis towards teaching rather than learning as a focus for research.
- Greater hospitality towards research that is less directly concerned with improving practice than with sharpening thinking, directing attention to important issues, clarifying problems, encouraging debate and the exchange of views.

- ➢ The expansion of cross-national research collaboration, including comparative research in science education.
- ➢ The on-going development of the European Summer School Programme.
- ➢ The building of research partnerships with others active in science education but who work and/or publish within their own specialist fields, museums, the print and broadcast media.
- ➢ The further development of mechanisms to promote and sustain research training and collaboration among European countries and regions.

References

Adey, P. & Shayer, M. (1994). *Really raising standards: Cognitive intervention and academic achievement*. London: Routledge.

Baird, J.R. & Northfield, J.R., Eds. (1992). *Learning from the PEEL experience*. Melbourne, Australia: Faculty of Education, Monash University.

Bandiera, M., Caravita, S., Torracca, E., & Vicentini, M., Eds. (1999). *Research in Science Education in Europe*. Dordrecht, The Netherlands: Kluwer.

Bassey, M. (1995). Creating education through research. Paper presented to the British Educational Research Association, Edinburgh.

Bishop, A.J. (1992). International perspectives on research in mathematics education. In D.A. Grouws, Ed., *Handbook of research on mathematics teaching and learning, (pp.* 710-23). New York: Macmillan.

Boenig, R.W. (1969). *Research in science education, 1938 through 1947*. New York: Teachers College Press.

Curtis, F.D. (1926, reprinted 1971). *A digest of investigations in the teaching of science*. Philadelphia, USA: Blakiston & Co. Inc.

Curtis, F.D. (1931, reprinted 1971). *A second digest of investigations in the teaching of science*. Philadelphia, USA: Blakiston & Co. Inc.

Curtis, F.D. (1939, reprinted 1971). *Third digest of investigations in the teaching of science*. Philadelphia, USA: Blakiston & Co. Inc.

Fraser, B.J. & Tobin, K.G., Eds. (1998). *International handbook on science education*. Dordrecht, The Netherlands: Kluwer, 2 volumes.

Freudenthal, H. (1983). Major problems of mathematics education. *Proceedings of the Fourth ICME, Berkeley*. Boston, MA: Birkhäuser.

Holzner, B. & Marx, J.H. (1979). *Knowledge application. The knowledge system in society*. Boston, MA: Allyn and Bacon.

Hopmann, S. & Riquarts, K., Eds. (1995). *Didaktik and/or curriculum*. Kiel, Germany: IPN.

Lave, J. (1988). *Cognition in practice. Mind, mathematics and culture in everyday life*. Cambridge, MA: Cambridge University Press.

Lawlor, E.P. (1970). *Research in science education, 1953 through 1957*. New York. Teachers College Press.

Layton, D., Jenkins, E.W., Davey, A., & Macgill, S. (1993). *Inarticulate science ? The public understanding of science and some implications for science education*. Driffield, UK: Studies in Education.

Lijnse, P.L. (1993). *European research in science education: Proceedings of the first PhD summer school*. Utrecht, The Netherlands: CDB Press.

Matthews, M.R. (1998). *Constructivism in science education. A philosophical examination*. Dordrecht, The Netherlands: Kluwer.

Phillips, D.C. (1995). The good, the bad, and the ugly: The many faces of constructivism. *Educational Researcher* 24, 5-12.

Shorrocks-Taylor, D. & Jenkins, E.W., Eds. (2000). *Learning from Others: International comparisons in Education*. Dordrecht, The Netherlands: Kluwer.

Sierpinska, A. & Kilpatrick, J. (1999). *Mathematics education as a research domain: A search for identity*. Dordrecht, The Netherlands: Kluwer.

Wilson, S., Shulman, L., & Richert, A.E. (1987). '150 different ways' of knowing: representations of knowledge in teaching. In J. Calderhead, *Exploring Teachers' Thinking*. London, UK: Cassell.

Acknowledgements

I am grateful to the European Science Education Research Association for the invitation to give this lecture and to the many colleagues who responded so generously to my request for assistance about developments in science education in individual European countries or regions. I am, of course, solely responsible for any interpretation placed upon the documents sent to me and for any tentative conclusions that I have based upon them.

Science Content as Problematic – Issues for Research[1]

Peter J. Fensham
Faculty of Education, Monash University, Australia

There are many persons in science education who have been influenced by Ros Driver's published work which, for 25 years, contributed so much to defining the frontiers of our field. There will also be a number, like me, who had the good fortune to work with her in Leeds or elsewhere. Others, will be able to feel again, the pulse of walking with her and Geoff in those Yorkshire dales they loved so much, or, as we once did, on the moors of Derbyshire, ablaze with purple, despite a glowering sky. More still, will remember being with her at conferences and other professional meetings. In any of these ways of close encounters, we were all touched by her honesty, enthusiasm, warm humanity and amazing commitment to furthering science education.

Abstract

In this lecture in tribute to Rosalind Driver, you will hear me inevitably refer a number of times to aspects of her research work. This will not be at all in the sense of a retrospective review, but because her research provides vantage points from which to see some directions science education research could well take in the next few years.

Introduction

In the 1980s, five developments in science education research drew attention to Content as an issue in school science education. These were:

1. The emergence of student alternative conceptions as an important field (1980 -)
2. The concept of "curriculum emphasis" (1982, 1988).
3. Differential affective responses by students to particular science content (1981-1987)
4. Beyond Processes - the interdependence of conceptual and process content (1987 -)
5. The recognition of a general weakness in science content knowledge of many science teachers (1986 -)

Those who know Rosalind Driver's work will be aware that she was a leading figure in two of these developments and a significant one in a third.

I plan to review these five developments from the particular perspective of regarding the "science" of science education as problematic, rather than the much more common one of seeing the "education" of science education as the problem area for research. I do this for several reasons. Firstly, despite drawing our attention as researchers to issues of content (ie. the "science" as problematic), the

[1] Lecture in Honor of Rosalind Driver

H. Behrendt et al. (eds.), Research in Science Education – Past, Present, and Future, 27–41.

overwhelming amount of our subsequent research response in areas 1, 4 and 5 (where there has been ongoing research) has been essentially pedagogical in character. Secondly, I have been intrigued that developments 2 and 3 did not evoke similarly vigorous research responses. Thirdly, I am now conscious that some years ago I failed to promote Content as a research field when I could have done so. Finally, as a result of the positive research responses in the last decade, I believe there is now no shortage of pedagogical knowledge in science education, albeit much needs to be done in initial and subsequent teacher education before this knowledge is apparent in the practices in the science classrooms of our countries. I begin with the failed opportunity.

The content of science project

In 1991, as the tide of alternative conception and cognitive change studies rolled on, I, with Richard Gunstone and Richard White, friends and colleagues at Monash University, had money to produce a new book. We asked some wise, international advisers the question. *Which of Content or affect, as rather neglected aspects of the current research in science education, should we try to highlight in a book project we were about to launch?*

The unanimous answer was *Content* and in due course we did produce a book, *The Content of Science : A constructivist approach to its teaching* (Fensham, Gunstone & White, 1994).

With the hindsight provided by the intervening years, it is now clear that, despite the book's main title, we failed to produce a book about the problematic character of the Content of school science, as perhaps our advisers had meant us to do. Rather, nearly all the authors, including myself, simply summarised, in our chapters, as the sub-title aptly indicates, what a decade of cognitive pedagogical research had provided about the teaching/learning of the particular science topic each of us had chosen to write about.

White (1994) was an exception. He argued, in his chapter, that a *Theory of Content* was overdue and made a start by setting out some properties of science content that research had already shown to be important. His list of properties is *openness to common experience, abstraction, complexity, presence of alternative models with explanatory power, presence of common words, mix of types of knowledge, demonstrable versus arbitrary, social acceptance, extent of links,* and *emotive power*. Finally, he suggested some steps that this major research task could take. A Theory of Content for school science still awaits our attention as a research community. Chevallard's (1991) work in didactics of mathematics may have relevance for us.

Alternative conceptions

Rosalind Driver's doctoral study at the University of Illinois with Professor Jack Easley (see her book, *The Pupil as Scientist*, 1983) was one of two or three studies that inspired in the 1980s, a most striking shift in research interest among science educators. Namely, the study of students' alternative conceptions in science. The attention her early work gave to pupils' views of the content of science and the use she and the other pioneers made of clinical interviewing to elicit their research data, shifted not only the focus of research, but also changed its methodologies. The dominant styles of science education research had been experimental studies or large scale survey studies of science teaching, but now more qualitative research studies, involving relatively small numbers of students or case studies of individual science classrooms were to be increasingly reported. Some of us have tried to describe how our own research underwent this profound transition (Gunstone, White, & Fensham, 1988).

These early studies led to a rapid growth of studies around the world of students' views of isolated science concepts. Large numbers of learners were found to hold conceptions that were alternative to the intended science ones and often these were held surprisingly strongly. Despite all these studies having as their focus, learners' conceptions of content, the responses of both the research community and curriculum designers was overwhelmingly pedagogical. That is, the research community saw these findings as a challenge to study processes like conceptual change, conceptual addition, and metacognition (e.g. Driver et al., 1994; Gunstone, 1994) and to develop theories of the pedagogical procedures that facilitated them.

These research outcomes have been very positive for teaching and learning but did nothing about the problems of the content of school science that the alternative conceptions research had also so directly raised. In a few isolated papers authors did draw conclusions that were directed at how the content of school science could be redefined in ways that would be more responsive to the difficulties their existing form presented for learners; e.g. Osborne (1982) *momentum as dissipating*; Wiser (1986) *heat as substance*; Duit (1981) *energy not being conserved*; De Vos and Verdonk (1985) and Andersson (1990) *chemical reactions*; and Russell et al. (1995) *dark before light.* I know, however, of no official curriculum that has made use of such redefinitions of its content.

Pfundt and Duit (1994), the bibliographers of alternative conceptions research, have now recorded more than 4000 studies. The quantity of this research is impressive indeed. Nevertheless, in terms of the range of science content that has been addressed, it is not unfair to say that almost all the science concepts that have been studied are ones that belong in the very academically defined science curricula that dominated school science after the 1960/70s reforms. Only a tiny fraction have been concerned with concepts that are associated with the environmental, technological, and socio-scientific content, that was beginning to be tried in the 1980s in STS-types of science curricula Thus, there are few, if any studies of students' conceptions of *green revolution, endangered species, bio-diversity, ozone hole, greenhouse effect, noise pollution, shelf life, radiation risk, and toxic level.*

The primary reason for researchers' concentration on traditional concepts rather than on the concepts of everyday relevance was, no doubt, because the former concepts still dominated the curriculum of school science in the countries where this research was most popular. Similarly, for those researchers who moved on to study pedagogies for conceptual change etc., such traditional science classrooms offered readily available contexts to try these new pedagogies for teaching and learning.

The response by curriculum bureaucrats (supported by academic scientists) to our research evidence that these traditional concepts are so poorly learned was to reaffirm them as the content for school science and to expand the opportunities in the primary and secondary years for students who had to learn them. They have also encouraged teachers, through in-service education, to make use of the pedagogies found by research to be effective.

It is interesting to speculate whether the cause of STS science in the curriculum for schooling would have received an equal fillip, if we had (as we surely would have done), produced evidence of similarly poor understanding of the technological and socio-scientific concepts (like those above) that bear so strongly on society and the lives of citizens.

Another weakness in the range of alternative conceptions research is that the focus in most of the studies is on isolated concepts of science, rather than on the contexts and processes of conceptualisation and nominalisation that led to their invention in science. Yet a large number of us, who helped to create this impressive library of research outcomes, encountered in our clinical interviewing many students who held very different perspectives from us on the *applicability of Laws of Nature, the role of empirical evidence, the tentativeness of scientific explanation, etc.*.

Since the tools for uncovering alternative conceptions are now so well developed, it will be a pity if they are not used in the years immediately ahead to extend the range of the science content being explored in these two directions.

Curriculum emphases

In 1982, Roberts in Canada published a paper based on his analyses of a large number of North American curriculum materials for school science, in which he introduced the concept of *curriculum emphasis*. This concept relates closely to the purposes for learning science in school that explicitly or implicitly determine a science curriculum's content, its styles of teaching and learning and its manner of assessment. Roberts was able to define seven different curriculum emphases. Their names are reasonably self explanatory - *Solid Foundation, Self as Explainer, Scientific Skills, Correct Explanations, Science/Technology Decisions, Nature of Science,* and *Everyday Coping.*

He used the word "emphasis"(rather than "purpose") quite deliberately, because he argued that, if students are to become aware and confident that their learning of science does have a coherent and meaningful purpose, and is not just isolated pieces of information, this purpose must be given explicit and repeated emphasis over a reasonably sustained period of learning. In other words, in any one semester or year of school study there should be only two or three emphases, with the other purposes of science education having less prominence during that time. The 1982 paper appears to have had little research follow up, and no direct impact on school curricula, even in the provinces of Canada.

Roberts (1998) developed the concept further, indicating among other things how the choice of an emphasis should influence the choice of content, and change the roles of science teachers and students. By this time, for a variety of reasons, some science curricula were beginning to lay more stress on content that was consistent with some of Roberts' emphases. For example, *Personal Coping* is present in the rationale for the sequence of the *Salters' Chemistry* project, *Science/Technology Decisions* was central in *Logical Reasoning in Science and Technology*, a secondary text used in some Canadian provinces and in some *PLON* units. *Self as Explainer* was quite explicit in *The Science Framework* in Victoria, Australia and in the *Science Plus* materials from the Atlantic Science Project in eastern Canada.

In a further extension of Roberts' conception in 1995, I identified three additional emphases, *Science in Application*, *Science as Nurturing*, and *Science through Technology*, as ones that were now evident in some of the science curriculum materials since 1982 (Fensham, 1997).

The idea of *curriculum emphasis* can be directly studied by introducing the characteristics of the ten emphases to teachers and then asking them to associate not more than three emphases (from the list of ten) with the years of schooling divided into 3+3+3+3, where the first 3 is the first years of primary schooling, etc. and allowing for a "non-science" and a "science" stream of students in the last three years. The criterion for their choice of emphasis is which one or ones they believe most coincide with the needs and interests of learners in these year blocks of schooling. I have usually found a remarkable agreement about at least the first two emphases for each set of 3 years of schooling. For example, *Everyday Coping* and *Science as Nurturing (environment)* are regularly chosen for the first three years, *Self as Explainer* in the next three years, *Science as Application* in the third block and *Science , Technology Decisions* in both cases of the later secondary years, with *Solid Foundation* as the other obvious choice for the science stream. The good sense this task has for teachers, and their general agreement about meaningful purposes for school science, are encouraging indications that more extensive studies of this could provide a new and radically different basis for designing a science curriculum and choosing its content. Roberts (1995), himself, describes a new junior secondary curriculum for lower secondary science in Alberta, Canada, that very explicitly set out to match its content with a limited number of emphases.

More recently, the curriculum emphasis concept has developed in the new and potentially powerful direction of Curriculum and Meaning. Roberts and Östman (1998) explain their new conception of *companion meaning* as being a development from curriculum emphasis. They argue that *companion meaning* has more substantial capacity, than *curriculum emphasis* had, of enabling the "collateral learnings" (Dewey, 1938) and "meta-learnings" (Schwab, 1962) that are hidden in a curriculum to be exposed. They, thus, see companion meanings such as *different views of nature, instrumentalism, dogmatism, etc.,* revealed by their discourse analysis of school science texts, as being both the context and the sub-text of the subject matter or content in science subjects. When these underlying meanings are exposed in this way, they can be criticised, rationally debated and used to revitalise a curriculum's purpose and choice of content. The very recent paper by De Vos and Reiding (1999) about the development of the new mandatory science subject in the Netherlands, *Public Awareness of Science*, is a fascinating example of how the companion meanings of this subject slipped and changed through the three stages of conception, describing, and becoming texts.

Affect and the content of science education

The Girls and Science and Technology (GASAT) movement in its biennial conferences beginning in 1981, regularly drew attention to affective aspects of school science. The initial studies of students' affect towards differences in science content simply used the various school science subjects as the variable. In many countries it was found that boys were more attracted to physics, girls to biology, and that the attitude to chemistry was less clearly gender biased. As some STS-type science curricula, like PLON in The Netherlands and Salters' Science and Chemistry in England began to be tried in schools, a few studies were reported concerning the attractiveness of their broad topics to students (e.g. Jorg & Wubbels, 1987; Lazonby,

1987). The former authors found gender differences in the response to some topics, but these did not follow the stereotypical differences associated with the future employment intentions of the students. A number of topics like *Weather, Traffic* and *Music* appealed equally to both sexes. Lazonby found students responded negatively to indiscriminate use of industrial applications as content in STS-science, and argued for a more cultural approach to this type of content rather than an instrumental ("implicit industrial recruiting") one.

One might have thought the drives in the 1980s for more *socially relevant* and *more personally meaningful* science education for all students, would have encouraged a great deal more research that seeks to associate science content with these two affective criteria of the *Science for All* movement. The curriculum victories in England and Wales, the USA and a number of other countries in the late 1980s and early 1990s, of the academic and elitist ideology (what Ball, 1994, described as the "cultural restorationist" ideology) over the democratic ideologies that underpinned STS science put a stop to this promising line of research. If the content to be learnt is essentially defined because of its importance in science, the motivational issue shifts from: *What content is intrinsically attractive to more students?* to *How (pedagogically), can this prescribed content be made attractive to more students?*

Sjøberg (1999) has now, however, revived the former question with his comparative, large scale survey research into what girls and boys are interested in learning about. For example, within the topic *Acoustics and Sound*, there were large gender differences; favouring boys among, for instance, the Norwegian sample. Within the topics *How can the ear hear?, Music, Instruments and Sound,* or *Sound and Music in Birds and other Animals*, the gender differences were eliminated or reversed.

Beyond processes

A number of new lines of research have flowed from Millar and Driver's (1987) critical paper entitled *Beyond Processes*. In the following years, Millar (1988, 1991) developed from this critique a three-fold classification of practical skills - *(i) general cognitive processing, (ii) practical science skills* and *(iii) inquiry tactics*. The first, including *observing, classifying, hypothesising* and *inferring*, he suggests, are developed very early without the aid of formal instruction. The second, *measuring specific properties,* such as temperature and mass, need to be taught and learnt. The third, like *identifying* and *controlling relevant variables* are associated with the conceptual content knowledge of the phenomenon concerned. Hence, students, whose knowledge of the science content in skill tasks of the third type varies, should perform differently when given such tasks. Erickson (1994) reported data, from assessment studies with students in grades 4, 7 and 10 in British Columbia, that seem to be consistent with how the learning of science conceptual content appears in Millar's classification. Some of the country analyses of the TIMSS results in the *Performance Expectation* and *Performance Task* dimensions of its testings seem also to be supportive (Tamir & Zuzovsky, 1998; Harris, 1998). There is, however, a need for more deliberate and direct studies of the distinctions Millar has made between content dependent and content independent skills in school science.

The notion that skills like *observing, classifying, hypothesising* and *inferring* are generic, and not content dependent was obscured for science educators for a long time, because of the way they had been identified in school science curricula with scientific method. For example, in the *Science A Process Approach* (USA) and *Science 5-13* (England) materials and their very pervasive influence in derivative curricula in other countries. Researchers in other areas of education have, however, been interested in the generic of skills for a longer period, and there is a rich body of findings to be mined in the cognitive psychology literature. For example, Chi, Feltovich and Glaser (1981), nearly twenty years ago, reported that experts are better problem solvers not because they have mastered a set of generic thinking skills, but because they know more about certain things than novices.

In the 1990s, another research development has occurred that probably should also be associated with the issues raised by *Beyond Processes.* This development changes the content in school science from extensive conceptual science knowledge to the procedures of *higher order scientific reasoning*. For example, Kuhn (1993) argued that the weight of detailed conceptual knowledge is so great that its acquisition by most students is impossible. As an alternative learning outcome, she proposes *scientific argumentation* as a fundamental and powerful capability for all students to acquire. This is a particular example of the claim made by Olson (1994) who stated that to understand a discipline like science requires the ability to participate in its higher order discourses. The number of reports in the literature about attempts to implement this claim in classrooms is increasing, but they are still at the case study stage. Furthermore, although the theoretical underpinnings of this emphasis on discourse are substantial, they are also in a flux of debate (e.g. Nuttall, 1997).

Some of the more socially-oriented theories give so little weight to individual cognition, that a shift in the classroom language in the direction of the intended scientific discourse is the only evidence that is usually presented. In this way, these research studies have tended to side-step the following important question for which educational systems and science educators will require answers, if the content of the science curriculum is, indeed, to be changed from detailed conceptual content to these discourses of science: *How much conceptual knowledge does an individual student need to have in order to meaningfully engage in these higher order discourses?*

There is a chance that the *OECD/PISA Science Project*[2] (1998) will provide some answers to this question. Ohlsson (1995) asked another critical question: *Will practice and participation in higher order reasoning lead to and/or deepen individual student's conceptual understanding?*

Only when these two questions have been answered, will we know whether these, so-called, *higher order processes of science* are a real advance for the Content of school science or a 1990s version of those earlier "science processes" that Millar and Driver so properly criticised in *Beyond Processes*.

Science content knowledge and teachers

The recognition, in the mid 80s, of the weakness of the science content knowledge of so many teachers in the USA led to the Holmes Group of universities agreeing to tighten their science content requirements for intending teachers and to the very large research project at Stanford University directed by Professor Lee

[2] see the contribution of Wynne Harlen in this volume

Shulman (1986), who identified content knowledge as the 'missing paradigm in research on teaching'. He asked (in the context of science education) questions such as :

What kind of science knowledge does a science teacher need? and *Is the science knowledge teachers need different from the knowledge held by scientists?*

Such questions, together with the concept of *pedagogical content knowledge (PCK)* he introduced in the following year (Shulman, 1987), sparked a series of research studies by the team at Stanford, some of which were with science teachers.

Shulman's original attempts to define PCK, included understanding how particular content will be constructed by learners, knowing what alternative conceptions will need to be dealt with, and what representations (models, metaphors, analogies, etc.) will be most effective for getting the content across. This diffuse and complex description did not provide a simple variable, or an obvious design for research studies. Neither did it suggest how a teacher's *science content knowledge* is to be combined with his/her *pedagogical knowledge* to become *pedagogical content knowledge.*

Since the Stanford studies, the concept of PCK has been the subject of research by science educators in other countries. In a number of these studies, the researchers have, in practice, focussed on the pedagogy used by selected teachers to teach particular content, - the interpretation of Content mentioned earlier in respect of *The Content of Science* book.

A number of studies have used an *exemplary teacher* design. That is, the research reports how science teachers, with good reputations for content knowledge and for effective teaching, taught particular topics. Too often these reports have little to say about how the content had been construed by the teacher, why the pedagogies for its development in the lessons were chosen and how anticipated alternative conceptions were recognised, etc. The relation between the nuances in the science content and the pedagogical decisions remain obscure.

Likewise, the strong current research interest into the use of models, metaphors and analogies in science teaching has tended to focus more on how teachers use these as pedagogical aids for students' learning of a science topic, and rather less on how well, and in what ways, the chosen model, metaphor, or analogy represents key aspects of the science content.

Osborne and Simon (1996) used a case study approach to compare the science teaching of several primary teachers. As a result, they made what may well be an important contribution to the particularly difficult issue of the science content knowledge needed by primary teachers. Based on case studies they distinguish between *A. What we know about a topic*, *B. How we know it,* and *C. In what alternative ways can we describe it.* There is a suggestion that the knowledge of the second kind may be a key to the confidence problem that these teachers have, even about the first kind of knowledge they do have. Furthermore, it may be that if these teachers' knowledge of the second kind is strengthened, their lack of first type knowledge, when challenged by students, or by new curriculum demands, will not be so threatening because they now have a sense of how various types of science knowledge are established.

Kennedy (1998) also recommends developing the second kind of knowledge in teachers, since without it they will inevitably misrepresent the character of science in their teaching. She includes, PCK within a wider concept and *subject matter knowledge*, that embraces, as well as PCK, *conceptual understanding of the subject, beliefs about the nature of work in science,* and *attitudes to science* and *actual teaching practices.*

The Dutch science educators, Van Driel, Verloop and de Vos (1998), describe PCK as (i) *the transformation of subject matter knowledge so that it can be used effectively and flexibly in the communication processes in classrooms,* (ii) *knowledge of comprehensible representations of subject matter* and (iii) *knowledge of content-related learning difficulties.* This description is the outcome of their review of the literature and of their analysis of the discussions of chemistry teachers, who agreed to teach chemical equilibrium in a prescribed manner and subsequently to participate in workshops to discuss this experience. As an empirical result, they conclude that, while subject matter knowledge was prerequisite, teaching experience was the major source of PCK.

Loughran and Gunstone (1999) draw, for their studies of PCK, on a medical analogy - comparing the knowledge of the human body that medical practitioners acquire from treating it with the physiology they learn in their training. So, for them, an important aspect of CK is the way a teacher's knowledge of content is altered by the extended experience of teaching it. That only a few studies have reported or attempted this aspect of PCK is not surprising, since it requires a longitudinal design that is difficult to achieve with teachers. Baird et al. (1987) studied the changes in science graduates' knowledge of the Cartesian Diver during their year of education for teaching, while Arzi (1988), more ambitiously, followed another group's knowledge of energy from graduation in science, through teacher education and beyond for two years of teaching.

The Didaktik vs. Curriculum conversations

Shulman's notion of PCK did, however, have another outcome that is much less well known among science education researchers. It was a series of meetings and conversations between small groups of researchers, who were interested in exploring two different traditions in education – the *Curriculum tradition* in which most Anglo-American research is set and a *Didaktik tradition,* which is strongly represented in research in some parts of continental Europe, and which relates albeit with differences to research in other parts. The appearance of the words, "subject content", in Shulman's (1987) account of the Stanford project was the stimulus for these meetings. The interest in North America in combining "subject content knowledge" and "pedagogical knowledge" in a major educational research project attracted the interest of those in the Didaktik tradition for whom subject content had always been integral to their research on teaching.

The major sources for the conversations are the report of the Kiel meeting, *Didaktik and/or Curriculum* (Hopmann & Riquarts, 1995) and the first number of volume 27 of the *Journal of Curriculum Studies (1995).*

Although one or two science education researchers from each tradition participated in the meetings, these reports are not well known in our community. Furthermore, there has been no discussion yet as to the implications that these conversations may have for the mainstreams of research in science education. Accordingly, in this final part of my lecture, I want to share with you the implications I have begun to draw from these meetings of two great traditions in education.

I must begin by saying that almost my whole career has been within the Anglo-American tradition of *Curriculum*, since Australia's education systems are derivative from these sources. I have, nevertheless, been aware on a number of occasions at Conferences where English is the medium of discourse, of a discontinuity and breakdown of communication when the word, "Didaktik", is introduced in a question or in discussion by European participants. Furthermore, in my contacts over the years with European colleagues, I have been very intrigued, from time to time, to encounter features of education and schooling that are quite other than how we express, organise, or do these things in Australia.

Didaktik

To begin to understand the Didaktik tradition it is evident that an appreciation of the German word, Bildung, is important, but that it is not simple to translate into English. "Bildung und Erziehung" makes sense in German; and carries a differentiated meaning that my elementary dictionary's "Education and Education" failed to capture. The metaphors associated with "Bildung" were, however, helpful. They include *the formation of the learner as a whole person* and *the cultivation or nurturing of a plant as it develops through the stages, from seedling to full fruit or flower.* These are ideas about Education that are not usually recognised very strongly in the Curriculum tradition in which I had been socialised for so long.

As a result of the ideas in Bildung, there is in the Didaktik tradition an analytical process that turns (transposes or transforms) sources of human knowledge, like the scientific disciplines, into knowledge for schooling so that it can contribute to the Bildung of young learners. In other words, the knowledge of biology, chemistry, or physics, as it exists in these scientific disciplines is not automatically in a form that makes it worthy of a place in schooling committed to education as Bildung. After all, the contexts of scientists in which this knowledge is constructed are not the contexts of young learners in school. This notion of a transformation of content was a particularly exciting one for me, because of my interest in the notion of *Science for All.* Elsewhere, I have tried to develop this aspect of the ideas of Bildung and Didaktik further (Fensham, 2000).

Wolfgang Klafki (1958) in a text book that was part, I am told, of the education the great majority of German secondary school teachers for a quarter of a century, challenged student teachers with a series of questions (reported by Uljens, 1995) that indicate the sort of analysis of content that illustrates Didaktik in a clear fashion.

I What wider or general sense or reality does this content exemplify and open up to the learner?
What basic phenomenon or fundamental principle, what law, criterion, problem, method, technique, or attitude can be grasped by dealing with this content as an "example"?

II What significance does the content in question, or the experience, knowledge, ability, or skill, to be acquired through this topic, already possess in the minds of the children in my class?
What significance should it have from a pedagogical point of view?

III What constitutes the topic's significance for the children's future?

IV How are the contents structured (which have been placed in a specifically pedagogical perspective by questions I, II, and III)?

V *a) What facts, phenomena, situations, experiments, controversies, etc., i.e. what intuitions, are appropriate to induce the child to ask questions directed at the essence and structure of the content in question?*

b) What pictures, hints, situations, observations, accounts, experiments, models, etc., are appropriate in helping children to answer as independently as possible, their questions directed at the essentials of the matter?

c) What situations and tasks are appropriate for helping students grasp the principle of the content by means of the example of an elementary "case", and to apply and practise it so that it will be of real benefit to them?

Questions in IV and V are familiar in both *traditions* as the content of subjects called *Methods of Teaching* or *Pedagogics*. The questions in I, II and III are much less commonly addressed in the *Curriculum tradition,* where the content of the intended curriculum in a reasonably detailed form is usually provided to the teacher from the educational system, through its curriculum body or examination board.

In the context of this paper, the questions in I, II and III assume somebody has the task of rating traditional science topics as only justifiable, if they exemplify *a general sense of reality, a principle, etc.* of importance for the learner's yesterday, today and tomorrow. As another difference between the traditions, it emerges that in the *Didaktik tradition*, the somebody with responsibility for this task is the teacher, and in the *Curriculum tradition*, if anybody does recognise this responsibility, then it must lie with the curriculum authority, for it is not the teacher.

It is instructive and challenging to insert various familiar school science topics into the questions in I.

What are chemical equations an example of in science?

What wider sense of physics do series and parallel circuits or the laws of reflection open up to students?

Of what fundamental principle or problem tn biology is the flow of energy through food chains an example?

Science teachers or student teachers with basic degrees in one or more of the scientific disciplines do not find these questions easy to answer.

In which of the following senses is a chemical equation an equation?	*e.g.* $2H_2 + O_2 = 2H_2O$
(a) 2 plus 4 is the same as 6	i.e. $2 + 4 = 6$
(b) 2 apples plus 4 oranges cannot be added, because only like things can be added	i.e. $2 + 4 =$
(c) 2 apples plus 4 oranges equals 6 parts of fruit salad	i.e. $2 + 4 = 6$ (a mixture of 2 and 4)
(d) adding 4 to 2 leads to a new number 6, different from 2 and from 4	i.e. $2 + 4 \rightarrow 6$

Although I oriented the teachers to this topic, and particularly to its conservation aspects, by asking them before the interview to balance some simple equations and calculate the yield of two other reaction equations, given the masses of the reactants, more than half of them overlooked the conservation comparison with (a) above and saw the connection between (d) and the synthetic aspect of the reaction as the only essential comparison. Instead of seeing chemical equations as an example, in chemistry, of the conservation of matter, (Klafki's first question) they saw balancing equations as teaching algorithmic rules.

These interviews reminded me of my bewilderment many years ago on a visit to IPN, when I was given a research paper by Werner Dierks (1980) on the topic of chemical equations that included no data on the teaching or on learning this topic. How was this, I thought at the time, considered as a research study in science education? At last, I now recognise it as addressing profound educational questions about this elementary topic in chemistry, as a source for '*a wider or general sense of reality*' for learners. It was, indeed, an excellent example of the first steps in didactical analysis.

I am not in any sense suggesting that the differences I am highlighting between the traditions are necessarily reflected in all classrooms of countries, where the educational system belongs to the respective traditions. Nevertheless, I believe there are fruitful new research questions we can each draw from the differences in tradition these conversations have opened up.

For example, the acceptance of the idea that the disciplinary knowledge of the sciences is not automatically appropriate for school science would open a field of studies in which the focus would be on exploring what worth, for school learning, the alternative representations any topic in science in fact do have. I mentioned much earlier that some studies of alternative conceptions pointed to alternative representations of some familiar topics in school science. In this new research I am now suggesting such findings would immediately no longer be isolated, but be contributions to a mainstream of research in science education.

It is interesting to revisit the original and subsequent PCK literature to see if there is any recognition at all of the first steps of Klafki's didactical analysis as part of PCK. A European member of the Stanford team, Gudmonsdottir (1991), commented that an important point is misunderstood when exemplary teachers are chosen for study by the grades their students achieve, and not by "an evaluation of their didactical interpretation and implementation of the study plan's intentions". The PCK concept has been further developed by researchers with a general interest in teaching and also by others within science education itself. Prawat (1989), not a science educator, saw PCK as: "knowledge that allows teachers to transform what they know into something meaningful for their students" - very akin to Klafki's first steps.

One emerging group of researchers in South Korea, straddles the two traditions in an interesting way. It is common practice for graduate students in this group to undertake, in step with designing their empirical study, an historical investigation of the earlier scientific thinking that has led to the contemporary account in science of the phenomenon concerned. Jongwon Park and his colleagues (1999) provided an example how this incorporation of the history of the content influenced the questions the researchers were addressing. They have recognised the immense contribution Galileo made, when he introduced *idealisation* into the thinking about physical natural phenomena. Hence, they are beginning to explore the implications for learning of the way idealisation is included in school physics.

The research questions they see flowing from this recognition include (i) What types of ideal conditions are included in physics concepts, in the processes of deriving physics laws? (ii) What ideal conditions are included in the problems posed for problem solving in physics? (iii) How are ideal conditions used to understand the relation between the world of physics and the real world? and (iv) How are ideal conditions used in creative inquiry in physics?

Already, they have obtained data on students' understanding of some of these ideal conditions, and their relation to the real world. This appears to be another fruitful direction for alternative conceptions research to take, since it takes seriously the contexts within science of the concepts, rather than simply going on asking students about them as if they exist in science in isolation.

Conclusion

In this lecture I have tried to demonstrate that there are a number of new frontiers for researchers in science education to explore if they are willing to take its Content as problematic. In a last great paper at the Rome meeting of ESERA, Rosalind Driver presented the case for, and a design for a major study of one of these new issues of Content (Driver, Newton & Osborne, 1999). Her colleagues at Kings are now embarking on that study. It will not, I trust, be the only one frontier of Content to be reported at the next ESERA meeting in 2001.

References

Andersson, B. (1990). Pupil's conception of matter and its transformations (aged 12-16). *Studies in Science Education 18*, 53-85.

Arzi, H.J. (1988, April). On energy in chocolate and yoghurt or On the application of school science concepts to everyday life and their integration across the curriculum. Paper presented at the meeting of the American Educational Research Association, New Orleans.

Baird, J., Fensham, P.J., Gunstone, R., & White, R. (1987). Individual development during teacher training, *Research in Science Education 17*, 182-191.

Ball, S.J. (1994). *Education Reform: A critical and post-structural approach*. Buckingham, England: Open University Press.

Chevallard, Y. (1991). La transposition didactique. 2cd ed.,Grenoble: La Pensée Sauvage.

Chi, M.T.H., Feltovich, P., & Glaser, R. (1981). Categorization of physics problems by experts and novices. *Cognitive Science, 5,* 121-152.

De Vos, W., & Verdonk, A.(1985). A new road to reactions, Parts1&2. *Journal of Chemical Education* 63(2), 238-240, and 62(8), 648-649.

De Vos,W. & Reiding, J. (1999). Public Understanding of Science as a separate subject in secondary schools in The Netherlands. *International Journal of Science Education* 21(7), 711-720.

Dewey, J. (1963). *Experience and Education*. New York: Collier (Original work published 1938).

Dierks, W. (1980). Das Verwenden der Anzahl beim stöchiometrischen Rechnen mit Grössenwertgleichungen und bei der Symbolisierung quantitativer Reaktionen. *Der mathematische und naturwissenschaftliche Unterricht* 34, 29-41.

Driver,R. (1983) *The pupil as scientist.* Milton Keynes, England: The Open University.

Driver R. (1988). Theory into Practice II: A constructivist approach to curriculum development. In P.J. Fensham, Ed., *Developments and dilemmas in science education* (pp. 133-149). London: Falmer.

Driver, R., Asoko,H., Leach, J. Mortimer, E., & Scott, P. (1994). Constructing scientific knowledge in the classroom. *Educational Researcher* 23, 5-12.

Driver, R. Newton, P., & Osborne, J. (1999). Establishing the norms of scientific argumentation in classrooms. *Science Education* (in press)

Duit, R. (1981). Understanding energy as a conserved quantity. *European Journal of Science Education* 3(3), 291-301.

Erickson, G. (1994). Pupils understanding magnetism in a practical assessment context. In P.J. Fensham, R.F. Gunstone & R.T. White, Eds., *The Content of Science: A constructivist approach to its teaching and learning* (pp. 80-99). London: Falmer.

Fensham, P.J (1999). Science for All: The issue of content. Chapter to be published in R.Millar, J.Leach & J.Osborne, Eds., *Improving Science Education: The contribution of research.* Milton Keynes: Open University Press.

Fensham, P.J., Gunstone, R.F., & White, R.T., Eds., (1994). *The content of science: A constructivist approach to its teaching and learning.* London: Falmer.

Gudmonsdottir, S. (1991). Pedagogical models of subject matter. In J. Brophy, Ed., *Advances in Research on Teaching, Vol.2* (pp. 265-304), Greenwich:

Gunstone, R.F., White, R.T., & Fensham, P.J. (1988). Developments in style and purpose of research on the learning of science. *Journal of Research in Science Teaching* 25(7), 5-13.

Gunstone, R.F. (1994). The importance of specific science content in the enhancement of metacognition. In P.J. Fensham, R.F. Gunstone, & R.T. White, Eds., *The content of science: A constructivist approach to its teaching and learning* (pp. 131-146). London: Falmer.

Harris, S. (1998, April) TIMSS Performance Assessment: Strengths and weaknesses of English students. Paper presented at the American Educational Research Association Meeting, San Diego.

Hopmann, S. & Riquarts, K., Eds. (1995). *Didaktik and/or Curriculum.* Kiel, Germany: IPN.

Jongwon Park & Ikgyun Kim (1999, July). Classifying students observational activities. Paper presented at ASERA Meeting, Rotorua, New Zealand.

Jorg, T. & Wubbels, T. (1987). Girls and physics. *International Journal of Science Education* 9(3) 296 -307.

Kennedy, M.M. (1998) Education reform and subject matter knowledge. *Journal of Research in Science Teaching* 35(3) 249-263.

Klafki, W. (1958). *Didaktische Analyse als Kern der Unterrichtsvorbereitung.* Weinheim, Germany.

Kuhn, D. (1993). Science as argument; implications for teaching and learning scientific thinking. *Science Education* 77(3), 319-337.

Lazonby, J. (1987). Do students want to learn about industry? In D. Waddington, Ed., *Education, Industry and Technology* (pp. 39-40). London: Pergamon.

Loughran, J. & Gunstone, R.F. (1999*). Science cases in action: Developing an understanding of teachers' pedagogical content knowledge.* Australian Research Council Project. Clayton, Victoria: Faculty of Education, Monash University.

Millar, R. & Driver, R. (1987). Beyond processes. *Studies in Science Education* 14, 33-62.

Millar, R. (1989). What is scientific method and can it be taught? In J.J. Wellington, Ed., Skills and Processes in science education (pp. 47-62). London: Routledge.

Millar, R. (1991). A means to an end: The role of processes in science education. In B. Woolnough, Ed., Practical Science (pp. 43-52). Milton Keynes: Open University Press.

Nuttall, G. (1997). Understanding student thinking and learning in the classroom. In B.J. Biddle, T.C. Good & I. Goddson (Eds.), *The international handbook of teachers and teaching.* Dordrecht: Kluwer Academic Publishers.

OECD/PISA (1998). *Science Framework. November 1998 Draft*, Camberwell, Victoria, Australia: Australian Council for Educational Research.

Ohlsson, S. (1995). Learning to do and learning to understand. A lesson and a challenge for cognitive modelling. In P.Reimann & H.Spada, Eds., *Learning in humans and machines: Towards an interdisciplinary learning science.* New York: Pergamon.

Olson, D.R. (1994). *The World on Paper: The conceptual and cognitive implications of writing and reading.* Cambridge: Cambridge University Press.

Osborne, R. (1982). *Private communication.*

Osborne, J. & Simon, S. (1996). Primary science: Past and future directions. *Studies in Science Education* 27, 99-147.

Pfundt, H. & Duit, R. (1994). *Bibliography: Students' alternative frameworks and science education*, 4th edition. Kiel, Germany: IPN.

Prawat, R.S. (1989). Teaching for understanding. *Teaching and Teacher Education* 5(4) 315-328.

Roberts, D.A. (1982). Developing the concept of "curriculum emphasis". *Science Education* 66, 243-260.

Roberts, D.A. (1988). What counts as science education? In P.J. Fensham, Ed., *Developments and Dilemmas in Science Education*, Chap.2 (pp. 27-54). London: Falmer.

Roberts, D.A. (1995). Junior high school science transformed: Analysing a science curriculum policy change. *International Journal of Science Education* 7(4), 493-504.

Roberts, D.A. & Östman, L., Eds. (1998). *Problems of meaning in science curriculum*, New York: Teachers College Press.

Russell, T., Qualter, A., & McGuigan, L. (1995). Reflections on the implementation of National Curriculum science policy for the 5-14 age range: findings, and interpretations from a national evaluation study in England. *International Journal of Science Education* 17(4), 481-492.

Schwab, J.J. (1962). The teaching of science as enquiry. In J.J. Schwab & P.F. Brandwein, Eds., *The Teaching of Science* (pp. 32-103). Cambridge, MA: Harvard University Press.

Shulman, L.S.(1986). Paradigms and research programs in the study of teaching: A contemporary perspective. In M.Wittrock, Ed., *The Handbook of Research on Teaching, 3rd edition* (pp. 3-36). New York: Macmillan.

Shulman, L.S. (1987). Knowledge and teaching: Foundations of a new reform. *Harvard Education Review* 57(1) 1-22.

Sjøberg, S.(1999, June). Is there a science curriculum that can serve the interests of children in different countries: Results from a study among 10000 children in 21 countries. In M. Keogh & S. Naidoo, Eds., *Proceedings of the 9th Symposium of IOSTE, Vol.2* (pp. 624-5), Durban.

Tamir, P. & Zuzovsky, R. (1998, April). The "knowledge" aspect in the practical performance assessment of elementary and junior high school students in Israel. Paper presented at the American Educational Research Association, San Diego, Ca.

Uljens, M. (1995). A model of school Didactics and it role in academic teacher education. In S. Hopmann & K. Riquarts, Eds., *Didaktik and/or Curriculum* (pp. 301-322). Kiel, Germany: IPN.

Van Driel, J.H. Verloop, N., & De Vos, W. (1998). Development of science teachers' pedagogical content knowledge. *Journal of Research in Science Teaching* 35(6), 673-695.

White, R.T. (1994). Dimensions of content. In P.J. Fensham, R.F. Gunstone, & R.T. White, Eds., *The Content of Science: A constructivist approach to its teaching and learning* (pp. 255-262). London: Falmer.

Wiser, M. (1986, April). Using computer based labs to induce students' differentiation of heat and temperature. Paper presented at the AERA Annual Meeting, San Francisco.

Science Education Versus Science in the Academy: Questions - Discussion - Perspectives

Helmut Dahncke and Reinders Duit, University of Kiel, Germany
John Gilbert, University of Reading, UK
Leif Östman, University of Uppsala, Sweden
Dimitris Psillos, University of Thessaloniki, Greece
David B. Pushkin, Wilmington College of Delaware, USA

Abstract

The paper seeks to address the troubling issues we, as science educators, encounter when coexisting with scientist colleagues. Significant disagreement exists regarding what constitutes an appropriate science learning experience, what constitutes legitimate research, how science should be structured and taught and even who should be allowed to teach or learn science. From a cultural perspective we should try to understand better ourselves and those we work with, from a historical perspective: identify the origins of the aforementioned issues, from a philosophical, epistemological, and ontological perspective: determine where the common ground exists between conflicting faculty, so that we may make more significant progress in our efforts to improve science teaching and learning as well as teacher education programs. We think that we should try to reflect on our own professional contexts within science departments or education departments. Ultimately, we should probe the institution of higher education, the very place where future teachers, scientists, and scientific literates learn science. This is where the most significant impact potentially resides for education reform. For this is where the foundations begin. If conflicts cannot be meaningfully resolved at the university level, the products of that environment will inevitably perpetuate the dysfunction we observe in schools, where the children of the present and future will be no different than of the past.

Initial situation, problems and fundamental questions

The interrelation of science and science education as a research discipline has increased in many universities and colleges during the last decades in connection with training science teachers. Reports on the character of this interrelation differ from one country to another and even from place to place within one country. This interrelation almost certainly depends on vested parties and on their characteristics, although several structural features can be detected regardless of this. However, despite differences in local prerequisites it is important to remember that under no circumstances is this relationship easy. Assessment ranges from „conflicts within the academic setting" and „field of conflict" to a report on constructive collaboration.

A significant degree of conflict and tension exists between those teaching discipline-based science (e.g. scientists) and science educators (e.g. teacher trainers and researchers in science education). Such conflicts and tensions occur regarding views on pedagogical practice, curricula designed to prepare future teachers and the definition of research (Psillos, this volume). Although differing opinions have the potential to broaden and enrich an academic community, discipline scientists and

H. Behrendt et al. (eds.), Research in Science Education – Past, Present, and Future, 43–48.

science education researchers, in particular, often view each other as contemptuous threats, all on the basis of incongruous paradigms.

Perhaps the root of the paradigmatic conflict is how each group views science. While discipline scientists may consider science as a body of canonical knowledge, science education researchers may consider it as a process, by which we come to understand natural phenomena. While one group may consider science as hierarchical and logical, the other may consider it as part of a broader context for literacy. While one group may consider science a speciality for the intellectually gifted, the other may consider it as exclusionary and alienating. Ultimately, the question both groups need to resolve is: are we dispensing science or are we educating about science?

The Second International Conference on Research in Science Education staged by ESERA, presented an opportunity to pose several questions on this significant relationship, for discussion and to reach a solution more easily. In order to do this, the six authors of this paper got together at a symposium with twenty other participants. The following questions formed the starting point: Do different countries within the ESERA community observe a degree of conflict between scientist faculty and science education faculty? If so, what are the possible manifestations of such conflict and what might be the potential causes? We are ultimately dealing with a cultural phenomenon in our academic institutions that permeates into the way we teach science courses, prepare future teachers, support current teachers, design curricula, assess learning, conduct research and participate in scholarly discourse. To understand this phenomenon better, we must ask the following fundamental questions: What are one's beliefs of the nature of science? What are one's beliefs of the practice of science? What purpose does one see in learning science? What purpose does one see in teaching science? Does conflict exist between science educators and scientists? Is the conflict a natural / inherent phenomenon? What might be the source of the conflict? Is this concern universal or locally / nationally limited? Or, expressed in other ways: Who does the research? What research is done? Who does the teaching? Who does the learning? What is the curriculum? Who is teaching / learning? These are broad questions, but some statements made on those taking action and on examples taken from the project such as university science staff were taken as a starting point:

For example, discipline scientists focus on nature, produce knowledge about nature, practice science education (potentially). Their work is referred to citation index: natural sciences. University science education staff focus on scientific knowledge about nature, produce knowledge about education, practice science education. Their work is referred to citation index: social sciences.

Science education is interrelated to (at least) eight disciplines as illustrated in figure 1:

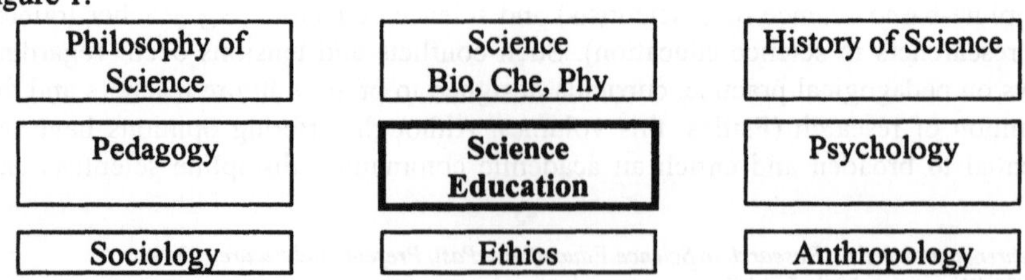

Fig. 1

Science education is an interdisciplinary domain. Science education research and development includes, of course, competence in the science reference domain. However, it also comprises competencies in a substantially large number of other reference domains as illustrated in figure 1. In short, if science education research and development is mainly oriented towards the science reference domain it will not be able to provide effective teaching and learning aids. What is necessary is a balance between science issues and educational issues in a broad sense (Duit, 1999). The interrelations between the reference domains determine important viewpoints relating to research in science education (e.g. as follows):

- Major concerns of science education research: *subject matter clarification / educational analysis (educational reconstruction)/ empirical research on teaching and learning / development and evaluation of media, methods, curricula*
- Research in science education is distinct from research in the sciences.
- Science education research methods have to merge issues of several reference domains.
- Science education researchers are amateurs in most of what they do. They fall between two stools (clashes with science and also with other key reference domains).
- There are certain misunderstandings (clashes): with biologists, chemists and physicists, but also with educational psychologists and with teachers.
- Evaluation of university teaching challenges scientists, at present, and may make them more aware that science education has something to offer.
- Science education as a university domain on its own has been challenged.

This paragraph may become a statement and an example of co-operation. Desirable goals are: collaboration between science and science education researchers, educational researchers, teachers from all levels and teachers at institutes for teacher training. Example: a long term project involving doctoral students, science education researchers and partners from science departments at teacher training colleges (e.g. several projects at Uppsala University).

Common viewpoints and perspectives

The beginnings of a common viewpoint emerge from discussion on such issues, on basic statements and goals, as well as from further examples of joint projects between science and science education. There is, however, a common awareness of the problem, even though national and local viewpoints vary considerably. The interrelation of science education and science is extremely significant not only for practical work and day-by-day work in research and teaching, but also for training science teachers. The almost historic aspect that researchers in science do not like to accept the fact that any research beyond their own specific methodology might exist is part of the problem. On the other hand it is a fact that research is carried out in the field of science education and that there are overall accepted quality standards regarding normal university work, as for instance in accepting or rejecting dissertations.

The references given in internationally referred journals, published during the last decades, also form part of this standard. An analysis of all this material may possibly still create an impression of diversity, yet it forms the basis of how researchers in science education see their own set of tasks and their own role and the perspectives of our present and future research. We have summarized this in six general statements on science education research, adding three more specialized aspects on the addressees, the prerequisites and the aims of science education:

1. Science education as a research discipline has its roots in several corresponding special science disciplines and in educational science. However, it is not a partial discipline, belonging to one of these disciplines, nor can it be solely derived from them or from a combination of them.

2. Although science education as a discipline also requires sound knowledge of special science disciplines, it cannot be limited to these alone. Former comprehension of science education as a depiction process or depiction outcome or as an application of the appropriate science discipline is not acceptable from an epistemological point of view, nor is it suitable for teacher education and lessons.

3. As defined in the first two statements, science education as a research discipline has its own areas of topics, independent goals, an independent system of secured knowledge, a (still limited) independent amount of research methods and (different) institutional allocations. This is occasionally summarized in a nutshell as „research in science education as a discipline sui generis".

4. Although practice in science education has been carried out from early times (e.g. Galileo's works still make good reading today for researchers in science education), it is quite young as an established research discipline and has been a teaching subject at university level in many countries only since the sixties.

5. Science education, as a research discipline, covers the goals, prerequisites and patterns of teaching and learning special subject matter. This teaching and learning takes place mostly in institutions, in particular in schools, but basically goes beyond this too. To put it in more general terms, science education as a discipline confronts people with objects and subject matter which could basically, be relevant to teaching and education and takes this as a starting point. Even the decision on whether certain subject matter is relevant to education, to prepare and to influence it is one of the scientific tasks within science education. With this in mind, all the questions which crop up when a person, a group of people or society as a whole are confronted with special subject matter are dealt with.

6. Even though we, understandably, insist on science education being scientifically independent, our fifth statement confirms that it overlaps special science disciplines and educational science and is also dependent on them.

To complement these common statements, which are, of necessity, rather general, we have to add three specific positions regarding the addressees, prerequisites and goals of science education:

I. Teaching and lessons in schools, universities and in adult education, as well as all forms of independent learning using texts, teaching programs and television programs etc. are the subject matter of research in science education.

II. Studies on the prerequisites and principles of teaching and learning can not, generally, be carried out successfully using the methods of a particular science. In science education as a research discipline, different methods have to be applied from those used in science in biology, chemistry and physics. The development of a number of specific research methods has been started, but is still in the initial stages. It is also necessary for advanced students to tackle these methods. It can, generally, be said that science education is regarded as an integral part of the curriculum for teacher education. These studies should therefore concern subject matter, epistemological and ontological viewpoints, methods and goals and tasks involved in science education as a research discipline.

III. The goals and tasks of science education as a research discipline include the following:

- the subject-specific process of discovery and its significance for teaching;
- analysis and structure of the scientific discipline according to subject matter, problems, basic terms, methods, theories and principles against the background of its historical development from a didactic point of view (structuring the discipline);
- assessment of the significance of science and its application in culture and society and for the present and future of the learner;
- analysis of goals in science lessons regarding their origin, roots and change according to the relevant conditions, as well as the development of principles for deciding on, assessing and establishing teaching and learning goals;
- development and application of criteria for deciding on the choice and implementation of the content of teaching;
- development and application of logical and psychological principles for transferring specialized knowledge to the appropriate level (didactic reduction, elementarization);
- interests, approach, motivation and extracurricular conceptions on scientific subject matter;
- previous experience gained by the learner from extracurricular experience with regard to the subject and its integration in his surroundings;
- subject-specific teaching, learning and communication processes including prerequisites;
- development of a subject-specific theory of teaching relating to teaching principles, types of lessons and procedure during lessons in connection with teaching experience;

- development, testing and revision of subject-related and interdisciplinary curricula and parts of the curricula (decisions on goals, the choice and implementation of subjects, planning lessons);
- analysis and development of subject-related media;
- development and testing of procedures for diagnosing learning difficulties and for assessing successful learning.

Different aspects of the way our didactics see themselves are dealt with here and the elements of scientific work carried out in research and teaching stated here produce long lists which are nevertheless open-ended.

References

Coppola, B.P. & Paerson, W.H. (1998). Heretical thoughts II - On lessons we learned from our graduate advisor that have impacted on our untergraduate teaching. *Journal of College Science Teaching* 27, 416-421.

Dahncke, H. (1994). Bilanz und Perspektiven fachdidaktischer Forschung zur Chemie und Physik. [Results and perspectives of research on chemistry and physics education]. In H. Behrendt (Ed.), *Zur Didaktik der Physik und Chemie - Probleme und Perspektiven* (pp. 17-25). Alsbach: Leuchtturm.

DeBoer, G. (1991). *A history of ideas in science education – Implications for practice*. New York: Teachers College Press.

Duit, R. (1999). Merkmale fachdidaktischer Lehr-Lern-Forschung aus Sicht der Didaktik der Physik [Characteristics of domain-specific research on teaching and learning - A physics education point of view]. Paper presented at the DFG-Rundgespräch "Domänspezifische Lehr-Lern-Forschung: Merkmale, Ertrag und Anforderungen an Forschungsprojekte".

Pushkin, D.B. (1999). Cookbook classrooms: Cognitive capitulation. In J. Weaver, P. Applebaum & M. Morris (Eds.) *(Post) Modern Science (Education)*. New York: Peter Lang. In press.

Roberts, D. A. & Östman, L. (Eds.) (1998). *Problems of meaning in science curriculum*. New York: Teachers College Press.

Trowler, P.R. (1998). *Academics Responding to Change – New Higher Education Frameworks and Academic Cultures*. Buckingham, UK: Open University Press.

Part 2: Scientific Literacy – Conceptions and Assessment

The Assessment of Scientific Literacy in the OECD/PISA Project[1]

Wynne Harlen
Visiting Professor, Bristol University, UK

Abstract

In this paper, the focus is upon the rationale for and the nature of the framework for assessing scientific literacy in the OECD/PISA project. Although pilot trials of test materials were conducted in May 1999, at this point it is only possible to report the procedures for analysis and selection of items and not their outcome. The paper begins with a brief overview of the intentions of the programme as a whole, since these set the parameters for the assessment in each of the domains selected for the surveys: reading literacy, mathematical literacy and scientific literacy. This first section attempts to answer questions such as: why another international survey? how does PISA differ from TIMSS? which countries are participating? The second section discusses the interpretation of scientific literacy put forward by the Science Functional Expert Group (SFEG) which was set up to advise on what to assess and how it might be assessed. This interpretation has been agreed by the committee, comprising representatives of the participating countries, which steers the project on matters of policy and ensures adherence to policy decisions as the programme proceeds. The third section describes the different aspects of scientific literacy that are being used to develop assessment units. Finally, the nature of the units is mentioned briefly and examples are given.

The significance of PISA

The purpose of PISA is to provide information that is internationally comparable and useful for informing educational policy decisions about the outcomes of the educational systems in the 32 participating countries. It is a collaborative project, steered by the countries involved, and undertaken, under the direction of the OECD Secretariat, through an international consortium led by the Australian Council for Educational Research[2]. The essential features of the programme and the frameworks for assessing educational outcomes have now been published, giving answers to the questions: what? when? and how? the surveys will be conducted. The key features are summarised in Box 1 (from OECD, 1999).

The 32 countries taking part are all members of the OECD apart from Latvia, the

---

[1] Plenary Address

[2] As from October 1999, the other members of the consortium are: the Netherlands' National Institute for Educational Measurement (CITO), the Educational Testing Service (ETS), the National Institute for Educational Research, Japan (NIER) and Westat.

H. Behrendt et al. (eds.), Research in Science Education – Past, Present, and Future, 49–60.

Russian Federation, Brazil and China. These represent more than a quarter of the world's population, which is greater than for any previous international survey of educational outcomes. This is not the reason for carrying out another international survey.

Basics
- An internationally standardised assessment, jointly developed by participating countries and administered to 15-year-olds in groups in their schools
- Administered in 32 countries, of which 28 are members of the OECD
- Between 4,500 and 10,000 students will be tested in each country.

Content
- PISA covers three domains: reading literacy, mathematical literacy and scientific literacy
- PISA aims to define each domain not merely in terms of mastery of the school curriculum, but in terms of important knowledge and skills needed in adult life. The assessment of cross-curriculum competencies is an integral part of PISA
- Emphasis is placed on the mastery of processes, the understanding of concepts and the ability to function in various situations within each domain.

Methods
- Pencil and paper tests are used, with assessments lasting a total of 3 hours for each student
- Test items are a mixture of multiple-choice test items and questions requiring the students to construct their own responses. The items are organised into groups based on a passage setting out a real-life situation
- A total of about 7 hours of test items is included, with different students taking different combinations of the test items
- Students answer a background questionnaire which takes about 20 – 30 minutes to complete, providing information about themselves. School principals are given a 30-minute questionnaire asking about their schools.

Assessment cycle
- The first assessment will take place in 2000, with the first results published in 2001, and assessments will continue thereafter, in three-year cycles
- Each cycle looks in depth at a 'major' domain, to which two-thirds of the testing time is devoted; the other two domains provide a summary profile of skills. Major domains are reading literacy in 2000, mathematical literacy in 2003 and scientific literacy in 2006.

Outcomes
- Basic profile of knowledge and skills among students at the end of compulsory schooling
- Contextual indicators relating results to student and school characteristics
- Trend indicators showing how results change over time.

Box 1

The points that distinguish PISA from the TIMSS study and justify a further survey are the following:

- It comprises a programme of surveys, not a single one-off event; the ability to provide comparable data from one survey to another is built in.
- It is concerned with the outcomes of the whole of basic education given to students during the years of compulsory education; therefore it assesses students at the end of this period of education, aged 15.
- What is assessed is not restricted to the common core of what is taught in participating countries but rather to a common view of what the education system should provide to prepare its future citizens for adult life and for life-long

learning.

- It assesses skills and knowledge in the context of extended units designed to reflect real-life contexts rather than performance in isolated test items.
- It will go beyond providing indicators of educational outcomes to supporting policy analyses.
- As the programme develops it will attempt to assess cross-curricular competencies and explore how self-regulated learning can be included.

These are ambitious aims and it will not be possible to evaluate the success in achieving some of them until the programme is in full operation, when there has been time to learn from trials and time for the development of suitably innovative instruments and procedures. The aims were not agreed without considerable debate, since the patterns of surveys set by TIMSS and its predecessors are deeply ingrained and not willingly relinquished, especially in countries with very 'traditional' school science curricula.

The first step was to interpret the intentions of the programme into statements of what knowledge and skills would be assessed. The three domains for testing were described in terms of 'literacy' to signal the concern with widely applicable skills and knowledge. For each domain a group called a Functional Expert Group (FEG) was set up to translate the notion of 'literacy' into operational terms that could be used for test item development and as a basis of reporting findings. The groups also oversee the subsequent creation of survey instruments.[3] So how are these features of the programme represented in the plans for assessment of scientific literacy?

The framework for assessing scientific literacy

It was both an advantage and a disadvantage that the term scientific literacy was already part of the vocabulary for discussing aims of science education; an advantage because there were plenty of starting points; a disadvantage because supporters of various perceptions expected their views to predominate. Of course various existing descriptions and models of scientific literacy were considered. In the end the SFEG members evolved their own. With the benefit of hindsight, the main points of the rationale were roughly as follows:

- In relation to the dual purposes of science education - to produce future scientists and to provide all students with understanding that will improve their future lives (Fensham, 1985) – there was no doubt that we were concerned with the latter.
- Whilst scientific literacy is something to be aimed for and developed throughout life, it is essential that it is begins in school.
- Scientific literacy is not to be equated with vocabulary; the term 'literacy' was interpreted metaphorically to mean general competence or being 'at ease' with scientific ways of understanding things.
- A key feature, is the use of evidence, which also includes knowledge of how evidence is collected in science, what makes some evidence more dependable than other, its short-comings and where it can and should be applied.

[3] Members of the science FEG are Wynne Harlen (Bristol University, UK, chair), Peter Fensham (Monash University, Australia), Raul Gagliardi (Geneva,Switzerland), Donghee Shin (Korean Institute of Curriculum and Evaluation), Svein Lie (University of Oslo, Norway), Manfred Prenzel (IPN, Kiel, Germany), Senta Raizen (NCISE, Washington, US), Elizabeth Stage (University of California, US).

Although the SFEG did not articulate the kinds of school experience that would lead to a firm foundation to scientific literacy what we had in mind is probably what is well expressed by Bybee and Ben-Zvi (1998):

> *Students should begin early with observing and describing the world around them and moving towards progressively more elaborated scientific explanations of phenomena. By the end of high school, students should be able to provide comprehensive explanations for the most obvious and compelling events that they experience, such as the seasons, day and night, disease, heredity and species variation, and dangers of hazardous substances.*
>
> *With respect to the methods of science, students should learn a disciplined way of asking questions, making investigations and constructing explanations of a scientific and technological nature. The latter can certainly be developed in a personal/societal context. Students should learn that scientific inquiry is a powerful, but not the only, route for progress in our world. Inquiry should not be taught in isolation but as a tool for finding answers to questions about the world in which students live.....*
>
> *Concerning the applications of science, students should confront contemporary and historical examples of how scientific knowledge is related to social advances and how society influences scientific advances. Once again, the focus should not be on learning about science and society for their own sakes, but to bring students to an appreciation of the complexity of the scientific/technological enterprise and to provide contexts and explanations for important science-related and technology-related societal challenges which they confront.*
>
> (Bybee and Ben-Zvi, 1998, p 491-2)

We were trying to capture the outcomes of such expences. The result was the PISA definition of scientific literacy as follows:

> *Scientific literacy is the capacity to use science knowledge to identify questions and to draw evidence-based conclusions in order to understand and help make decisions about the natural world and the changes made to it through human activity.*

The SFEG stated explicitly the view that scientific literacy is not polar. That is, that people cannot be categorised as being either scientifically literate or scientifically illiterate. Rather, there is a progression from less developed to more developed scientific literacy. However this progression was not conceived as the progress identified by Bybee (1993). In Bybee's framework, scientific and technological literacy proceeds from what is described as *Nominal* to *Functional* to *Conceptual and Procedural* to *Multidimensional*. The first two of these relate to the use of vocabulary that, as noted, did not reflect the SFEG's view of the meaning of scientific literacy. It will be possible to describe development with more certainty with the help of empirical data from trials and surveys. Meanwhile, the SFEG's view is that the student with less developed scientific literacy might be able to identify some of the evidence that is relevant to evaluating a claim or supporting argument or might be able to give a more complete evaluation in relation to simple and familiar situations. A more developed scientific literacy will show in giving more complete answers and being able to use knowledge and to evaluate claims in relation to evidence in less familiar and more complex situations.

There is a great deal packed into this definition and, so, it is important to look at it in a little more detail. The following notes on the meaning of phrases in the above definition are given in OECD (1999, 60-1).

- 'Science knowledge' is used to mean far more than knowledge of facts, names and terms. It includes understanding fundamental science concepts, the limitations of science knowledge and the nature of science as a human activity.
- The questions referred to in 'to identify questions' are those questions that can be answered by scientific enquiry, implying knowledge *about science* as well as about the scientific aspects of specific topics.
- Drawing 'evidence-based conclusions' means knowing and applying processes of selecting and evaluating information/data, whilst recognising that there is not often sufficient information to draw definite conclusions, thus making it necessary to speculate, cautiously and consciously, about the information that is available.
- 'Understand and help make decisions' indicates first, that to understand the natural world is valued as a goal in itself as well as being necessary for decision-making and, second, that scientific understanding can contribute to, but rarely determines, decision-making. Practical decisions are always set in situations having social, political or economic dimensions and science knowledge is used in the context of human values related to these dimensions. Where there is a consensus of values in a situation, the use of scientific evidence can be non-controversial. Where values differ, the selection and use of scientific evidence in decision making will be more controversial.
- The phrase 'the natural world' is used as a short-hand for the physical setting, the made world, living things and the relationships among them. Decisions about the natural world include science-linked decisions about self and family, community and global issues.
- 'Changes made through human activity' refer to planned and unplanned adaptations of the natural world for human purposes (simple and complex technologies) and their consequences.

Aspects of behaviour indicating scientific literacy

Although the framework, as published, identifies three aspects of units in which scientific literacy is assessed, this forces two rather separate aspects into a single one. Mainly for the sake of uniformity with the description of the other domains, of mathematical literacy and reading literacy. It is more straight-forward to identify four aspects;

- *science processes* that, because these are *scientific,* will involve some knowledge of science, although in the assessment this knowledge must not form the major barrier to success
- *science concepts* of which understanding is assessed by application to certain content
- *areas of application* in which the processes and concepts are applied
- *situations* within which the assessment units are presented (this aspect is often referred to in common usage as the 'context' or 'setting' of a unit).

Scientific processes

What the SFEG means by processes are mental and physical actions used in conceiving, obtaining, interpreting and using evidence or data to gain knowledge or understanding. Processes have to be used in relation to some subject matter; there is no meaning to a content-free process. They can be used in relation to a wide range of subject matter; they become *science processes* when the subject matter is drawn from scientific aspects of the world and the outcome of using them is to further scientific understanding.

Points that led to the selection of process to be included in the framework were:

- Whilst processes relating to *collecting* evidence through investigation in practice - planning and setting up experimental situations, taking measurements and making observations using appropriate instruments, etc. - are important aims of school science education, few will require these practical skills in life after school. What they will need is the *understanding of processes and concepts developed* through practical, hands-on enquiry.
- Scientific literacy, as identified above, gives higher priority to *using* scientific knowledge to 'draw evidence-based conclusions' than to the ability to collect evidence for oneself. It is being able to relate evidence or data to claims and conclusions about it that is seen as central to what all citizens need in order to make judgements about the aspects of their life that are influenced by science. It follows that every citizen needs:
    - to know when scientific knowledge is relevant, distinguishing between questions which science can and cannot answer.
    - to be able to judge when evidence is valid, both in terms of its relevance and how it has been collected.
    - to be able to relate evidence to conclusions based on it and to be able to weigh the evidence for and against particular courses of action that affect life at a personal, social or global level.

These arguments lead to giving priority to processes *about* science as compared with processes *within* science and this is how the list in Box 2 should be interpreted.

The science processes selected for inclusion in PISA
1. Recognising scientifically investigable questions
2. Identifying evidence needed in a scientific investigation.
3. Drawing or evaluating conclusions
4. Communicating valid conclusions
5. Demonstrating understanding of science concepts

Box 2

Process 1 can involve identifying the question or idea that was being (or could have been) tested in a given situation. It may also involve distinguishing questions that can be answered by scientific investigation from those that cannot, or more openly suggesting a question that it would be possible to investigate scientifically in a situation.

Process 2 involves identifying the information that is needed for a valid test of a given idea. For example, what things should be compared, what variables changed or controlled, what additional information is needed or what action should be taken so that relevant data can be collected.

Process 3 involves drawing, or critically evaluating, conclusions that have been drawn from given data. It also includes giving reasons for or against a given conclusion in terms of the data provided or identifying the assumptions made in reaching a conclusion.

Process 4 means communicating valid conclusions to a specified audience from available evidence or data and involves the production of an argument based on the situation or data given or on relevant additional information.

Process 5 means explaining relationships and possible causes of given changes, or making predictions as to the effect of given changes, or identifying the factors that influence a given outcome, using scientific ideas and/or information which have not been given.

All of these processes involve knowledge of science concepts. In the first four processes this knowledge is necessary but not sufficient since knowledge about collecting and using scientific evidence and data is essential. In the fifth process, the understanding of science concepts is the essential factor.

Scientific concepts

Major scientific themes (with examples of related concepts) for the assessment of scientific literacy

Structure and properties of matter	(thermal and electrical conductivity)
Atmospheric change	(radiation; transmission; pressure)
Chemical and physical changes	(states of matter; rates of reaction; decomposition)
Energy transformations	(energy conservation; energy degradation; photosynthesis)
Forces and movement	(balanced/unbalanced forces; velocity; acceleration; momentum)
Form and function	(cell; skeleton; adaptation)
Human biology	(health, hygiene, nutrition)
Physiological change	(hormones, electrolysis, neurons)
Biodiversity	(species; gene pool; evolution)
Genetic control	(dominance; inheritance)
Ecosystems	(food chains; sustainability)
The Earth and its place in the universe	(solar system; diurnal and seasonal changes)
Geological change	(continental drift; weathering)

Box 3

The concepts were selected according to the following criteria:

- relevance to everyday situations and to life in the 21st century,
- relevance to the situations identified as being ones in which scientific literacy should be demonstrated,
- use in combination with selected science processes. This would not be the case for concepts which are labels for a particular group of characteristics or events.

Given that the major assessment of science is planned to take place in the year 2006, it was emphasised that 'relevance' should be judged, as far as possible in relation to life throughout the next decade and beyond. Thus concepts chosen must be ones likely to remain important in science and public policy for a number of years. Clearly, a very large number of concepts would meet these criteria and the SFEG has not attempted to identify them comprehensively. Instead the group identified themes or groups of related concepts and some examples within each, as in Box 3. The scope of the survey will not allow, even when scientific literacy is the major theme, coverage of all concepts, even if these could be agreed.

Areas of application

Box 4 gives the list of those areas of application of science that raise issues that citizens of today and tomorrow need to understand and to make decisions about. The ten areas identified are grouped under three headings. These are helpful in ensuring that, there is a broad balance across the bank of items and units. The aim is, eventually to ensure a good representation of each of the issues under each heading. It is these applications that guide the selection of content for units and items within them. Box 4, therefore, indicates the areas of application in which the understanding of the concepts in Box 3 will be assessed.

Areas of application of science for the assessment of scientific literacy

Science in life and health
Health, disease and nutrition; Maintenance of and sustainable use of species; Interdependence of physical/biological systems

Science in Earth and environment
Pollution; Production and loss of soil; Weather and climate

Science in technology
Biotechnology; Use of materials and waste disposal; Use of energy;Transportation

Box 4

Situations

To complete the description of the framework before providing some examples to illustrate what it means in practice, the fourth aspect to be discussed is the situation in which the issues are presented. This is often called the *context* or *setting* of the units, but here the word 'situation' is used to avoid confusion with other uses of these words. The particular situations are known to influence performance and so it is important to decide and control the range of situations intended for the assessment units. It is not intended to report performance in relation to particular situations. Nevertheless the situations in which the units are set need to be identified in order to ensure a spread of units across those felt to be important.

In selecting situations, it is important to keep in mind that the purpose of the assessment in science is to assess the ability of students to apply in their lives as citizens the skills and knowledge that they have acquired by the end of the compulsory years of schooling. For this reason we have excluded the 'school learning context' from the items and focused on real life situations.

Real world situations involve problems which can affect us, for example
- as individuals (e.g. about food and energy use),
- as members of a local community (e.g. about treatment of the water supply or siting of a power station),
- as world citizens (e.g. global warming, diminution of biodiversity).

All of these should be represented in the range of assessment units used in the surveys. A further type of situation, appropriate to some topics, is the historical one, where understanding of the advances in scientific knowledge can be assessed.

The knowledge that is applied in these situations is knowledge most likely to have been gained in the science curriculum, although some may be gained from other subjects and from non-school sources. However, in order to find out if this learning has gone beyond knowledge of isolated facts and is serving the development of scientific literacy, PISA is assessing the application of that knowledge in items reflecting real-life situations.

The situation is the aspect which is most difficult to select in the construction of units which aim to be relevant to students' interests and lives in all countries.

Sensitivity to cultural differences has a high priority in unit development and selection, not only for the sake of validity of the assessment, but to respect the different values and traditions in participating countries. Feedback from field trials is important in ensuring that situations chosen for the survey units are relevant and appropriate across the different countries, whilst involving the combination of scientific knowledge and the use of science processes. PISA has set up a Cultural Review Panel to look for, and hopefully to eliminate, cultural bias in question in all three domains.

Assessment units

The assessment units in PISA comprise a series of questions about some stimulus material that presents the situation. Between them the items within a unit may assess more than one process and science concept, whilst each item assesses one of the science processes.

One reason for this structure is to make units as realistic as possible and to reflect in them to some extent the complexity of real-life situations. Another reason relates to the efficiency of use of testing time, cutting down on the time required for a student to 'get into' the subject matter of the situation, by having fewer situations in which several questions can be posed rather than separate questions about a larger number of different situations. However, it inevitably means that there is more reading to do at the beginning of the unit than is the case for conventional stand-alone items that require recall only.

Points taken into account in the design of units include:

- the necessity to make each scored point independent of others within the unit,
- the need to minimise bias which may be due to the situation (all the more important when there are fewer situations used).

All the science assessment units, at least in the 2000 and 2003 surveys, are in written form. The following are examples of items taken from longer units.

Examples of test items assessing science processes

Examples of the items being considered for assessing some of these processes will help in conveying their operational meaning. The first two processes are assessed in two questions within a unit entitled 'Stop that Germ!' The students are asked to read a short text which includes this extract about the history of immunisation. The extract on which two example questions are based is shown in

Box 5.

Example One

As early as the 11th century, Chinese doctors were manipulating the immune system. By blowing pulverised scabs from a smallpox victim into their patients' nostrils, they could often induce a mild case of the disease that prevented a more severe onslaught. In the 1700's, people rubbed their skins with dried scabs to protect themselves from the disease. These primitive practices were introduced into England and the American colonies. In 1771 and 1772, during a small pox epidemic, a Boston doctor named Zabdiel Boylston scratched the skin on his six-year-old son and 285 other people and rubbed pus from small pox scabs into the wounds. All but six of his patients survived.

Box 5

Example item 1: What idea might Zabdiel Boylston have been testing?

Item 1 requires a constructed response which is scored as 2, 1 or 0 according to the amount of relevant detail given in the answer. (A score of 2 would be given to an idea along the lines of 'breaking the skin and applying pus directly into the blood stream will increase the chances of developing immunity against small pox'). The item assesses Process 1 - *Recognising scientifically investigable questions* using knowledge of *human biology* applied in the area of *science in life and health,* set in an *historical* situation.

Example item 2: Give two other pieces of information that you would need to decide how successful Boylston's approach was.

This item is also scored as 2, 1 or 0 according to whether one or both pieces of information are mentioned (the rate of survival without Boylston's treatment and whether his patients were exposed to small pox apart from within the treatment). It assesses the Process 2 - *Identifying evidence needed in a scientific investigation* using knowledge of *human biology* and applied in the area of *science in life and health,* set in an *historical* situation.

The following four items are part of a unit for which the stimulus material is a passage about Peter Cairney who works for the Australian Road Research Board.

Example Two

.....*Another way that Peter gathers information is by the use of a TV camera on a 13 metre pole to film the traffic on a narrow road. The pictures tell the researchers such things as how fast the traffic is going, how far apart the cars travel, and what part of the road the traffic uses. Then after a time lane lines are painted on the road. The researchers can then use the TV camera to see whether the traffic is different in any way after the lines are there. Does the traffic now go faster or slower? Are the cars closer together or further apart than before? Do the motorists drive closer to the edge of the road or closer to the centre now that the lines are there? When Peter knows these things he can give advice about whether or not to paint lines on narrow roads.*

Box 6

Example item 3: If Peter wants to be sure that he is giving good advice, he might collect some other information as well as filming the narrow road. Which of these things would help him to be more sure about the advice he could give about the effect of painting lines on narrow roads?

a) Doing the same on other narrow roads. Yes/No
b) Doing the same on wide roads. Yes/No
c) Checking the accident rates before and after painting the lines. Yes/No
d) Checking the number of cars using the road before
and after painting the lines. Yes/No

Scoring: Yes of a) and c) No to b) and d) (score 2)
Yes to a) No to b), c) and d) (score 1)
Any other combination (score 0)

This item assesses Process 2 - *Identifying evidence needed in a scientific investigation,* using knowledge of *forces and movement* in the area of *science in technology,* set in a *community* situation.

Example item 4: Suppose that on one stretch of narrow road Peter finds that after the lane lines are painted the traffic changes as in this table:

Speed	Traffic moves more quickly
Position	Traffic keeps nearer edges of road
Distance apart	No change

On the basis of these results it was decided that lane lines should be painted on all narrow roads.

Do you think this was the best decision?
Give your reasons for agreeing or disagreeing.
Agree Disagree

Reason:...

This item assesses Process 3 - *Drawing or evaluating conclusions*, using knowledge of *forces and movement* in the area of *science in technology.* No credit is given for agreeing or disagreeing but for the reason which is consistent with either and the given information (e.g. Agree because there is less chance of collisions if the traffic is keeping near the edges of the road even if it is moving faster; if it is moving faster there is less incentive to overtake. Or, disagree because if the traffic is moving faster and keeping the same distance apart this may mean that they don't have enough room to stop in an emergency.)

Example item 5: Drivers are advised to leave more space between their vehicle and the one in front when they are travelling more quickly than when they are travelling more slowly because faster cars take longer to stop. Explain why a faster car takes longer to stop than a slower one.

Reason:...

This assesses Process 5 - *Understanding of science concepts* about *forces and movement* in the area *of science in technology.* It requires a constructed response, scored 2, 1, 0 according to whether one or both of the significant points are mentioned, Reference to: a) greater momentum of a vehicle when it is moving more quickly and the consequent need for more force to stop it b) at a higher speed a vehicle will move further whilst slowing down than a slower vehicle in the same time.

To answer all of these questions, the student is required to use knowledge that would be gained from the science curriculum and apply it in a novel situation. Where assessment of conceptual understanding is not the main purpose of the item the knowledge required is not the main challenge (or hurdle) and success should depend on ability in the particular process required. Where assessment of conceptual understanding is the main aim, as in Example Item 5 the process is one of demonstrating this understanding.

These items have been chosen to illustrate the framework and not the distribution of items across the various components. This distribution has to meet the requirements of the Board of Participating Countries, which is that 40 – 50% of the items will assess Process 5 (demonstrating understanding of scientific concepts), whilst being spread evenly across the three areas of application (Box 4). The Board has also placed a restriction of 35% on the proportion of items that require an extended constructed response, as opposed to a selected response or an unambiguous short answer.

In conclusion

International surveys take place in a political and not just an educational and social context. At all stages, compromises have to be struck among the claims and priorities of different countries with regard to what is assessed and how it is assessed. What the OECD/PISA study assesses, as this paper has indicated, does not related directly to what is actually being taught in most countries and in many not even to what is included in the goals of education. This raises serious questions of 'fairness' in relation to making comparisons between countries, some of whom might embrace scientific literacy (in some form or other) as a goal whilst others do not. There is the further issue of the form of test items that are used in assessing application of scientific ideas in real-life situations, and the items most often used in tests – and familiar from TIMSS.

Moreover the unfamiliarity of the items, raises the question of motivation for students to attempt questions of a very different form and content than they have met before. Is it 'fair', or even moral, to place students in situations where they feel unprepared to cope with the tasks put in front of them? Of course information will be collected by questionnaire directly from students and from schools about exposure to relevant learning experiences, but the impact on validity and reliability remains and is unknown.

Whilst the participating countries have 'signed up' to the framework and item types, PISA raises more issues even than the IEA surveys about the role of international surveys and of international organisations such as the OECD. In particular, what is assessed by the OECD/PISA derives from a view outside the countries of what is important in science education, rather than being based on what is currently being taught. This is a big step in assessment leading the curriculum.

References

Bybee, R.W. (1993). *Reforming science education: social perspectives and personal reflections.* New York: Teachers College Press.

Bybee, R. W. & Ben-Zvi, N. (1998) Science curriculum: transforming goals to practices. In B.J.Fraser & K.G. Tobin (Eds.), *International Handbook of Science Education* (pp. 487–498). Dodrecht, The Netherlands: Kluwer Academic Publishers.

Fensham, P.J. (1985). Science for all. *Journal of Curriculum Studies* 17, 415-435.

OECD (1999). *Measuring Student Knowledge and Skills: A New Framework for Assessment.* Paris: OECD.

Scientific Literacy: From Theory to Practice

Wolfgang Gräber and Peter Nentwig
Institute for Science Education (IPN), Kiel, Germany
Hans-Jürgen Becker, University of Paderborn, Germany
Elke Sumfleth and Anja Pitton, University of Essen, Germany
Kai Wollweber, Integrierte Gesamtschule Eckernförde, Germany
Doris Jorde, University of Oslo, Norway

Abstract

Gerhard Schaefer (1998), in his keynote address at the second International IPN-Symposium on Scientific Literacy, thought about educational aims for citizens in a changing world: "Taking into account the growing complexity of our world, caused by the opening of national borders and by almost infinite electronic communication and the increasing speed of global ecological, economic and political changes, number one of the educational challenges of the next century seems to be: high flexibility, both in storing and using knowledge and in international communication." Thus, he argues for "life-competence" as a goal of school education. This ESERA symposium focussed on the part the sciences have to play to contribute to this goal. Three examples from educational research and classroom teaching illustrate the theoretical references of the second International Symposium on Scientific Literacy.

The Background

There is worldwide consensus that our societies, regardless of any potential cultural differences, need scientifically literate citizens. The US National Research Council (1996), in its Science Education Standards, states for example: "All of us have a stake, as individuals and as a society, in scientific literacy. An understanding of science makes it possible for everyone to share in the richness and excitement of comprehending the natural world. Scientific literacy enables people to use scientific principles and processes in making personal decisions and to participate in discussions of scientific issues that affect society. A sound grounding in science strengthens many of the skills that people use every day, like solving problems creatively, thinking critically, working cooperatively in teams, using technology effectively and valuing life-long learning. In addition, the economic productivity of our society is tightly linked to the scientific and technological skills of our work force."

The BLK (1997), a governmental commission for the coordination of research activities and educational matters in Germany, in its expertise for the preparation of a just recently installed program for the enhancement of science education claims: "Biology, chemistry and physics provide basic scientific concepts for the interpretation of nature, humanity and a world that is formed by science and technology. The various epistemic methods of the sciences serve as basic tools for understanding oneself as part of the world."

Millar (1996) summarizes the legitimization for science teaching in schools:

H. Behrendt et al. (eds.), Research in Science Education – Past, Present, and Future, 61–70.

- because it is practically useful in everyday contexts;
- because it is needed to participate in debating controversial issues;
- because science is a major achievement of our culture.

If one accepts that scientific literacy is a must, then what could the term mean? There are bookshelves full of definitions. International concern about what constitutes appropriate and adequate science education for citizens has received ever greater attention in recent years (Shamos, 1996; Bybee, 1997; Gräber & Bolte, 1997).

Rodger Bybee (1997) has proposed a sequence of steps for all students to achieve higher levels of scientific literacy. He distinguishes nominal, functional, conceptual, procedural, multidimensional scientific and technological literacy and describes scientific literacy as a process of life-long learning in different science domains. Koballa, Kemp and Evans (1997) also combine levels and domains, but add "values" as a third dimension to form a three-dimensional landscape representation of the scientific literacy spectrum. An interesting definition was recently added by the Science Functional Expert Group of the OECD PISA project (1999), who propose: "By scientific literacy, we mean being able to combine science knowledge with the ability to draw evidence based conclusions in order to understand and help make decisions about the natural world and the changes made to it through human activity."

Morris Shamos (1995) has taken all this into account and has suggested that the science education community has deceived itself into thinking that a definition of scientific literacy, including both broad and in-depth content knowledge as well as process competence is possible. He proposes a more realistic definition what challenges science educators to help students become competent consumers and to trust the real issues to science experts.

Summarizing the various proposed definitions and models and taking into account the skepticism of Shamos and others (Millar, 1996; Hamill, 1997), we would like to suggest a competency based model of scientific literacy, going back to Schaefer's request for "life-competence" as a goal for education in school. In this approach the question is not the causal "why" do we teach science to children but rather the final "what for" and, consequently, the answer is not "because our societies need a scientifically literate workforce" (or other justifications, as indicated above). The answer is "for the individual to cope with our complex world." Competencies are needed for this task and some specific competencies can be acquired in the science domain better than in others.

Subject competence includes declarative and conceptual knowledge: a continuum of science knowledge and understanding throughout the various domains of science. When combined, depth and breadth provide an individual profile of science knowledge and understanding

Epistemological competence includes insight into (an overview of) the systematic approach of science as one way of seeing the world, as compared with technology, the fine arts, religion, etc.

Learning competence includes the ability to use different learning strategies and ways of constructing scientific knowledge.

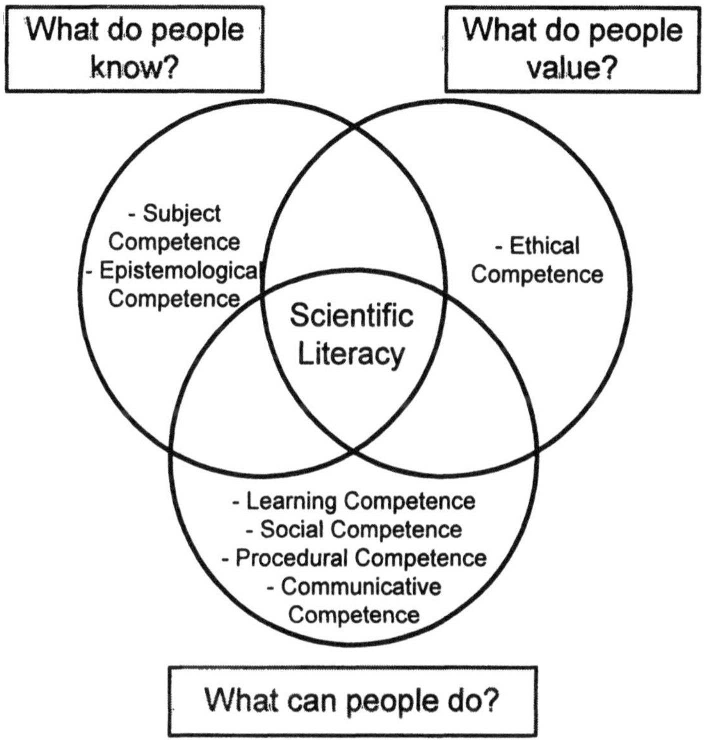

Social competence includes the ability to cooperate in teams in order to collect, produce, process or interpret – in short, to make use of - scientific information.

Procedural competence includes abilities to observe, experiment, evaluate; ability to make and interpret graphic representations, to use statistical and mathematical skills, to investigate literature. It also includes the ability to use thought models, to analyze critically, to generate and test hypotheses.

Communicative competence includes competence in using and understanding scientific language, reporting, reading and arguing scientific information.

Ethical competence includes knowledge of norms, an understanding of the relativity of norms in time and location and the ability to reflect norms and develop value hierarchies.

Science teaching traditionally concentrates on the knowledge aspect, adding perhaps a few of the procedural skills, but usually neglecting the other competencies. The view proposed here may help to reconsider the balance between the various competencies and to reflect what specific contribution science education can make.

An analysis of science-classroom videos at the Second International IPN-Symposium on Scientific Literacy shows that science teaching can be described in three dimensions. Each dimension can be characterized by two extremes on a scale:

teacher centered - student centered

It is either the teacher governing the classroom activities, steering the students' learning processes and dominating the communication process or the students taking responsibility for their own progress, initiating their learning processes in a more or less autonomous way.

teaching facts - teaching processes

The teaching/learning activities aim either at science facts, laws and formulae or at the acquisition of problem solving strategies and skills of processing information and interpreting data.

discipline oriented - daily-life oriented

The aim of the lessons can be either to delineate the structure of a scientific discipline and to reproduce research findings on a reduced level or to provide means to understand daily-life phenomena, including their social, technological and economic implications.

Traditionally, science teaching leans towards the left side of each of these scales. Teachers tend to dominate the teaching/learning process with fact oriented lesson contents and the aim to reproduce (at least part of) the structure of their scientific discipline in the heads of their students. The results of these efforts have been disillusioning world wide.

A look at the list of competencies that, each with their own weight, constitute "scientific literacy", makes it evident that such science teaching must fall short of reaching the goal. We are not suggesting now that all engines should be turned on reverse and that the opposite approach - student autonomy, acquisition of process skills, daily-life orientation - is the recipe for successful science teaching. Obviously, often enough, students do not take responsibility for their own learning processes, facts are necessary for applying skills and daily-life phenomena are often too complex to understand with regular students' limited knowledge.

However, to reach some extent of scientific literacy for all students, more of these ingredients must be introduced to science classrooms and in order to do that, teachers must be able to act expertly on both sides of the scale in all three dimensions.

If we take scientific literacy seriously as a goal of science education and if we want science classes to contribute to the general education of emancipated citizens, we must re-balance the orientation of our teaching. Careful guidance of children towards self-regulated learning must have priority. This can be justified with terms such as self-determination, self-responsibility and self-activity.

Self-determination, as a basis of all emancipative efforts, makes the learner independent of a teaching person. *Self-responsibility,* for ones own learning process, is the prerequisite for life-long learning, which depends on *self-activity*, if it is meant to be fruitful and lasting.

Example 1: The "daily-life" approach to scientific literacy

What do a judge, a musician, a businessman remember when they think about their chemistry classes? Redox processes, delocalized electrons, inductive effects? Hardly! If they have any memory at all, it will probably not be without a negative aftertaste. They might remember that water can be hard or soft, how glass can be made and how steel is tempered. We would hope so, but we cannot be certain. Such topics have played only a marginal role in the discipline oriented chemistry teaching of the past twenty years. The "real issue" was the elaboration of the scientific discipline. Applications, technological contexts, societal relevance, environmental problems related to the chemical content matter were, if at all, mentioned as a byline. These, in reality, are the issues that will concern our students in their life after school. For such issues, it is Science Education's responsibility to provide necessary orientation and decision-making aids. Freise, for instance, describes a daily-life teaching concept as a chance to relate science and technology to the understanding and the lives of the students in our society and thus as a possibility for "understanding life in a world made scientific".

Only few students benefit from the current theory-oriented chemistry teaching. The competency to act appropriately in daily life is only acquired if facts, theories and phenomena are taught in daily life contexts. In life after school, knowledge of chemistry is necessary to understand and evaluate technical aspects (including their effects on society), environmental problems and questions about health, the energy discussion, law making procedures, substances at home and in everyday life. Knowledge about materials, machinery, production processes, or social conditions at the workplace will be important in the students' future crafts and businesses. Students need to be prepared for independent and critical behaviour in such contexts. Science teaching has to have in mind the students' activities in their future adult life. Chemistry and daily-life phenomena are then not viewed as opposites but as a unit.

We do not deny the problems caused by this orientation. Daily-life situations are most complex and of diverse structure. They are perceived differently by different individuals. The question is still open as to, how the structure of a discipline can be communicated with this approach. Apparently incompatible aspects have to be brought together in the discussion about daily-life approaches. Its high motivational power is an asset, its complexity is a problem. Both arguments are relevant from a psychological point of view. Seldom are themes like washing space powder, soda, steel, glass, polyvinylchloride or glues obligatory parts of a syllabus. Usually it is left to the methodological and experimental skill of the teacher to liven up the discipline-oriented course with daily-life phenomena. Science education has offered help for that task, however:

In simulation games the development and solution of complex problems is acted out in a reality-related way. Among simulations, role-playing is probably the best way to liven up science lessons. Role playing can easily and quickly be incorporated into the lesson, whereby the students' interests can be taken into consideration. It is also a chance to train students' competence of judgement, even though role playing is more a way of rethinking an issue, a game set down by role cards. Teaching experience shows that role playing in chemistry lessons promotes motivation, activates participation and can be effectively built into lessons. Students as actors (role players) can experience in which daily-life situations scientific (chemistry) knowledge can be important. Systematic fundamental knowledge, that can be imparted in various methodological ways, can be applied, practiced and expanded. For younger children the teacher prepares the various roles for the play. Older students can be involved in this as well. Points of view and arguments are written on role cards for each role.

We do not have much feedback yet about role playing in daily teaching practice. We, therefore, feel that it is a most important task for teacher training to provide adequate methodological foundations in connection with the "change in legitimation".

Example 2: Instructional discourses for the acquisition of definitional knowledge and thinking skills

Summarising the results of research in science education, there is general agreement that it is important to improve the quality not the quantity of reproducable elements of knowledge. With respect to learning, Duit (1991, p. 68) summarises major trends: „Learning is not seen as a process of simply storing pieces of knowledge provided, for instance, by the teacher. On the contrary, it is seen as a process of active construction of knowledge on the part of the learners themselves on the basis of their already existing conceptions.“ Additionally Brody (1994, p. 423) formulates: „Children have existing conceptions that influence their understanding of the world and their acquisition of new knowledge. Learning cannot be considered as simply the acquisition of a set of correct responses. Alternatively, meaningful learning is considered as aprocess of conceptual change. Learning is a process of active knowledge construction based on learners' previous knowledge. Problem solving requires that the learner pick up his previous knowledge to integrate new elements of knowledge in existing individual cognitive structures."

Often learning is related to verbal interactions between two or more individuals, as for example learning at school. Does a student at school get this chance to construct his knowledge actively? In many cases the students remain passive, listening. They comply with the request of the teacher to answer and adopt the teacher's words. Roth (1994, p. 437) says very pessimistically: „In contrast, it is a common practice in today's schools that students sit, listen and respond to teacher-investigated question-answer-evaluation sequences.“ Teachers that they are supporting the students in problem solving, but actually, they are focusing on imparting and examining definitional knowledge without realising this discrepancy. Teaching should mean helping the students to construct knowledge actively, to activate previous knowledge and to relate new structures to known ones (Kluwe & Spada, 1981; Sumfleth, 1988; Stachelscheid, 1990). Teachers must influence students' processes of problem understanding by structuring the objective task environment as clearly as possible. They have to try to recognise the individual thinking and acting of the students and to judge them with regard to a meaningful use of prior knowledge (Fleer, 1992).

Communication is mainly verbal. In classes verbal acting is adapted to the institutional framework 'school'. In order to identify students' difficulties, teachers have to take students' statements seriously. An instruction based on students' preconceptions is possible only when teacher and students exchange their individual arguments and compare them in relation to their meaning in explaining the facts discussed (Gramm, 1992). During lessons, the teacher cannot pay attention to each student individually. Communication between teacher and certain students, as well as among students has to provide impulses to the individual learning processes. The teacher must refer to students' audible and visible signs and he must make sure that he interprets the signs in the way they were meant (Dierks & Weninger, 1988).

According to Geißner (1994) the communication structure in lessons is obvious:. The teacher is the leader. The students are leader-oriented. Thus, the course of the discussion is asymmetrical. In contrast, during a symmetrical discussion, the „leader" participates in the discussion. The group determines the velocity of learning and searches for a solution together. The students are partner-oriented and content-oriented. These patterns of communication are determined by the type of teacher questions and the aims pursued by the teacher. Therefore students' problem solving ability depends on the type of teacher instruction.

The results of lesson-observation prove that only the minority of students show active problem solving. More often, the students accumulate reproducible pieces of knowledge. The bulk of unstructured information inhibits students' problem solving. The students do not learn to use their previous knowledge in class but rather depend on reproduction of definitions. This is sufficient to reply correctly to a teacher question or to pass a written examination. Students store reproducible but meaningless words, that can seldom be integrated into a context. This system hardly allows the construction of knowledge structures. Misunderstanding remains unidentified because teacher and student believe they are speaking the same professional language – which is not the case (Sumfleth & Pitton, 1997).

Example 3: Learning scientific contents in cooperative groups

It is well known that science is predominantly taught in a teacher centered way, focussed on learning science facts according to structured disciplines. This has led to students' decreasing interest, development of negative attitudes toward the sciences and particularly low learning outcomes. These are the reasons that have convinced us to, not only focus on selecting and structuring suitable contents, but also to consider forms of organization and social aspects of learning. One appropriate method seems to be the cooperative learning model, which not only promotes the acquisition of social skills, but also helps to enhance subject learning processes and thinking skills. Cooperative learning causes students to communicate their thinking in an understandable way, to debate, to take different perspectives and to handle discrepant opinions and judgements. This should help to promote students' scientific literacy through facilitating mainly subject competency, communication competency and reasoning competency.

"Cooperation is working together to accomplish shared goals and cooperative learning is the instructional use of small groups so that students work together to maximise their own and each other's learning. ...Thus, a student seeks an outcome that is beneficial to himself or herself *and* beneficial to all other group members.

In order to be productive, cooperative learning groups must be structured to include the essential elements of *positive interdependence*, in which each member can succeed only if all members succeed, *face-to-face promotive interaction*, during which students assist and support each other's efforts to achieve, *individual accountability* to ensure that all members do their fair share of the work, the *interpersonal and small group skills* required to work cooperatively with each others and *group processing* (in which groups reflect on how well they are working together and how their effectiveness as a group may be improved)." (Johnson & Johnson 1992)

The aim of the lessons presented in the symposium is to show how students can learn the chemical content of "aminoacids and proteins" in cooperative groups. The chosen method of teaching is the project oriented approach according to Bastian and Gudjons (1990). The content of the lessons, the arrangement of the working groups and a set of rules is given by the teacher. Students are free to choose particular topics and methods. They work autonomously over several lessons, they are responsible for the organisation of the learning process, and they have to prepare a product (e.g. a poster, demonstration experiment, report, video, etc.), thus demonstrating the results of their efforts. The unit is completed with a test of basic knowledge and understanding of the theme.

The observed group is a basic course in chemistry, grade 12 from an integrated comprehensive school, consisting of 3 male and 15 female students.

From grade 5 to 10 the students had been taught integrated science, from grade 11 upwards they had to choose two of the basic science disciplines biology, physics or chemistry. All students on this chemistry course had decided to take biology as their second science discipline. As they had all taken biology at grade 11, the content of aminoacids and proteins can be taught in liasion with biology. At level 11 they studied the basics of aminoacids and proteins in biology: Formulas of the aminoacids and the combination into peptides and proteins were taught. The secondary and tertiary structures of proteins were discussed. Experiments and the theoretical framework of enzymes were the main aspects in these lessons. Chemical properties regarding functional groups were left out at that time.

The grade 12 lessons in chemistry were now supposed to focus on the chemical properties and reactions of aminoacids and proteins and to leave out the biological issues. Nevertheless last year's biology knowledge was to be used as a starting point.

The topic of the parallel courses in biology is genetics and the biosynthesis of proteins is a main topic within the molecular genetics.

Results

In the final round-table discussion the students were asked about their motivation during the sequence, the contents and the teaching method. The majority agreed that:

- the theme was interesting;
- the teaching method offered the opportunity to choose content, learning style and speed autonomously;
- without cooperation in the group the unassisted (self-regulated) development of the theme would have been asking too much of them;
- in future, they would find it easier to work on a theme independently;
- they were uncertain what the teacher expected of them, particularly with respect to the anticipated written test.

The teachers' observation of the students during the sequence and the assessment of students' performance (mind-maps, records, oral reports, written test) show that:

- students were quite insecure about their own procedures;
- in the beginning they tended to stick to the textbook and organized their activities according to the proposed order in the textbook with little reflection;
- after overcoming this initial insecurity they soon arrived at well planned procedures;
- they elaborated the theme in various directions, different groups developing different strategies;
- results were presented on a rather high intellectual level;

- the groupwork helped the higher achieving students to adopt leading roles and to show better results than in traditional teaching modes, whereas lower achieving students hardly benefited at all from this method.

References

Bastian, J. & Gudjons, H. (1990). *Das Projektbuch II.* Hamburg: Bergmann + Helbig

Baumert, J. et al. (1998). *TIMSS/III. Schülerleistungen in Mathematik und den Naturwissenschaften am Ende der Sekundarstufe II im internationalen Vergleich.* Berlin: MPI für Bildungsforschung. Studien und Berichte 64.

Becker, H.-J. et al. (1992). *Fachdidaktik Chemie (2. Auflage).* Köln: Aulis Verlag Deubner.

Becker, H.-J. (1990). *Verbraucherdialoge im Chemieunterricht. Waschmittel-Reinigungsmittel-Pflegemittel.* Berlin, speziell, 135-153

Brody, M. .J. (1994). Student science knowledge related to ecological crises. *International Journal of Science Education* 14, (4), 421-435.

Bund-Länder Kommission für Bildungsplanung und Forschungsförderung (BLK) (1997). *Gutachten zur Vorbereitung des Programms "Steigerung der Effizienz des mathematisch-naturwissenschaftlichen Unterrichts".* Materialien zur Bildungsplanung und zur Forschungsförderung, Heft 60. Bonn: BLK.

Bybee, R. (1997). *Achieving scientific literacy: From purposes to practices.* Portsmouth , NH: Heinemann Publishing.

Dierks, W. & Weninger, J. (1988). *Stoffe und Stoffumbildungen. IPN-Lehrgang, Teil 3: Auf dem Weg zu einer Chemie der Aggregate: Kompositheneden und Henadenaggregate. Lehrgangsbeschreibung, 1. Auflage.* Stuttgart: Klett.

Duit, R. (1991). Students' conceptual frameworks: Consequences for learning science. In S.M. Glynn, R.H. Yeany & B. Britton (Eds.), *The psychology of learning science* (pp. 65-85). Hillsdale, New Jersey: Lawrence Erlbaum Associates, Publishers.

Erdmann, T., Gräber, W. & Nentwig P. (1999). *Second International IPN-Symposium on Scientific Literacy (CD-ROM).* Kiel: IPN.

Fleer, M. (1992). Identifying teacher-child interaction which scaffolds scientific thinking in young children. *Science Education* 76(4), 373-397.

Gräber, W. (1998). Schooling for life-long attention to chemistry issues: The role of interest and achievement. In L. Hoffmann, A. Krapp, K.A. Renninger & J.Baumert (Eds.), *Interest and learning: Proceedings of the Seeon-conference on interest and gender* (pp. 290-300). Kiel: IPN-Schriftenreihe.

Geißner, H. (1994). Klären und streiten. In M. Beck, *Unterrichtsgespräche: Zwischen Lehrerdominanz und Schülerbeteiligung; eine sprachwissenschaftliche Untersuchung didaktischer Ansätze zur Unterrichtskommunikation.* St. Ingbert: Röhrig.

Gräber, W. & Bolte, C. (Eds.) (1997). *Scientific literacy – An international symposium.* Kiel: IPN.

Gramm, A. (1992). Kommunikation und Chemieunterricht. *Der mathematische und naturwissenschaftliche Unterricht* 45, 362-366.

Häußler, P. (1987). Measuring students' interest in physics — design and results of a cross-sectional study in the Federal Republic of Germany. *International Journal of Science Education* 79-92.

Hamill, A. (1997). Science education for the new millennium. *School Science Review* 79, 9, 21-26.

Hoffmann, L. & Häußler, P. (1995, April). Assessment of students' interest in physics as a means to improve instruction. Paper presented at the Annual NARST Meeting in San Francisco.

Johnson, D. W. & Johnson, R. (1992). Encouraging thinking through constructive controversy. In N. Davidson & T. Worsham (Eds.), *Enhancing thinking through cooperative learning* (120-137). New York: Teachers College Press.

Kluwe, R.H. & Spada, H. (1981). Wissen und seine Veränderung: Einige psychologische Beschreibungsansätze. In K. Foppa. & R. Groner (Eds.), *Kognitive Strukturen und ihre Entwicklung* (pp. 284-327). Bern, Stuttgart, Wien: Huber.

Koballa, T., Kemp, A. & Evans, R. (1997). The spectrum of scientific literacy. *The Science Teacher* 69 (10), 27 – 31.

Millar, R. (1996). Towards a curriculum for public understanding. *School Science Review* 77, 3, 7-18.

Miller, J. D. (1997). Civic scientific literacy in the United States: A development analysis from Middle-School through adulthood. In W. Gräber & C. Bolte (Eds.), *Scientific Literacy – An International Symposium* (pp. 121-142). Kiel: IPN.

National Research Council (1996). National science education standards. Washington, DC: National Academy Press.

OECD-PISA, (1998). Framework for assessing scientific literacy. Unpublished paper from the Science Functional Expert Group.

Roth, W.-M. (1994). Science discourse through collaborative concept mapping: new perspectives for the teacher. *International Journal of Science Education* 16 (4), 437-455.

Schaefer, G. (1998). Scientific literacy for general competences – Teaching "subject-transcending subjects" at School. Keynote address at the *Second International IPN-Symposium*, Strande, Germany, www.ipn.uni-kiel.de/projektea4_5/sympos2/allgem2.htm.

Shamos, M. (1995). *The myth of scientific literacy.* New Brunswick, NJ: Rutgers University Press.

Sjöberg, S. (1997). Scientific literacy and school science. In S. Sjöberg & E. Kallerud (Eds.), *Science, Technology and Citizenship* (pp. 9-28). Norway, Oslo, NIFU – Norsk institutt for studier av forskning og utdanning.

Stachelscheid, K. (1990). *Problemlösender Chemieunterricht in der Sekundarstufe I – Gymnasium. Eine empirische Untersuchung zur Förderung von verknüpftem Wissen.* Essen: Westarp Wissenschaften, Naturwissenschaften und Unterricht, Bd. 7.

Sumfleth, E. (1988). *Lehr- und Lernprozesse im Chemieunterricht.* Frankfurt am Main, Bern, New York, Paris: P. Lang.

Sumfleth, E. & Pitton, A. (1997). Learning chemistry today - Examples related to different groups of learners. In W. Gräber & C. Bolte (Eds.), *Naturwissenschaft und Allgemeinbildung / Scientific Literacy* (pp. 349-376). IPN-Schriftenreihe, Kiel.

Making Formative Use of a National Summative Assessment Regime

Terry J. Russell and Linda McGuigan
Centre for Research in Primary Science and Technology, University of Liverpool, United Kingdom

Abstract

While the impact of *formative* assessment practices on learning outcomes is receiving increasing attention, the dominant function of the statutory assessment system in England and Wales is *summative*. As authors of the statutory end of Key Stage 2 tests between 1995 and 1999, we are aware of a volume of test performance data generated in the course of the summative regime. Summative performance data are available on the cohort of 600,000 pupils, at age 11, assessed annually. We have further illuminated these data by an annual qualitative re-marking of a sub-sample stratified by three overall achievement levels. We suggest that many test items are analogous to concept probes within the constructivist paradigm. When pupils' assessed understanding can be mapped onto lines of progression, assessment can have a powerful formative capability in informing classroom teaching and learning practices. The characteristics of test items which may combine summative and formative utility are discussed.

Introduction

The education system in England and Wales has invested huge resources in measuring educational outcomes using summative assessment. A recent review (Black & Wiliam, 1998) pointed to the educational gains associated with formative assessment practices. We suggest that when certain conditions in test construction are met, summative instruments can play a dual role, summative and formative, in informing education systems. The adoption of these principles link our work as constructivist researchers with our activities in the test development process. Our discussion is illustrated with test material and pupil responses relating to the end of Key Stage 2 standard tests, administered annually to a cohort of about 600,000 pupils at age 11, at the end of the primary phase of education in England and Wales.

There are two particular conditions, which we believe enhance the possibility of bridging opportunities between the summative and formative functions of the tests:

- the tests should incorporate 'ecologically valid' items, illustrating desirable practices in primary science education;
- test items should be capable of being located within a framework of conceptual progression.

The summative measurement remit, within the national testing regime requires a variety of test presentation and response modes. The first of our two principles cannot, therefore, apply invariably to all test items. The second can apply more extensively, in theory, but in practice requires the backing of concept development research. Because of space constraints, the second condition will be discussed more fully than the first.

H. Behrendt et al. (eds.), Research in Science Education – Past, Present, and Future, 71–76.

The ecological validity of the test items

'Ecologically valid' test items sample knowledge set in contexts which would be recognisably 'good practice in primary science.' A teacher looking at such an item might react by deciding that the scenario presented is a good and effective way to approach the particular part of the Programme of Study concerned. Not all tests attempt this or regard it as helpful or necessary. For example, our test items do not resemble those used in the TIMSS survey, which favoured decontextualised presentations.

In questions which we would deem to have high ecological validity, pupils are often illustrated engaging in the kind of activity which would constitute good classroom practice. Pupils will be shown to be active, dealing with quantified data and reflecting on the outcome of their experience. We endeavour to make the links between assessment and teaching and learning as transparent as possible for teachers. Our assumption is that the tests, because of their 'high stakes' nature, will, inevitably impact on practice. Our objective is that this impact should be as positive as possible. (Wiliam, 1996, discusses test validity and its relationship to teaching). Test questions, which are ecologically valid, have face validity for teachers.

The selection of items located within a framework of progression

Locating ideas on a sequence of progression can help the teacher to determine what has already been learned, as well as suggesting possibilities for subsequent learning. Viewing assessment data in this way helps to integrate assessment and teaching/learning. To inform the process of conceptual change as it happens in schools, constructivist research must describe pupils' conceptual change in the relevant time scales involved – perhaps five to eight years, or even longer. This is usually accomplished by cross-sectional rather than longitudinal research. (The progression inherent in micro-developmental conceptual change is another issue addressed by a different research paradigm. See e.g. Karmiloff-Smith, 1992).

Pupils' understanding, at a particular level, has to be accommodated in the form we refer to as 'intermediate understandings'. We are not using 'progression' in the sense of 'logical sequence' or 'progressive complexity'. Our idea is one of developmental progression in understanding. There are science educators who find this notion difficult to accommodate, preferring a fixed, positivist definition of what is 'correct' understanding and what is not.

Diagnostic analysis

Each year since 1995, a sub-sample of national standard test scripts has been qualitatively re-marked shortly after the administration of the tests. The sample comprises a minimum of 450 pupil scripts, stratified by overall national curriculum Level achieved and by gender. On the basis of the total score achieved and 'cut-scores' defined by the Qualifications and Curriculum Authority (QCA – the statutory body) pupils are defined as operating at one of three Levels: Level 3, Level 4 or Level 5. Level 4 corresponds to the 'expected' criterial level of performance for the age group - not the norm; Level 3 is lower and Level 5 higher than criterion for the Year 6 (age 11) cohort. Approximately equal numbers of each of these achievement bands is included in the re-marking sample.

A code scheme is developed to reflect the range and quality of both correct and incorrect responses. The data generated by the re-marking are then analysed by gender and Level sub-groupings.

We illustrate the outcomes of the qualitative re-marking process by reference to a question called *Plumbline* (see Fig. 1) to exemplify the quality of data which re-marking can make available. We characterise *Plumbline* as a concept probe, low on the dimension of ecological validity but high in its capacity to provide data relating to progression in understanding. (The 'ecologically valid' classroom activity from which this item is abstracted might involve children working with a three-dimensional globe, attaching miniature model people to the different continents of the Earth in the correct orientation). The summative statistics reveal only that about two-thirds of the age 11 cohort gained credit for drawing the plumbline towards the centre of the Earth.

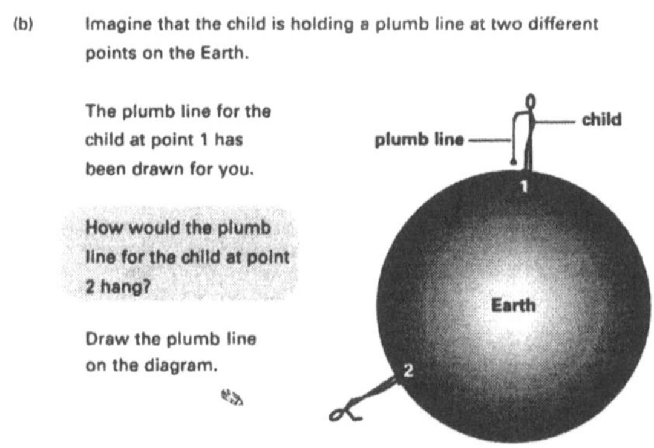

Fig. 1

Qualitative re-marking used a rational framework (see Fig. 2) to analyse children's actual responses. The mark-scheme adopted by QCA (the commissioning agency) accepted any response in what we have defined as regions a and b as being creditworthy. Regions b and c we would be inclined to regard as lower level responses, consistent only with the understanding of gravity acting 'downwards' or 'towards the ground' rather than towards the centre of the Earth. Analysed in this way, less than half the sample meet our stricter criterion of *understanding* as contrasted with two thirds *scoring* (see Table 1).

Looking at the more unambiguously incorrect responses, region e groups those which appear to use the edge of the page as the reference point. Such responses might be described as 'egocentric' in the sense that their reference is the child's own very local position, rather than any reference to the Earth's role in a global sense.

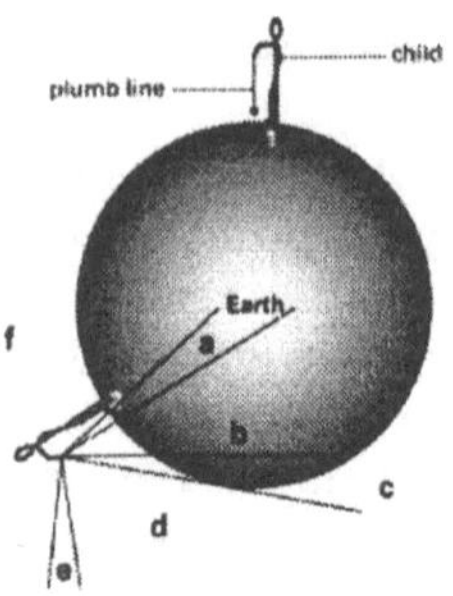

Table 1. 'Plumbline' - Diagnostic analysis of responses (%)

Scoring responses	
region a	region b
48	15

non-scoring responses			
region c	region d	region e	region f
3	4	10	3

Fig. 2

We have also routinely analysed data by reference to the performance of sub-groups defined by Level. Such data might be thought of as sharing some of the qualities of cross-sectional developmental information. Table 1 shows the Plumbline performance data analysed by Level, confirming that the Level 3 pupils dominate response type **e**, while the Level 5 pupils are clearly in the majority in response type a.

		Scoring responses		non-scoring responses			
		region a	region b	region c	region d	region e	region f
Overall Level achieved on the test	L3	19	9	7	5	15	4
	L4	49	19	2	4	11	3
	L5	82	14	-	1	2	1

Tab. 1: 'Plumbline' diagnostical analysis by overall level achieved (%)

In the case of gravitational attraction, following a review of the literature (e.g. Nussbaum & Novak, 1976 and Vosniadou & Brewer, 1994) and some work with teachers of children in the 5-14 age range, we have proposed a sequence of progression in the development of understanding, (Russell et al., 1998, see table 2). Coupled with data such as those emerging from the Plumbline question, we suggest that sequences of this nature are what teachers require in order to get a sense of progression in understanding. This in turn is likely to support and promote the practice of formative assessment. Clearly, those pupils who have no understanding of the force of gravity acting in a line through the centre of the Earth are not in a position to understand the distinction between mass and weight. Teachers could be alerted to the possibility of looking for opportunities to establish confident understanding of weight as that force acting on a body as the result of gravitational attraction, and how this differs from the fixed amount of 'stuff' in a body which constitutes its mass.

1. Gravity keeps things on the ground, stops them floating away.
2. Gravity is a property of the Earth, so is a pull from beneath.
3. The pull of gravity is directed towards the centre of the Earth.
4. The size of the force of gravity depends on the mass of the object being pulled by the Earth.

5. The size of this force is the weight of the object.
6. The Moon is smaller than the Earth and gravity on the Moon is less than that on Earth.
7. The size of the gravitational force is determined by the mass of the object and the mass of the Earth/Moon/planet.
8. There is a gravitational force between any two objects.
9. The size of this gravitational force depends on the distance between the objects.

Table 2: Suggested progression in pupils' acquisition of understanding of concept of gravitational attraction

Implications for formative assessment by teachers

Diagnostic assessment provides illustrative responses indicative of the nature of pupils' understanding, or difficulty with understanding. Formative assessment must locate that analysis in a sequence suggesting what preceded the pupil's current state and what is likely to come next.

The utility of the kind of qualitative data described above for classroom practice resides in their capacity for validating the sequence of progression, sensitising teachers to the kinds of difficulties in understanding experienced by pupils and suggesting ways forward in teaching. Such feedback can be useful in combating the sense that many teachers seem to have that they must teach all that is in the curriculum and pupils must understand it. We offer teachers insights into pupils' intermediate understandings and show that there can be evidence of progress in thinking which might nonetheless fall short of complete understanding.

The kind of practical consequence that we have in mind is that teachers consider the kind of diagnostic data summarised in Fig. 2 against the sequence summarised in table 2. One test item clearly does not generate data to address the entire sequence (which spans probably ten or more years of development in this instance). All the 'Plumbline' responses in regions a, b and c are consistent with statements 1 and 2 but only a is consistent with statement 3. Responses in regions d, e and f are not yet on the scale - though e reveals a particular idea relevant to the first level. Perhaps of particular interest to a teacher are those children who are offering responses in region b, accounting for about two-fifths of the sample. Are children in this group limited to a view of gravity as acting 'downwards' relative to their own local body position? Or do they have some inkling of the sphericity of the Earth and the fact that continents around the globe are populated by people whose sense of 'downwards' is relative? What manner of intervention might provoke each group respectively to shift in the desired direction?

Future possibilities

Educational assessment in the form of the summative national tests are consistently referred to as being 'high stakes' (Daugherty, 1995). There is a gulf between the summative uses to which the tests are put and any attempt to inform curriculum and classroom practices. Assessment is seen as the tool by which politicians make teachers accountable for their pupils' performance, rather than informing the system as it operates day by day in classrooms. The failure to make connections between the performance of a cohort and the teaching and learning going on in individual schools is distressing, incomprehensible, even irrational.

We argue for the gradual, cumulative and iterative assembly of a bank of assessment items which map onto a curricular matrix of lines of progression, representing the accumulated wisdom of educational research and assessment. We would like, at the very least, to establish an agreed agenda for such a programme of action to take place. It is notable that this will be a long slow process as the result of the decision that every pupil in the age cohort should be measured against the national tests. This means that the number of items which might be exposed each year is strictly limited, in contrast to the possibilities which a matrix sampling offer.

Historically, the government's Assessment of Performance Unit (Department of Education and Science, 1989) offered the opportunity to set up an item bank – albeit based on a process model of assessment – but the political priorities for education shifted and the APU baseline was abandoned. Educational research has been criticised for its failure to accumulate wisdom (Hargreaves, 1996). Is the development of assessment materials another such instance, or can something more positive be said?

The future holds possibilities of computerised assessment; with greater bandwidth, our principle of ecological validity could be realised in the form of incorporated multimedia images to set contexts and sophisticated scoring algorithms for open responses. The matrix of lines of progression could be realised as a computerised item bank operating as an intelligent system in its interactions with pupils' responses.

In turn, the same principle of cumulative and iterative development and research should lead to the accumulation of evidence to inform multimedia teaching and learning sequences to support the acquisition of the agreed curricular building blocks. Such multimedia programmes could and should be of the highest quality, developed using extensive trials and feedback to generate a highly polished, professional product. In the context of support for core concepts agreed as central to a national curriculum, we are thinking of very high volume usage. Such volume should permit low unit cost of expensively produced professional quality software informed by accumulated high-quality national and international data linking research and assessment.

References

Black, P. & Wiliam, D. (1998). Assessment and classroom learning. *Assessment in Education: principles, policy and practice.* Vol 5. no 1, 7-74.

Daugherty, R. (1995). *National Curriculum Assessment: A review of policy 1987-1994.* London, UK: Falmer Press.

DES (1989). *National Assessment: The APU approach,* Her Majesty's Stationery Office.

Hargreaves, D.H. (1996). *Teaching as a research based profession: possibilities and prospects.* Teacher Training Agency Annual Lecture U.K.

Karmiloff-Smith, A. (1992). *Beyond modularity: A developmental perspective on cognitive science.* MIT Press.

Nussbaum, J. & Novak, J.D. (1976). An assessment of children's concepts of the earth using structured interviews. *Science Education* 60, 4, 535-500.

Russell, T., McGuigan, L., & Hughes, A. (1998). *Forces SPACE research report.* Liverpool University Press.

Vosniadou, S. & Brewer, W.F. (1994). Mental models of the day/night cycle. *Cognitive Science* 18, 123-183.

Wiliam, D. (1996). National curriculum assessment and programmes of study: Validity and impact. *British Educational Research Journal* Vol 22, no 1, 129-142.

A Comparison of STS-Teaching and Traditional Physics Lessons – On the Correlation of Physics Knowledge and Taking Action

Helmut Dahncke and Helga Behrendt
University of Kiel, Germany
Priit Reiska
Tallinn University of Educational Sciences, Estonia

Abstract

In this paper one of a total of three components of the research project "Learning Physics and Taking Action" is described (Dahncke, 1996, 1997). We report in key words on the frame of reference, aims of the study, structure of the three components, hypotheses and the design of the study. Then, we briefly explain the basic concepts which underlie our hypotheses and the design: knowledge as propositionally represented knowledge, taking action according to Edelmann's criteria (1996), lessons as traditional physics lessons or STS-teaching (Science-Technology-Society), the effect of teaching as part of a control group design with six secondary school classes (130 subjects / 9th graders in Germany and Estonia) and a design of repeated measuring with data collection before and after the corresponding lessons. The acquisition, processing and evaluation of data as part of a multifactorial analysis are briefly mentioned and presented together with the findings of the study.

Frame of reference and research methods

This project has two interrelated main objectives. On the one hand, we examine the relation of learning and taking action on the basis of instruction - which is constantly postulated didactically, but rarely substantiated. On the other hand we examine this aim on the basis of several learning matters which differ with relevance to learning and taking action. Thus, we assume that learning and taking action take place specifically for each field. The relevant fields are represented in the three project components A, B and C. The concept of energy is the interlinking element between the three components. Component A deals with "The turnover of heat energy in a one-family house", B deals with "The turnover of electrical energy in the home" and C deals with "Energy supply in a dynamic system of different power stations". The systems, which we study in computer simulations, combined with the method of thinking aloud, progressively become more general and more comprehensive. A and B are closer to the subject's direct everyday surroundings than C. On the other hand, the amount of previous physics knowledge required to solve the problems in C is smaller than in B and this again is smaller than in A. The interlinking of the components A, B, C is described in Dahncke (1996).

The findings of extensive research carried out during the last 20 years on alternative conceptions and on phenomena in our natural and technical environment can be interpreted in such a way that structured networks, which are created at a very early age and remain consistent through things which are experienced

H. Behrendt et al. (eds.), Research in Science Education – Past, Present, and Future, 77–82.

correspond to students' alternative conceptions; they are networks, which are extremely difficult to link to the networks constructed during lessons. This is one of the most important aims of our work group, i.e. drawing up concept maps in the field of energy and energy supply and observing changes which occur during lessons. We describe this as propositionally represented knowledge of science.

According to Edelmann (1996), taking action, as we understand it, means that the subject makes a plan and follows it. In addition to this, actions must be elective (i.e. there are alternatives), must be able to be carried out arbitrarily (i.e. actions have a subjective meaning for the person taking action) and must be controlled by a plan which anticipates further action. We use these characteristics as criteria in our study and accordingly regard intervention in computer simulation as actions.

Thus, we make observations on the subjects' intervention and behaviour. In addition to this, we make observations on verbal (oral) statements, since the subjects are requested to explain their intervention. Questions that we think are important are: whether the subjects fall back on their knowledge of physics, whether they give explanations relating to physics or to other fields, whether they make use of subjective theories (as set down in concept maps, for example), whether they show exploratory, anticipating or reactive behaviour in dealing with a dynamic system and how many times they have to intervene in order to solve the problem and whether this is influenced by different approaches to teaching (STS, traditional physics lessons).

The computer simulation, that we developed for the field of energy supply, required the subjects to solve problems for which a physicist could draw on his knowledge, but would not necessarily be forced to. Observations carried out following the process of thinking aloud, make it clear whether the subjects base their actions on their knowledge of physics or whether they use a different approach. In the three project components, students are requested to use their knowledge of physics in varying degrees. In component C the requirement is deliberately smaller than it is in components B and A, in order to be able to examine the presumption that the majority of subjects do not put forward arguments relating to physics of their own accord when thinking aloud and that they do not do so even if physics lessons have taken place.

We report on data collection, processing and evaluation and on the findings of the project component C. The main issues based on our three hypotheses are : "The comparison of STS and traditional physics lessons regarding the effect of teaching on knowledge and taking action (H 1)", "The link between knowledge and taking action (H 2)" and "The reciprocal influence of data collection methods (H 3)".

Our design (Fig. 1), with three parallel subject groups, consists of three phases (before lessons, during lessons and after lessons). Thus, we use the same measuring instrument before and after lessons. In terms of data evaluation, this design of repeated measuring means that we use a multivariate analysis (MANOVA-procedure and LSD-test) (Bortz, 1985).

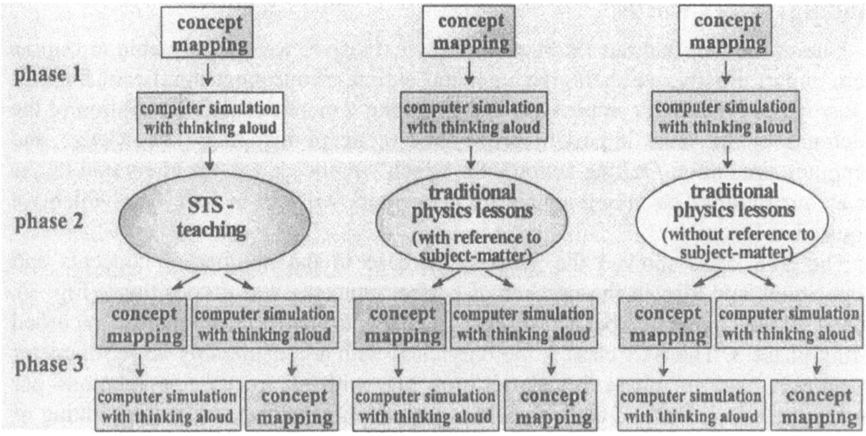

Fig. 1: Study design for three phases (teaching and data acquisition before and after lessons)

The methods used to acquire data are concept maps, computer simulation and thinking aloud; the content deals with energy and energy supply in a comparative study of STS-teaching and traditional physics lessons (Behrendt, 1999).

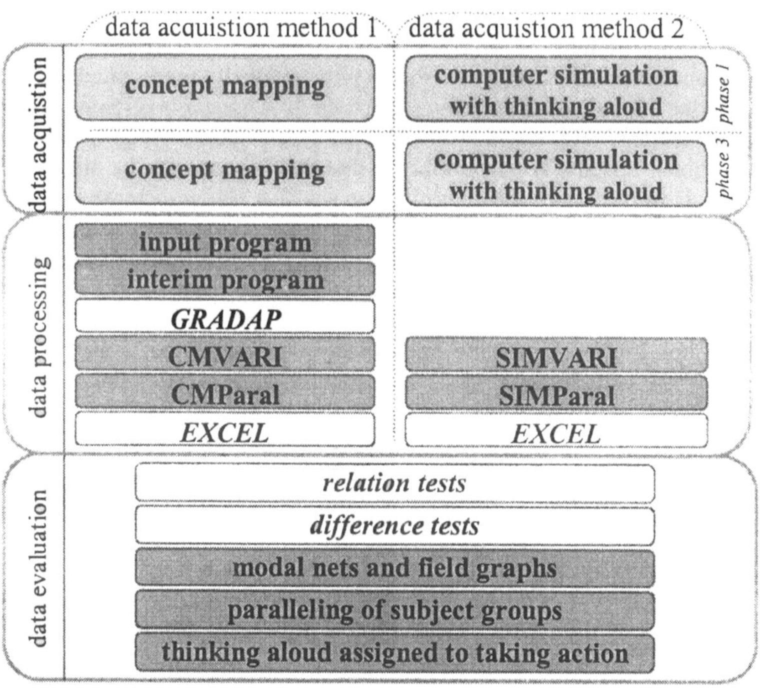

Fig. 2: Survey of the process of data collecting and data evaluation

- self developed
- adapted and further developed
- commercial programs

Findings in key words

Due to the fact that our findings are very extensive, we are only able to outline them in part. Firstly, we shall give a general outline of our most significant findings in key words. We shall complement this by giving a more detailed description of the outcomes of the most important procedures (concept mapping / knowledge and computer simulation / taking action), for which we choose two or one variable. As far as our findings on taking action are concerned, we also include data which we acquired in Estonia.

The STS class showed the greatest increase in the number of concepts and propositions and also in the number of correct concepts and propositions (Fig. 3). However, the largest number of mistakes in concepts and propositions was recorded during phase 3. The STS class is the only class with a considerably larger diameter of concept maps in phase 3 and the only one with an increase in relations per concept (density). Both of these variables provide information on the interlinking of concepts and on the size of the net. Thus, in the STS class the nets are better than those in the other classes with regard to size, correctness and density. In our study we cannot, however, explicitly determine special knowledge of the subject-matter, but using terms from the field of scientific terminology after lessons indicates whether the pupils have integrated new concepts regarding the subject-matter of the lessons into their network of knowledge. As it is the STS class which shows a greater increase in the number of concepts from the field of scientific terminology (Fig. 4).

In summary, it may be said that lessons relating to specialist subject-matter produces an increase in knowledge, that is propositionally represented. The increase in knowledge depends on the teaching approach. It is greater in STS-teaching than in traditional lessons. STS-teaching also led to greater competence in taking action than traditional lessons. We cannot prove that this increase can be attributed solely to lessons. However, since the control group shows a strong increase in competence to take action, the effect of practice from previous actions and a maximum ceiling effect also play a role. Thus, subjects who were successful in the first phase, were not able to increase their competence to take action as much as those subjects who were not as successful in phase 1 (Behrendt, 1998).

In addition to the main study in Germany, we carried out a study in Estonia following to the same design (Fig. 1). By using the same teaching units and the same methods of data acquisition, we were able to confirm the findings established in Germany (Fig. 5); this also provides us with additional information on the effects of the different teaching approaches (Reiska, 1999).

Here, we present further examples in more detail from the study carried out in Germany using concept mapping:

The findings illustrated in Fig. 3 and Fig. 4, clearly show a greater increase in concepts and propositions during STS-teaching than during traditional lessons. The results from the control class, during lessons which did not relate to the subject matter, remained more or less the same.

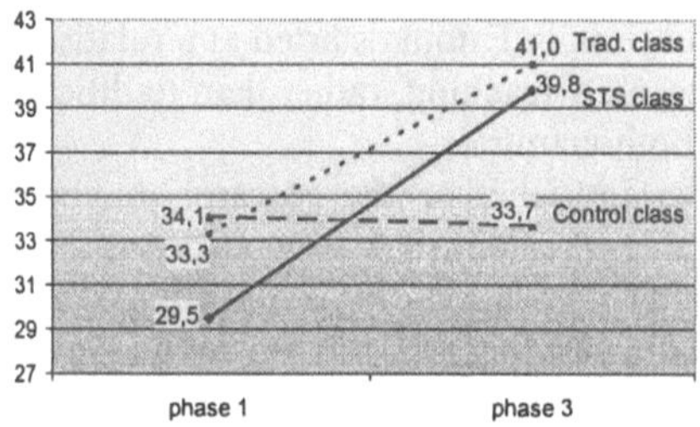

Fig. 3: Variable: Number of concepts (AlBe)

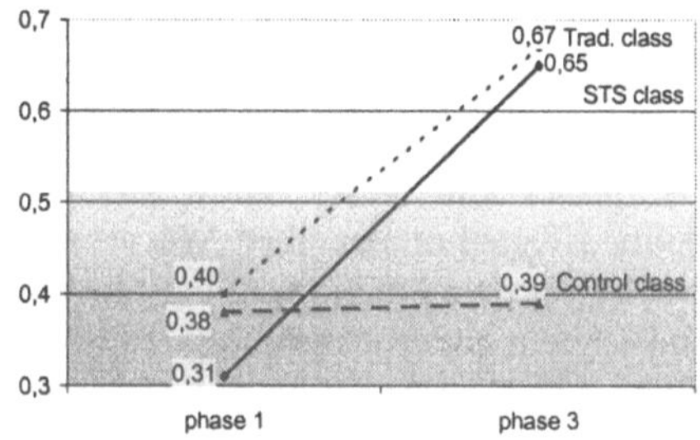

Fig. 4: Variable: Number of concepts from the field of terminology (ZaFa)

The study contributes to STS-research by providing us with quantatively reliable results, due to newly-developed research methods. Former research findings were, for the most part, merely able to determine a more positive attitude amongst teachers and learners - produced by STS-teaching - and this was only possible in qualitative studies.

The question of whether more knowledge produces more successful actions and whether there is a link between knowledge and taking action at all cannot be answered for certain using the data supplied by the German subjects. If we include the findings from Estonia, however, we can establish that scientific knowledge is useful for taking action. There is a link, not only between knowledge stated while thinking aloud and taking action during computer simulation, but also with knowledge represented by the concept maps. Although knowledge is useful for taking action successfully, it is, however, not the only and most essential prerequisite.

Here, we give a more detailed example of the effects of teaching observed during computer simulation:

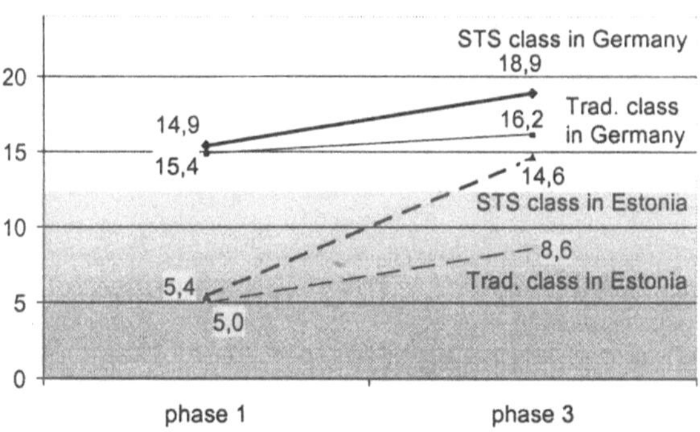

Fig. 5: Variable: Number of correct results in using power plants during computer simulation (ZaRL)

Fig. 5 indicates that, by using computer simulation, it is possible to observe differences in the actions subjects take. The number of correct actions taken during computer simulation (a maximum of 24) is considerably higher after lessons. A smaller increase in the classes in Germany is related to the high degree of success at the beginning (phase 1). In comparison, the subjects in Estonia started at a relatively low level and achieved a considerable increase. STS-teaching, rather than traditional physics lessons, leads to a greater increase in both countries.

A further aim of the study was to ascertain whether the research methods possibly affected one another. We were not able to establish that concept mapping had an effect on taking action, but that computer simulation did have a slight effect on subsequent concept mapping. However, the small increase in knowledge gained during computer simulation did not have any bearing on the findings of different effects produced by teaching approaches.

References

Behrendt, H. (1997). Physiklernen und Begriffsnetze. In H. Behrendt, Ed., *Zur Didaktik der Physik und Chemie - Probleme und Perspektiven (pp.* 299-301). Alsbach, Germany: Leuchtturm.

Behrendt, H. (1998). Einfluß unterschiedlicher Unterrichtsansätze auf deklaratives und prozedurales Wissen - Erste Ergebnisse einer Studie zum Thema Energieversorgung. In H. Behrendt, Ed., *Zur Didaktik der Physik und Chemie - Probleme und Perspektiven (pp.* 361-363). Alsbach, Germany: Leuchtturm.

Behrendt, H. (1999). *STS-Unterricht und traditioneller Physikunterricht – Empirische Untersuchung mit den Methoden Concept Mapping und Computersimulation.* University of Kiel, PhD-Thesis.

Bortz, J. (1985). *Lehrbuch der Statistik für Sozialwissenschaftler.* Berlin, Heidelberg, New York, Tokyo.

Dahncke, H. (1996). Physikkenntnisse und Eingriffe in Computersimulationen. In Deutsche Physikalische Gesellschaft e. V., Fachverband Didaktik der Physik, Ed., *Didaktik der Physik. 60. Physikertagung (pp.* 106-111). Jena.

Dahncke, H. (1997). Physiklernen - Physikunterricht - Handlungskompetenz. In H. Behrendt, Ed., *Zur Didaktik der Physik und Chemie - Probleme und Perspektiven (pp.* 293-295). . Alsbach, Germany: Leuchtturm.

Dahncke, H. (1998). Energieumsatz und Energieversorgung - Vergleich von Physikunterricht und STS-Unterricht mit den Mitteln von Computersimulation und Concept Mapping. In H. Behrendt, Ed., *Zur Didaktik der Physik und Chemie - Probleme und Perspektive (pp.* 358-360). Alsbach, Germany: Leuchtturm.

Edelmann, W. (1996). *Lernpsychologie - eine Einführung.* München, Weinheim: Urban & Schwarzenberg.

Fuhrmann, A. (1996). *Physikalisches Wissen und Alltagssituation.* University of Kiel, PhD-Thesis.

Reiska, P. (1999). *Physiklernen und Handeln von Schülern in Estland und in Deutschland. Eine empirische Untersuchung zu zwei unterschiedlichen Unterrichtskonzepten im Bereich von Energie und Energieversorgung mit den Methoden Concept Mapping und Computersimulation.* Frankfurt (Main): Peter Lang.

Part 3: Students' Conceptions

On the Quantum Thinking of Physics Undergraduates

Gren Ireson
Research In Science Education (RISE) Group, Loughborough University, UK

Abstract

Using a questionnaire for data collection and two multivariate techniques for analysis, this paper reports on the concepts undergraduate physics students in England hold regarding quantum phenomena. The analysis is by Cluster Analysis to find underlying groups of concepts and Multidimensional Scaling to map the concepts onto a two-dimensional epistemological space. A sample of 338 first and second year undergraduates, from six institutions, was surveyed during 1998. Cluster Analysis generated three distinct Clusters. Multidimensional Scaling allowed the responses to be mapped onto two dimensions. The results suggest that at the level of the group there is an underlying structure to the students' responses. Insight into a complex topic has been gained which is in agreement with earlier work with pre-university physics students in England and undergraduate physics students in the USA.

Introduction

Research into the concepts held by pre-university students in England and elsewhere is plentiful in virtually all topic areas. For surveys see, for example, (Pfundt & Duit, 1994; Eylon & Linn; 1988: Osborne & Freyberg, 1985). The exception being 'modern physics' - including quantum phenomena. The only study into this area in England being due to Mashhadi (1996). Mashhadi's work was supportive of that carried out by Niedderer, Bethge and Cassens (1990) with German pre-university students. The situation with regard to undergraduates is, again, such that little or no research is available within the European context. However European studies into how an individual student developed an understanding of quantum physics, Petri and Niedderer (1998), Niedderer, Bethge, Cassens & Petri (1997), focused on the important role played by models. Ingham and Gilbert (1991), when working with undergraduate and postgraduate chemistry students, took a similar line when investigating analogue models. Studies have been carried out into some aspects of quantum thinking in the USA (Ambrose, Shaffer, Steinberg & McDermott, 1999) and England (Ireson, 1999). This work points to a number of misconceptions in the quantum understanding of physics majors. The aim of this study was to investigate, at the level of the group, the concepts held by undergraduate physics students and to determine any underlying structure.

H. Behrendt et al. (eds.), Research in Science Education – Past, Present, and Future, 83–88.

Design

During 1998 a forty-item questionnaire was issued to first and second year physics undergraduates across six institutions in England. Of the forty items twenty-nine directly addressed the students' understanding of quantum phenomena with the remaining eleven items addressing their concept of 'a model'. The questionnaire was based on the fifty-four item version used by Mashhaddi (1996) when investigating the understanding of quantum phenomena amongst pre-university students. The reduction to forty items was based on two factors;

1. The removal of items addressing 'reality of entities'
2. Three further items when analysed under Factor Analysis showed high loadings onto more than one factor. This made their inclusion suspect and hence these were removed.

Procedure

The questionnaire was piloted on a group of thirty, second year undergraduates, prior to instruction, in order to ascertain whether the revised questionnaire would generate Clusters to which meaning could be attached. Given that three Clusters were generated with such a small sample, it was decided to continue with the full survey. The final sample was N=338. The questionnaires were administered, without further advice from the physics tutors, to first and second year students at the beginning of their taught course in quantum mechanics. The returned questionnaires were analysed, using SPSS version 7, by Cluster Analysis, using Ward's method, and Multidimensional Scaling.

Data Analysis

Cluster analysis generated three main clusters, which can be interpreted in terms of students' understanding of quantum phenomena as shown in table 1. Multidimensional Scaling allows the clusters to be mapped onto a two dimensional epistemological space which explains 98.9% of the variance for items dealing with *modelling* and 94.1% of the variance for items dealing with *quantum phenomena*. This is shown in figures 1 and 2.

Findings

Interpretation of Clusters

Cluster 1 is labelled *Quantum Thinking / Descriptive Models*. Students in this group can be regarded as having the 'accepted' understanding of quantum phenomena. Examples of quantum thinking can be drawn from statements B26 [When a beam of electrons produces a diffraction pattern it is because the electrons themselves are undergoing constructive and destructive interference] and B28 [Whether one labels an electron a 'particle' or 'wave' depends on the particular experiment being carried out]. The descriptive nature of models can be seen in statements B34 [Models are of a temporary nature. With the increase of knowledge a model may become obsolete or useless and either adapted or replaced by another model] and B38 [Models are aids in communication (e.g. in teaching)].

Cluster 2 is labelled *Conflicting Thinking / Conflicting Models*. Students in this group are still struggling with the notion of abandoning 'determinism.' Conflict of thinking is seen in B25 [Electrons move around the nucleus in definite orbits with a

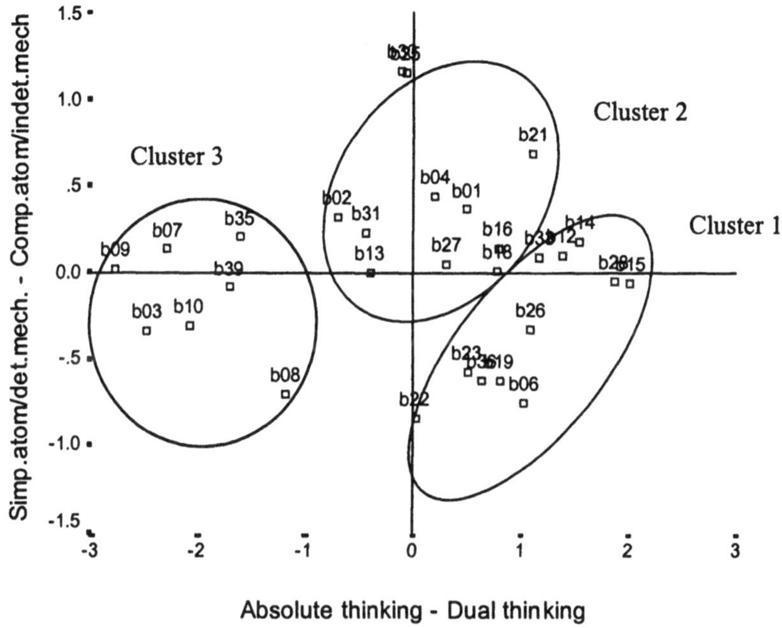

Fig. 1: Clusters (quantum phenomena only)

high velocity], B27 [Electrons move randomly around the nucleus within a certain region or at a certain distance] and B21 [Nobody knows the position accurately of an electron in orbit around the nucleus because it is very small and moves very fast]. Conflicting concepts of models are shown in B11 [A model is a scaled up or down version of something real, with a one to one correspondence between the model and the real thing it represents (e.g. the solar system or an atom)] and B32 [All models are mental images (existing only in the human mind)].

Cluster 3 is labelled *Mechanistic Thinking / Complete Models*. Students in this group still hold a mechanistic world view. The mechanistic thinking is typified by, for example, B35 [Electrons are fixed in their shells] and B07 [The electron is always a particle]. Whilst item B24 [A model always provides a complete description of the object, structure or process that it models] refers to a model providing a complete description.

Interpretation of the dimensions

Dimension 1 is labelled; Absolute thinking ⟵⟶ Dual thinking

This dimension maps the student thinking from Newtonian to Quantum in terms of the behaviour of 'light' and 'electrons'. The 'Absolute' end of the spectrum is typified by B09 [Light is always a wave], B07 [The electron is always a particle] and B03 [The energy of an atom can have *any* value]. The 'Dual' end is typified by B28 [Whether one labels an electron a 'particle' or 'wave' depends on the particular experiment being carried out] and B15 [How one thinks of the nature of light depends on the experiment being carried out].

Dimension 2 is labelled; *Simple atom/deterministic mechanics* ⟷ *Complex atom/indeterministic mechanics*

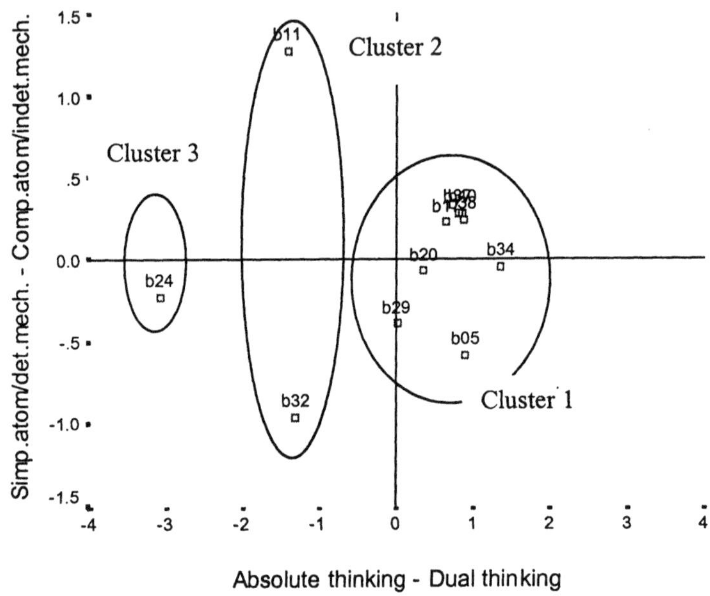

Fig. 2: Clusters (models only)

This dimension maps the student thinking from Newtonian to Quantum in terms of determinism or the lack thereof. The 'Simple atom' end of this spectrum is exemplified by B25 [Electrons move around the nucleus in definite orbits with high velocity] and B30 [If a container has a few gas molecules in it and we know their instantaneous positions and velocities then we can use Newtonian mechanics to predict exactly how they will behave as time goes by]. The 'Complex atom' end can be exemplified by B22 [It is possible for a single photon to constructively and destructively interfere with itself], B08 [An atom cannot be visualised] and B06 [Coulomb's law, electromagnetism and Newtonian mechanics cannot explain why atoms are stable].

Discussion

These findings are in accord with the previously quoted research regarding student misconceptions with quantum phenomena, suggesting that the misconceptions are real, stable over time and cross-cultural. A number of physics undergraduates do not, even after two years of university study, have an understanding of quantum phenomena which approximates to the accepted explanation. These misconceptions could, I feel, be addressed by a change in pedagogy which is informed by research such as quoted here. In England the usual route into University physics is via the General Certificate of Education (GCE) Advanced level in physics. These courses and much of the physics studied at undergraduate level present the students with a 'wave model' and a 'particle model' which leads to confusion for the student. A new pedagogy based on 'quantum objects' and which avoids 'duality' is needed. The Physicist turned science populariser, Professor Russell Standard, uses such an approach in his book aimed at junior school students, (Stannard, 1994). More recently Pospiech (1999) presents an approach to the EPR paradox which starts with the uniquely quantum phenomena of 'spin'.

Cluster	Item number	Statement
One	**B15**	How one thinks of the nature of light depends on the experiment being carried out.
	B28	Whether one labels an electron a 'particle' or 'wave' depends on the particular experiment being carried out.
	B34	Models are of a temporary nature. With the increase of knowledge a model may become obsolete or useless and either adapted or replaced by another model.
	B14	When an electron 'jumps' from a high orbital to a lower orbital, emitting a photon, *the electron is not anywhere in between the two orbits.*
	B12	The photon is a sort of 'energy particle'.
	B37	A model is formulated using facts obtained by experiment and/or observation.
	B38	Models are aids to communication (e.g. in teaching)
	B17	A model depicts or describes a theory.
	B40	Models can be used to predict phenomena, structures or processes that have not previously been observed.
	B29	Models are visual representations of the abstract and unseen
	B33	Individual electrons are fired towards a *very* narrow slit. On the other side is a photographic plate. What happens is that the electrons strike the plate one by one and gradually build up a diffraction pattern.
	B05	Models are constructions of the human mind.
	B20	Models do not represent the 'true picture' of atoms.
	B16	Electrons move along wave orbits around the nucleus.
	B18	The photon is a 'lump' of energy that is transferred to or from the electromagnetic field.
	B26	When a beam of electrons produces a diffraction pattern it is because the electrons themselves are undergoing constructive and destructive interference.
	B19	Electrons consist of smeared charge clouds which surround the nucleus.
	B36	Orbits of electrons are not exactly determined.
	B06	Coulomb's law, electromagnetism and Newtonian mechanics cannot explain why atoms are stable.
	B23	Since electrons are identical it is not possible to distinguish between them.
	B22	It is possible for a single photon to constructively and destructively interfere with itself.
	B01	The structure of the atom is similar to the way planets orbit the sun.
Two	**B04**	The atom is stable due to a 'balance' between an attractive electric force and the movement of the electron.
	B21	Nobody knows the position accurately of an electron in orbit around the nucleus because it is very small and moves very fast.
	B25	Electrons move around the nucleus in definite orbits with high velocity.
	B27	Electrons move randomly around the nucleus within a certain region or at a

		certain distance.
	B31	During the emission of light from atoms the electrons follow a definite path as they move from one energy level to another.
	B13	Electrons are waves.
	B32	All models are mental images (existing only in the human mind).
	B02	It is possible to have a visual 'image' of an electron.
	B11	A model is a scaled up or down version of something real, with a one to one correspondence between the model and the real thing it represents (e.g. the solar system or an atom).
	B30	If a container has a few gas molecules in it and we know their instantaneous positions and velocities then we can use Newtonian mechanics to predict exactly how they will behave as time goes by.
Three	**B03**	The energy of an atom can have *any* value.
	B07	The electron is always a particle.
	B09	Light always behaves as a wave.
	B10	In passing through a gap electrons continue to move along straight line paths.
	B24	A model always provides a complete description of the object, structure or process in nature that it models.
	B39	The photon is a small, spherical entity.
	B35	Electrons are fixed in their shells.
	B08	An atom cannot be visualised.

Tab. 1: Clusters listed by statement

References

Ambrose, B. S., Shaffer, P. S., Steinberg, R. N. & McDermott, L. C. (1999). An investigation of student understanding of single-slit and double-slit interference. *American Journal of Physics* 67, 146-61.

Eylon, B-S. & Linn, M. C. (1988). Learning and instruction: An examination of four research perspectives in science education. *Review of educational research* 58, No. 3.

Ingham, A. & Gilbert, J. K. (1991). The use of analogue models by students of chemistry at higher education level. *International Journal of Science Education* 13, No. 2.

Ireson, G. (1999). A multivariate analysis of undergraduate physics students' conceptions of quantum phenomena. *European Journal of Physics* 20, 193-199.

Mashhadi, A. (1996). *What is the nature of the understanding of the concept of 'wave particle duality' among Advanced level Physics students?* Unpublished DPhil thesis (St. Cross College / University of Oxford, UK).

Niedderer, H., Bethge, T. & Cassens, H. (1990). A simplified quantum model: A teaching approach and evaluation of understanding, P. L. Lijuse, P. Licht, W. de Vos & A. J. Waarlo (Eds.), *Relating macroscopic phenomena to microscopic particles - A central problem in secondary science education.* Utrecht: CD-ß Press.

Niedderer, H., Bethge, T., Cassens, H. & Petri, J. (1997). Teaching quantum atomic physics in college and research results about a learning pathway. *AIP conference proceedings.*

Osborne, R. & Freyberg, P. (1985). *Learning in science. The implications of children's science.* Portsmouth: Heinemann Educational.

Petri, J. & Niedderer, H. (1998). A learning pathway in high-school level quantum atomic physics. *International Journal of Science Education* 20, No. 9.

Pfundt, H. & Duit, R. (1994). *Students' Alternative Frameworks and Science Education.* Kiel: Institute for Science Education at the University of Kiel.

Pospiech, G. (1999). Teaching the EPR paradox at high school? *Physics Education* 34, 311-315.

Stannard, R (1994). *Uncle Albert and the quantum quest.* London: Faber and Faber.

Experiences with a Modern Course in Quantum Physics

Gesche Pospiech
Institut für Didaktik der Physik, Universität Frankfurt / Main, Germany

Abstract

Quantum Physics belongs to the most important physical theories. Its fundamental aspects show up, for example, in teleportation, which has high resonance in media. Consequently an educational access to quantum theory is highly desirable and should fulfill three conditions: to quickly reach the heart of quantum theory without distraction with too much mathematics, to focus on philosophical aspects and to open the door to future developments. In addition such a course should sharpen the ability to perceive presuppositions of classical physics by contrasting it with the different view thrown on nature by quantum theory. It conveys an enriched look on nature and, therefore, can be a building stone for general education. A course designed in this way was tested on a small group of students, mostly preparing to become teachers.

Subject and objective

Quantum theory has been heavily discussed since its invention in the beginning of the 20th century. It soon became obvious that the interpretation of its mathematical structure would have great influence on world view. The technological applications, e.g. the laser, developed more slowly, but nowadays are integrated into daily life. Hence, quantum theory, as a fundamental physics theory, should have a place in school. The first step to achieve this is the education and in-service training of teachers with the aid of a suitable course. There are three aspects that should be contained in such a course:

The mathematics of quantum theory: It should be made clear that the basic mathematical structures are important and build the basis of an interpretation.

The physical phenomena: Quantum objects behave quite differently compared to classical objects as shown by recent experiments concerning spin states.

Philosophical aspects: According to Heisenberg, the invention of quantum physics was one of the greatest achievements of the 20th century in the humanities and Weizsäcker called it "a fundamental philosophical advance".

Peculiarities in teaching quantum theory

Fundamental change of concepts

There are inevitable difficulties in understanding quantum theory, not only for students but in principle. These are founded in physics and nature itself. Hence, teaching quantum physics provides insight into the fundamental problems of any change of concepts, that also occur in teaching classical physics. The problems have to be taken as a unique opportunity for teaching science and scientific thinking.

H. Behrendt et al. (eds.), Research in Science Education – Past, Present, and Future, 89–94.

Absence of analogies from classical physics

Anschaulichkeit is a point of controversies in treating quantum theory: on the one hand quantum phenomena are different from every classical physical object by the very nature of the theory. On the other hand students need a familiar „hook" where they can fix new ideas and develop their understanding. Often attempts are made to explain the peculiarities of quantum theory in relation to the classical concepts position and momentum or with the so-called duality of wave and particle. But it is precisely these classical concepts, borne from daily experience that have to be overcome.

Prerequisites of teacher students entering university

According to the historical development of quantum theory, traditional education at school has certain fixed points, mostly from atom physics with some digressions to particle-wave dualism and the double slit experiment (Wiesner, 1989). This choice of curriculum has the effect that quantum physics is likely to be equated to atom physics. Therefore, most students come to university with a rather classical perception of quantum physics (if any). Many experts felt uneasy with the traditional focus and developed new ways for teaching quantum theory, e.g.:

- Lichtfeldt, (Lichtfeldt, 1995), tried to avoid misunderstandings by taking the preconceptions of pupils into account. He, therefore, aimed at generating cognitive conflicts by starting with diffraction of electrons, in contrast to the usual photo effect. His work is distinguished by thorough evaluation at schools.
- Zollman, (Zollman, 1998), concentrates on observable phenomena in atom physics. The students are supposed to acquire a working knowledge related to simple experiments (e.g. LED). In addition they gain experience by handling atoms in computer simulations. This is called „Hands-on quantum mechanics".
- Wiesner, (Wiesner, 1989), has a focus on preparation of quantum objects, thus bridging the gap between classical and quantum physics. He relates observations to the non-existence of fixed values for formal properties of quantum objects.

New possibilities

The analysis of the new results in research concerning the fundamentals of quantum theory - e.g. realizations of EPR thought experiment, quantum teleportation and GHZ-states - show new possibilities for didactical reconstruction of quantum theory. Hence, new ways in teaching arise by taking into account its three afore mentioned aspects as well as the needs of students for help in imagination as shown in table 1, (Pospiech, 1999c).

Aspects of quantum theory	Problems	Solution/Opportunities
Mathematics	abstract formalism	Reduction to most simple case: 2-dim Hilbert space
Physical phenomena	There are no analogies from other parts of physics	Use of intuitive images from psychology
Philosophical aspects	interpretation is not yet resolved.	Students gain insight that science is developing

Table 1: Difficulties and opportunities in quantum physics

However, to my knowledge so far, there has been no systematic attempt to teach quantum physics along the lines of the new experimental results, although there are some attempts, (Jaeckel, 1992; Audretsch, 1994). In my opinion they open a completely new view on quantum theory and their philosophical potential. But teachers then need special training at university, preparing them to teach this difficult subject without too much effort for lesson preparation. Most courses at university, however, generally focus on the mathematical formalism.

Therefore, I propose an elementary procedure, that quickly reaches the heart of quantum theory. Since there is no „royal street" in learning quantum theory, there are still problems to be mastered, but always connected to opportunities.

Guidelines for teaching concepts and design of the course

Motivation: good motivation can be easily achieved through hints towards fascinating phenomena that can also be found in newspapers: quantum cryptography, teleportation, quantum computing and, last but not least, the philosophical aspects mainly concerning the changing perception of "reality".

Reference example: It is important to have a point of reference that can be used as an analogy, preferably a known phenomenon and one which can serve as a basis for further development. The spin together with the EPR-experiment is such an example.

One single explanatory concept: Students have to gain experience with *one* single explanatory concept providing them with a method for explaining. We find the concept of information, information exchange and irreversibility a suitable concept. Perhaps this concept works on an abstract level but it can be „visualized" by a model of Schlüter, (Görnitz, 1999), paralleling information exchange to photon loss. The students are then able to understand phenomena by applying this concept. With this model, new phenomena can be explained in the same way as known phenomena previously dealt with. This can promote a feeling of competence.

„Hook" for building a conceptual net: In this category the consequent use of intuitive images serve as illustrations of strange quantum phenomena, allowing an intuitive understanding apart from mathematical formalism.

Structure of lesson plan

1. Some simple experiments with polarized light and calcite crystals.
2. Intuitive images in the spirit of Schrödinger's cat.
3. The spin and its mathematical description.
4. Philosophical context through a specifically designed dialogue between the philosophers Parmenides and Kant and a quantum philosopher.
5. The EPR - thought - experiment.
6. Original papers from Einstein and Schrödinger.
7. Bell Inequality and experimental tests in quantum mechanics.
8. The measuring process and its philosophical aspects (model of Schlüter).
9. Technological perspectives such as quantum computer, quantum cryptography and teleportation.
10. General concept of information (Example: which-way-information or quantum eraser).
11. Delayed - choice Experiments.

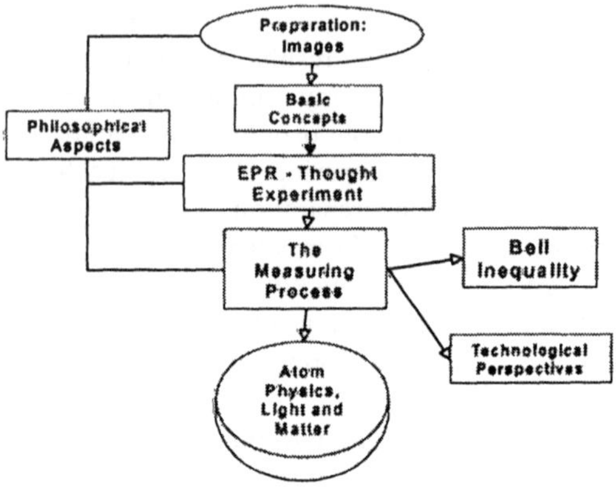

Fig. 1: Design of Course and Organisation of Contents

The course is organised in a modular manner. Details are described elsewhere, (Pospiech, 1999b).

Evaluation of the course

A course in teacher education along these lines has been conducted twice by myself. The first course (9 participants) was attended mostly by student teachers preparing for higher secondary education; in the second course there were six students with quite different backgrounds. Both groups were strictly selected according to their interest in interpretation of quantum theory. The main purpose of the courses was to show that the access described so far is feasible and to uncover its advantages as well as perhaps drawbacks. The subject matter „interpretation of quantum theory" for the most part precudes the use of the categories „wrong" or „true" in answering questions. Hence, to test the acceptance of the contents the evaluation stressed the subjective and intuitive understanding in addition to the mathematical formalism. In spite of the low number of some suggestions concerning the contents of the course, the main difficulties of students and ways to circumvent them could be ascertained. The following section describes the evaluation of the second course.

Method of evaluation

At the beginning of the course participants were given a pre-course questionnaire concerning prior knowledge, interest in the subject matter as well as expectations. A second questionnaire at the end of the course asked for an overall summary, any difficulties encountered and changes in competence and knowledge gain through subjective statements. Most questions were posed in open format. The answers to open questions were grouped and analysed in suitably derived categories. Due to the open atmosphere throughout the course the students posed their questions and discussed them broadly. A description follows of the results collected in the context of having had to course myself.

Results of the pre-course questionnaire

The formulation of interest and expectations in the course depended mostly on prior knowledge, whereas associations to the word "quantum" were mainly determined by interest in popular science literature. The participants were highly heterogeneous with regard to their university studies but all were quite concerned with questions about world view and the links between interpretation of quantum theory and their view of nature. The students wanted to discuss world view on a pragmatic, well-founded basis separately from esoteric attitudes, which sometimes might be found in connection with quantum theory. This common interest met the goals of the course and provided a common base.

Result of post-course questionnaire

In the post-course questionnaire I posed questions on how well expectations had been met, on changes and understanding of concepts, on remaining difficulties and on knowledge gain. Furthermore, the students were asked about the role of intuitive images used during the course. The participants stated that their overall expectations were met. The different aspects of the course - philosophy, mathematics, physics, didactics and experiments - were perceived as evenly balanced. Terms that very clearly signal differences to classical physics: non-locality, uncertainty, measuring process and information were deemed to be of most importance. Most difficulties arose with mathematical structure and in understanding the measuring process – is to be expected. In addition to mentioning these concepts, the students hinted to problems of perception of space and time. These problems had been addressed in a philosophical dialogue, especially designed for this course in order to build a bridge to philosophy. Moreover the students strongly felt that knowledge gains in some aspects of mathematical structure and concerning philosophical points of view in "nonlocality", mostly prompted by the presentation of several quantum mechanical phenomena such as teleportation and quantum eraser. Furthermore they stated that they had gained insight into the measuring process. On request, they expressed a feeling of better understanding the differences between the quantum world and the classical world.

In order to promote an intuitive understanding, intuitive images from psychology had been introduced. The participants found the images helpful to understanding. They remembered, particularly, one image (described in figure 2) that had been discussed very intensely with respect to whether it meets the relation between classical world and quantum world. The validity of this image was tested by developing and reflecting on the identification of elements of the image with physical and mathematical terms. Another intuitive image (Pospiech, 1999a) mentioned by nearly all students dealt with the EPR-experiment. Nevertheless the evaluation showed that classical concepts, such as wave-particle dualism or mathematical formulation for entanglement or uncertainty relation, still remain permanent.

Example of intuitive image: Woman and child talking

Imagine you observe a woman talking to a child. There are two possibilities:

1. They both met by chance. Then you could definitely describe their talk, quite precisely, by simply stating the phrases they use. This is the classical case.
2. They are mother and child. In this case, simply restating the phrases would probably not meet all the implications and hidden meanings of their talk, because there are relations between them that cannot be described in complete detail, without destroying them.

Fig. 2: Explanation of an example for "intuitive image"

In explanation of the EPR-experiment, all participants mentioned the twin photons (Diphoton) and nonlocality as essential concepts. They regarded as important the common history of both photons stemming from one single process.

Some also described the motivation of Einstein, the question as to whether the classical perception of reality could be conserved.

Résumé

The questionnaire showed a high degree of contentnent with the overall weight given to the different aspects of quantum theory. Constant use of one single concept - spin state of photons as building blocks of the theory - helped the students to see a link between different phenomena. However, by referring to concepts not mentioned in the course, the answers showed that the prior-knowledge still remained valid. The new concepts have been added and students with good knowledge tried to build them into an unifying image. A helpful ingredient for achieving understanding and memorizing were the intuitive images which allowed visualisation without interference with everyday experience.

Conclusion

I demonstrated a consistent way of teaching the basic structures of quantum theory, mainly for future teachers, on a modest mathematical level. Besides the fundamental mathematical structures, aspects of Weltanschauung (conceptions of the world) and recent experiments play a significant role. The courses conducted show that education in quantum theory might exert influence on how students view nature and that any conceptual change needs time and careful preparation. Quantum theory requires a deep change in reflecting on physics and its meaning for people as do other theories. Children start to learn science with an "Aristotelian" view of nature. Turning, for example, to the Newtonian view as required in science education, needs much effort and often is incomplete, as many studies show. I regard the conceptual change from Newtonian to quantum view as equally important. Therefore, teachers have to be carefully prepared and to become aware of different ways of looking at nature. Therefore, the proposed course can contribute to general education.

References

Audretsch, J. (1994). Die Unvermeidbarkeit der Quantenmechanik. In K. Mainzer & M. Schirmacher, *Eds.Quanten, Chaos und Dämonen*. Mannheim: BI.

Görnitz, Th. (1999). *Quanten sind anders.* Heidelberg: Spektrum Akademischer Verlag.

Jaeckel, K. & Pade, J. (1992). EPR-Paradoxon in der Schule - Spukhafte Fernwirkung und Bertlmanns Socken. In H. Fischler, Ed., *Quantenphysik in der Schule.* Kiel: IPN.

Lichtfeldt, M. (1992). Schülervorstellungen als Voraussetzung für das Lernen der Quantenphysik. In H. Fischler, Ed., *Quantenphysik in der Schule* . Kiel: IPN.

Pospiech, G. (1999a). Spukhafte Fernwirkungen in der Quantentheorie. *Physik in der Schule* 37, 56-59.

Pospiech, G. (1999b). Ein neuer Lehrgang für Quantenphysik in der Lehrerausbildung. In R. Brechtel, Ed. *Zur Didaktik der Physik und Chemie* (pp. 280-282). Alsbach: Leuchtturm Verlag

Pospiech, G. (1999c). Teaching the EPR-Paradox at High School? *Physics Education* 34, 311-316.

Wiesner, H. (1989). *Beiträge zur Didaktik des Unterrichts über Quantenphysik in der Oberstufe.* Essen: Westarp

Zollman, D. (1998). *Hands-On Quantum Mechanics.* Duisburg: Proceedings of GIREP-Conference, CD-ROM.

Learning Process Studies in the Field of Fractals

Michael Komorek, Reinders Duit, Nils Bücker and Barbara Naujack
Institute for Science Education (IPN) at the University of Kiel, Germany

Abstract

Our studies on understanding core ideas of fractals are part of a broader project on educational reconstruction of non-linear systems (Duit, Komorek & Wilbers, 1997). They focus on investigations into whether core ideas of chaos theory and fractals are worth teaching relating to the general aims of science instruction and whether they can be understood by students at the age of 15-17. The notion of fractals has already played a certain role in mathematics instruction in recent years. There are also ideas on how to employ this concept in explaining well known and new phenomena to students in science lessons. Studies on how students may learn and understand the new teaching materials (like experiments) are almost non-existent. The two studies address different core ideas of fractals. The first study lays its focal point on the concept of self-similarity (or self-affinity) and on the insight that very complex structures may evolve from simple rules. The second study focuses on employing the concept of fractals to explain certain phenomena. We investigate how students spontaneously understand the occurrence of dendrite structures in two quite different experiments and how they may be guided towards the science explanation. Both studies are designed as learning process studies. Nine groups of two students each are interviewed by an interviewer using the method of »teaching experiment« (Steffe & D'Ambrosio, 1996). The interviews last about one hour. In study, one the students were aged 16 and in study two they were aged 18. Data analyses follow category-based qualitative interpretation (Mayring, 1995; Bortz & Döring, 1995).

Study 1: Towards understanding of self-similarity

The students were presented 14 pictures (figure 1). Some of the pictures denote the principle of self-similarity, i.e., they show geometrical or real objects (like the cross-section-cut of a cauliflower, or a fern leave) where »zooming« into the objects reveals the same or similar structures. In the case of a cauliflower, for instance, a branch is similar to the whole object, a branch at the branch is again similar to a branch on the next level up. Students were asked to form groups of pictures that make sense to them and to write down the features that defined the categories. A short instruction on the concept of self-similarity followed. First, the cross-section-cut of a real cauliflower was investigated by the students. They were asked to describe the structure. The interviewer guided them towards seeing that, for a couple of steps, the branches on the next levels are similar to ones on the higher level. Second, they investigated how students understand the construction of the so called Koch-curve. The first three steps were given (figure 2). Their task was to find the fourth step. This task serves the purpose of further elaborating the idea of self-similarity and also to allow students to deal with an example where simple rules may result in complex structures if employed many times. After this, similarities and differences between structures of the cauliflower and the Koch-curve were

H. Behrendt et al. (eds.), Research in Science Education – Past, Present, and Future, 95–100.

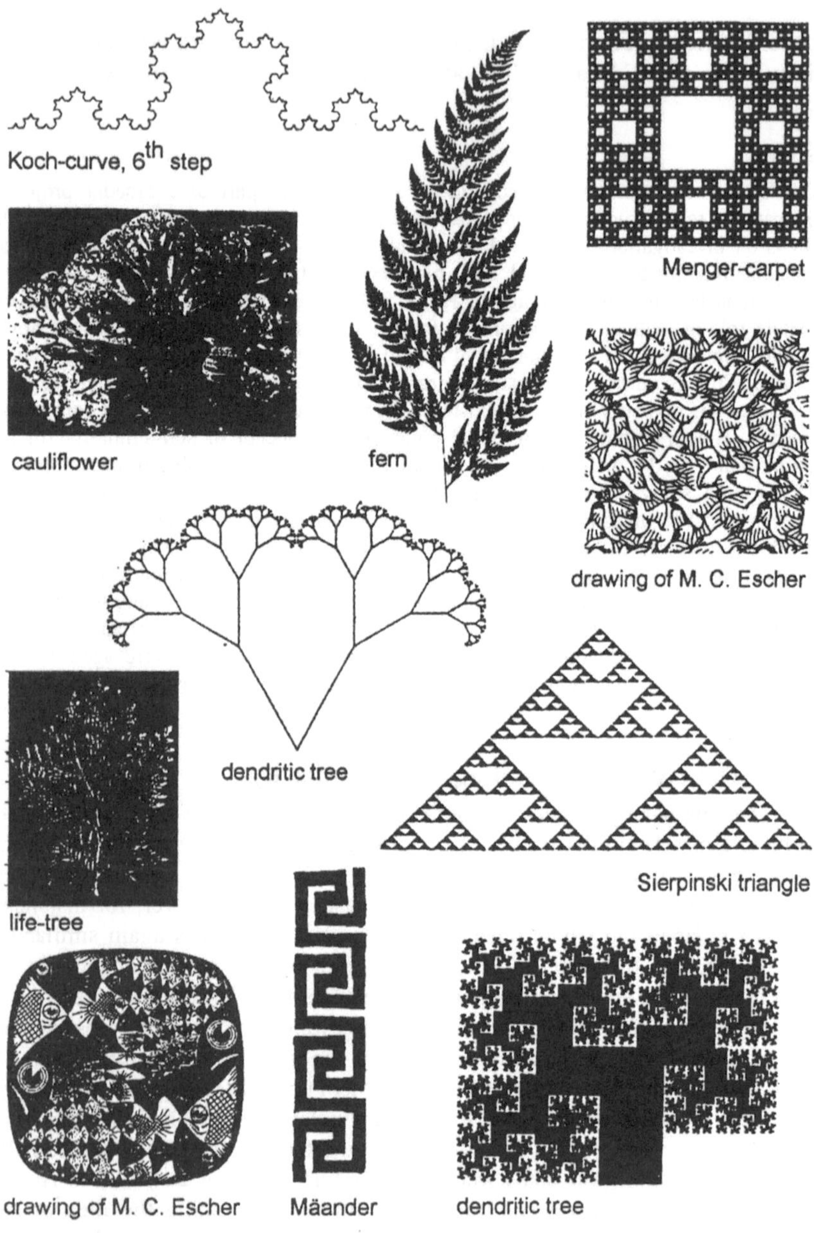

Fig. 1: Most of the 14 pictures employed in study 1

discussed. Finally, students were asked to re-group the 14 pictures they dealt with during the interview. Afterwards pictures of a human lung and a zinc-dendrite (figure 4) were presented. Students were asked to explain whether these objects are self-similar or not.

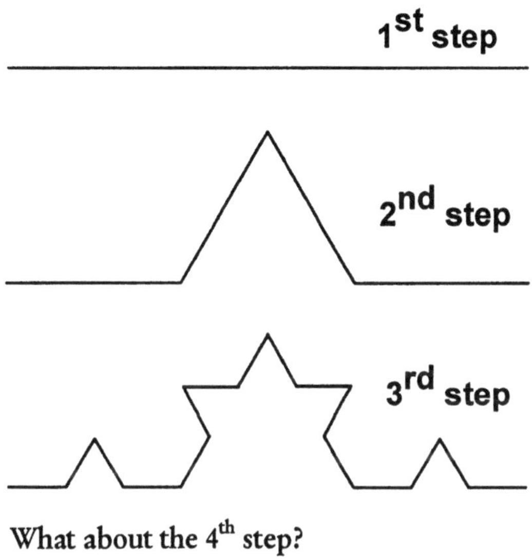

Fig. 2: Construction of the Koch-curve

Briefly summarized, the results from this study are as follows (for more details see Naujack, 1998). Most students do not spontaneously sort the 14 pictures presented to them in groups of self-similar objects. It does not come as a surprise that their major sorting categories were features of the geometric objects presented. The idea nearest to the idea of self-similarity is that certain objects are »branched« was employed by a number of groups. The idea that certain objects are repeatedly branched is also a view developed by many students when investigating the structure of the cauliflower. About half of the students interviewed in this case also develop preliminary ideas of self-similarity although the explanations they present are only partly in accordance with the science definition. The investigation of the Koch-curve turns out to be somewhat difficult for most students. It appears that they usually understand that a quite complex structure evolves from a simple rule permanently applied. They have severe difficulties however finding out the fourth step of construction. Katrin's and Silke's drawing of the fourth step (figure 3) is a quite telling example. Apparently the increasing number of indents leads them to draw the star.

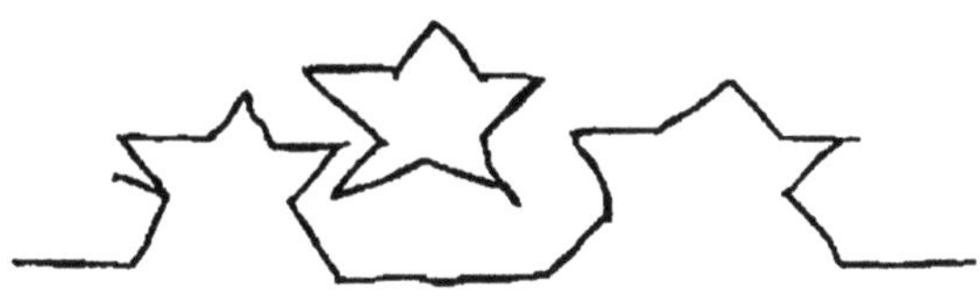

Fig. 3: Katrin's and Silke's idea of the fourth step

In the final task, all groups are able to arrange the initial 14 pictures in such a way that pictures depicting self-similarity are put in one group and that this grouping is adequately justified. There are also no serious difficulties to see that the lung and the zinc-dendrite are further examples of self-similar objects. It is therefore viable to conclude, that it is possible to guide students at the age of 15-16 towards the idea of self-similarity and to employ this idea to describe certain mathematical and real objects. There is also evidence in the collected data that the concept of self-similarity allows students to see similarities across different objects and phenomena. This includes also understanding of the biological function of self-similar structures of living objects like cauliflower, fern or the lung.

Study 2: Growth of dendrite structures

This study is based on two experiments that illustrate the growth of dendrite structures (for more details see Bücker, 1998). Electrolysis is the process underlying the first experiment. There is a solution of $ZnSO_4$ in a round container. There is a voltage between a zinc-ring and a center-cathode. The experiment allows to observe the growth of a »zinc-dendrite« (figure 4) (Bunde & Havlin, 1994; Witten & Sander, 1981). If the experiment is repeated, a similar global structure evolves exhibiting quite different details. Briefly put, the growth of this »fractal« structure may be explained in the following way. There is an interplay of random Brownian motion of the ions on the one hand and directed motion of the zinc-ions towards the center cathode due to the voltage applied on the other. Randomly the first ions arrive at the cathode forming small »hills«. As the cathode grows it becomes more likely that an ion arrives at the top of these »hills« than further down in the »valleys«. There are two reasons for that. First, for most ions moving towards the center-cathode it is nearer to the top of the hills, hence the probability that they end there is higher than ending down in the valleys. Second, the attractive forces towards indents (the tops) are slightly higher than for other places. Basically the same explanation holds for the further growth. The tops of the hills form the germs of the main branches. There are sub-branches developing for the reasons just outlined and so on. It is important to keep in mind that we have an interplay of random and deterministic processes here that resembles a similar interplay in the case of chaotic systems.

There is in addition, the principle of »self-amplification«: If a »hill« is forming, the further growth is supported by this occurrence. We would also like to mention that the fractal structure is to be expected if the principle of minimum energy dissipation holds. The fractal structure may be seen as the optimal structure that leads to minimal energy dissipation in this case. Thus, we have a key to the understanding of dendrite structures in general as »natures way to find optimal solutions«.

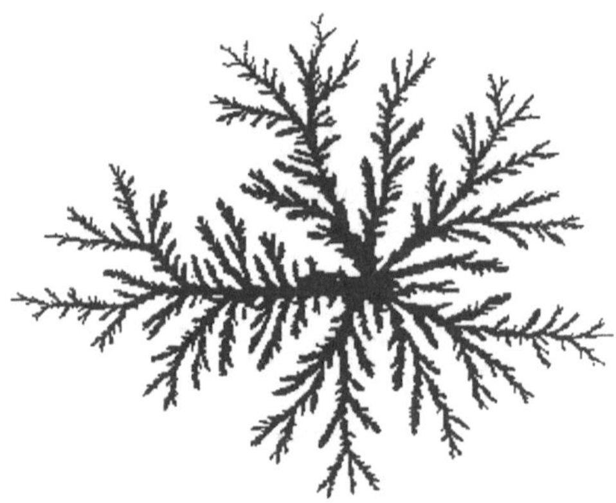

Fig. 4: »Zinc-dendrite«

The other experiment leads to similar dendrite structures (figure 5). Here we have a somewhat viscous liquid (glycerine) between two plates. A syringe is used to push air into the liquid. A dendrite-like structure results (Nordmeier, 1998). The explanation here draws on the same general principles as in the case of the zinc-dendrite. There is an interplay of random and directed processes together with the principle of self-amplification. As a rough outline: first a circle like front of air forms, but small instabilities result in a break through of the air at certain places; the main branches form. Then there is motion of air in an outward direction until the driving forces and the forces of the surface tension of the glycerine balance out. The same explanation holds for the

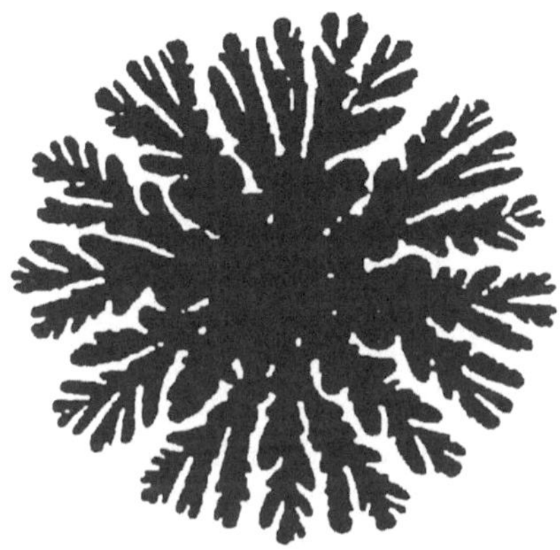

Fig. 5: Dendrite boundaries between two liquids of different vicosity

occurrence of sub-branches. Here also the fractal structure may be seen as the structure that necessarily has to occur due to the principle of minimal energy dissipation.

In total, 18 students were interviewed. Only one of the students predicted that in the first experiment a branched structure evolves. All the others thought that there would be some sort of circle like growth. Usually the symmetry of the set up is taken as argument in favour of the occurrence of a regular figure. If irregularities arise they will even out immediately. Wolfgang, for instance, argues:

It will expand in a circular way .... every other structure would not be logical as we have a circle like set up of the experiment.

It is interesting to see that the conceptions of what »random« denotes influences students attempts to make sense of the unexpected structure. One group of students holds that processes are determined by certain laws but that, due to a large number of variables the systems' behaviour appears to be random. Consequently they have severe difficulties in understanding that where the first ions reach the cathode is random. Random behaviour for them seems to be irrelevant to physics. They think that exactly the same structure (a structure with the same number of branches) will result when the experiment is repeated. For other students, random behaviour is closely related to irregularity. They think that such behaviour is not determined by principle. Hence, they have no difficulties in understanding that it may not be predicted where the first ions reach the electrode. As this process is random, for them the occurrence of »hills« and »valleys« is already sufficiently explained as something that looks like random results. Considerable effort is necessary to make these students aware that their argument actually does not (at least fully) explain the fractal structure.

Despite the above-mentioned difficulties caused by the two views of random behaviour, at the end of the interview all students receive at least some preliminary understanding of the fractal structure being explained by the interplay of random and deterministic influences. Already at the beginning of the second experiment, no student predicts that the same number of branches will occur when the experiment is repeated. It is also noteworthy that a number of students argue that the fractal structure is »optimal« in the sense of energy distribution. In other words, a number of students appear to hold or develop views that are in accordance with the physics concept of minimal energy dissipation mentioned above. Although, generally speaking the experiments and the interventions by the interviewer guide most students to the science view in the case of the first experiment, there are also some limitations to the view gained. Most of them have problems to taking the random Brownian motion into account when explaining the random motion of ions towards the cathode.

However, attempts to compare the quantities »driving« the processes (e.g., voltage in the case of the first experiment and pressure in the second) do not lead to a deeper understanding of common causes of the fractal structure. It appears that the most dominant idea to »explain« the similarities, is the conviction that »similar causes necessarily result in similar effects« - in this instance, however, the »nature« of the similar cause is not further elaborated.

Conclusion

The two studies, presented above open windows into understanding students' learning processes in a domain where almost no research is available. We conclude from these preliminary studies that it is worthwhile to investigate the educational significance of fractals and possibilities to make core ideas of this concept understandable to students. The findings of the studies presented encourage us in our attempts to make the core ideas accessible also to 15-16 year old students. We would like to mention that the method of the teaching experiment applied in the studies has proven a powerful means to investigating how new topics or new methods of teaching and learning work should be introduced into school practice.

References

Bortz, J. & Döring, N. (1995). *Forschungsmethoden und Evaluation* [Research methods and evaluation]. Berlin: Springer.

Bücker, N. (1998). Experimente, Elementarisierungen und Schülererklärungen zum fraktalen Wachstum [Experiments, elementarizations, and students explanations of fractal growth]. In M. Komorek, R. Duit & M. Schnegelberger, Eds., *Fraktale im Unterricht* [Fractals in school] (pp. 203-234). Kiel, Germany: IPN.

Bunde, A. & Havlin, S., Eds. (1994). *Fractals in Science.* Heidelberg: Springer.

Duit, R., Komorek, M. & Wilbers, J. (1997). Studies on educational reconstruction of chaos theory. *Research in Science Education* 27, 339-357.

Mayring, P. (1995). *Qualitative Inhaltsanalyse* [Qualitative content analysis]. Weinheim: Deutscher Studien Verlag.

Naujack, B. (1998). Eine empirische Untersuchung zum Lernen von Grundideen des Fraktalmodells [An empirical study on learning basic ideas of fractal model]. In M. Komorek, R. Duit, & M. Schnegelberger, Eds., *Fraktale im Unterricht* [Fractals in school] (pp. 151-182). Kiel, Germany: IPN.

Nordmeier, V. (1998). Fraktale in der Physik. In M. Komorek, R. Duit & M. Schnegelberger, Eds., *Fraktale im Unterricht* [Fractals in school] (183-202). Kiel, Germany: IPN.

Steffe, L. & D'Ambrosio, B. (1996). Using teaching experiments to understand students' mathematics. In D. Treagust, R. Duit & B. Fraser, Eds., *Improving teaching and learning in science and mathematics* (pp. 65-76). New York: Teacher College Press.

Witten, T. A. & Sander, L. M. (1981). Diffusion-limited aggregation - A kinetic critical phenomenon. *Physics Review Letters* 47, 1400.

Students' Understandings of their Internal Structure as Revealed by Drawings

Michael J. Reiss and Sue Dale Tunnicliffe
Homerton College, Cambridge, UK

Abstract

How do people develop their understanding of what is inside them? This study looks at students' understandings of their internal structure. A cross-sectional approach was used involving a total of 158 students in England from six different age groups (ranging from 4 year-olds to first year undergraduates). Students were given a blank piece of A4-sized paper and asked to draw what they thought was inside themselves. Repeated inspections of the completed drawings allowed us to construct a seven point scale of these representations. Our analysis shows the extent to which student understanding increases with age and the degree to which pupils know more about some organs and organ systems than others.

Introduction and background

As is widely acknowledged, there are many ways of gathering information about students' understandings of scientific phenomena (White & Gunstone, 1992). However, despite the richness and variety of the methods used by science educators, it remains a fact that most of these methods rely on students either talking or writing about science. Such methods include oral interviewing of students (Osborne & Gilbert, 1980), gathering students' written responses (Leach et al., 1995), recording students' spontaneous conversations (Tunnicliffe & Reiss, 1999a) and getting students to construct written concept maps (Novak & Musonda, 1991).

Each of these approaches has its own particular advantages and disadvantages but we wanted, in this study to use an approach that relied less on words. We hope that this approach is less likely to disadvantage students who are very shy in conversation, students who lack certain linguistic skills and students who speak a language (or languages) other than that used by the researcher. This last point means that drawings should be of especial value for international comparative studies.

In this study we report on students' understandings of their own internal structures. We decided on a cross-sectional approach, in which students of different ages would simply be asked to draw what they thought was inside themselves. There is perhaps a certain appropriateness in asking subjects to represent (albeit in two dimensions) anatomically their own anatomy. In the language of Buckley, Boulter and Gilbert (1997), such representations can be viewed as the expressed models - that is, representations of phenomena placed in the public domain - of the students. These expressed models relate to (but do not equate with) the mental models - i.e. the private and personal cognitive representations - held by the same students.

As far as students' knowledge, as revealed by drawings, of what is inside themselves goes, perhaps the most thoroughly studied organ system is the skeleton (Caravita & Tonucci, 1987; Guichard, 1995; Cox, 1997; Tunnicliffe & Reiss,

H. Behrendt et al. (eds.), Research in Science Education – Past, Present, and Future, 101–106.

1999b). Those research reports and papers that have looked at other organ systems have often reported valuable data (from Gellert, 1962, onwards) but there is very little work that systematically and quantitatively examines how knowledge, as revealed by drawings, of the various human organs and organ systems depends on student age.

Methodology

Fieldwork was carried out in the South of England in a primary school, a secondary school and a college of higher education. The primary school (for 4/5 to 11 year-olds) is a state Church of England aided school and is in a new town (established after the Second World War); the secondary school (for 11 to 16 year-olds) is a state comprehensive in a rural setting; the College of Higher Education contains mainly four year Bachelor of Education students training to be primary teachers. SDT carried out the primary fieldwork; MJR carried out the secondary and undergraduate fieldwork.

Cox (1989) discusses some of the ways in which children can be asked to do drawings. We simply asked our subjects, in a whole class setting, to draw what they thought was inside themselves. Students were not examined under formal examination conditions but were told not to copy one another's work. They were given as long as they wanted (up to about 10 minutes) to complete their drawing and were asked to write their name on it. A note was also made, by us, of the gender of each student.

The fieldwork was conducted in whole class settings. In all, data were obtained from 16 Reception children (aged 4 or 5), 21 Yr. 2 children (aged 6 or 7), 33 Yr. 3 children (aged 7 or 8), 32 Yr. 6 children (aged 10 or 11), 24 Yr. 9 children (aged 13 or 14) and 32 first year undergraduates (mostly aged 18 to 20). In the primary and the secondary school, all pupils were in mixed ability groups. The undergraduates who participated came from two separate student groups. One group of 12 were all English specialists, none of whom had studied biology after the age of 16. The other group of 20 were all biology specialists, all of whom had studied biology after the age of 16. The biology undergraduates all knew MJR as their lecturer; the other students in the study knew MJR or SDT only slightly, if at all.

Analysis

The 158 students made a total of 158 drawings, i.e. one per student. After we had collected all the drawings, we jointly and repeatedly sorted through them, attempting to arrange them in a ranked order which we felt reflected different levels of biological understanding. Our ranking was informed both by previous work in the field - especially Osborne, Wadsworth and Black (1992), Guichard (1995) and Cox (1997) - and by our own knowledge of anatomy. We were also extremely keen to provide a scoring system, that gave as little credit as possible to the 'artistic' quality of the drawing and was as unambiguous as possible to score. No notice was taken of the students' ages in determining the scoring system.

Eventually, we agreed on the following overall rank order for the biological quality of each drawing:

Level 1 No representation of internal structure
Level 2 One or more internal organs (e.g. bones and blood) placed at random

Level 3 One internal organ (e.g. brain or heart) in appropriate position

Level 4 Two or more internal organs (e.g. stomach and a bone 'unit' such as the ribs) in appropriate positions but no extensive relationships indicated between them

Level 5 One organ system indicated (e.g. gut connecting head to anus)

Level 6 Two or three major organ systems indicated out of skeletal, gaseous exchange, nervous, digestive, endocrine, urinogenital, muscular and circulatory

Level 7 Comprehensive representation with four or more organ systems indicated out of skeletal, gaseous exchange, nervous, digestive, endocrine, urinogenital, muscular and circulatory.

This scoring system requires a definition of organ systems. We used the following definitions for eight human organ systems:

Skeletal system
Skull, spine, ribs and limbs.

Gaseous exchange system
Two lungs, two bronchi, windpipe, that joins to mouth and/or nose.

Nervous system
Brain, spinal cord, some peripheral nerves (e.g. optic nerve).

Digestive system
Through tube from mouth to anus and indication of convolutions and/or compartmentalisation.

Endocrine system
Two endocrine organs (e.g. thyroid, adrenals, pituitary) other than pancreas [scored within digestive system] or gonads [scored within urinogenital system].

Urinogenital system
Two kidneys, two ureters, bladder and urethra *or* two ovaries, two fallopian tubes and uterus *or* two testes, two epididymes and penis.

Muscular system
Two muscle groups (e.g. lower arm and thigh) with attached points of origin.

Circulatory system
Heart, arteries and veins into and/or leaving heart and, at least to some extent, all round the body.

We then, separately and independently, scored all the drawings. Having agreed on the level (i.e. 1 to 7), we then, for each of the eight organ systems, decided whether or not the drawing met the criterion for that organ system. We also recorded whether or not at least one organ was present on the drawing for that organ system. We agreed on the great majority of scorings. In those cases where our views differed, we discussed each such case until we agreed. Data were entered into Minitab and Excel for analysis. All statistical tests are 2-tailed.

Results

The effect of student age on the level of the drawing

As one would expect, older students generally attain higher levels, on average, than younger ones (Table 1). In all, there are seven different 'age' categories:

Reception (Yr. 0), Yr. 2, Yr. 3, Yr. 6, Yr. 9, 1st year undergraduates who are English specialists (Yr. 14E) and 1st year undergraduates who are biology specialists (Yr. 14B).

Year	Mean level	Median level	sem	n
0	2.00	2	0.09	16
2	3.71	4	0.10	21
3	4.27	4	0.13	33
6	4.41	4	0.13	32
9	4.67	4	0.18	24
14E	4.67	5	0.19	12
14B	6.30	6	0.16	20

Table 1: The average levels of the drawings. sem is the standard error of the mean; n is the number of students in each age category.

However, what is also notable, is that while there is a very rapid increase between Yr. 0 and Yr. 2, subsequent increases (aside from the Biology specialists) are successively smaller.

The effect of student gender on the level of the drawing

There are no significant gender effects ($p > 0.05$).

Students' understandings of organ systems

Lumping together all the data and thus ignoring differences between the drawings resulting from student age, gender or degree of biology specialism, Table 2 shows, for each organ system, the percentage of students whose drawing displayed an organ system. For each of the eight organ systems, only a minority of drawings show the organ system drawn sufficiently completely to be classified by us as an organ system. In addition, there are significant differences between the eight organ systems in terms of how well they are represented ($c^2 = 82.1$, 7 df, $p << 0.001$).

Organ system	Percentage of drawings with this organ system
Digestive	22%
Gaseous exchange	18%
Urinogenital	13%
Skeletal	11%
Nervous	6%
Circulatory	2%
Endocrine	1%
Muscular	0%

Table 2: For each of the eight organ systems, the percentage of students whose drawing showed the organ system as defined in the text.

Students' understandings of organs

Again lumping together all the data, and thus ignoring differences between the drawings resulting from student age, gender or degree of biology specialism, Table 3 shows for each organ system the percentage of students whose drawing represented an organ (rather than the entire organ system). Students do much better at this.

Organ system	Percentage of drawings with an organ from this organ syste
Circulatory	93%
Skeletal	87%
Nervous	75%
Gaseous exchange	66%
Digestive	65%
Urinogenital	54%
Muscular	33%
Endocrine	4%

Table 3: For each of the eight organ systems, the percentage of students whose drawing included an organ from the organ system.

As was the case with whole organ systems, there are highly significant differences between the likelihood of students drawing organs from the different organ systems ($c^2 = 159.6$, 7 df, $p << 0.001$). There are also certain clear differences between the orderings in Tables 2 and 3, notably with respect to the circulatory system which is poorly represented as a whole system (Table 2), yet components of which are very frequently drawn (Table 3). Nevertheless, there is a high correlation between the rankings of how well represented whole organ systems and partial organ systems are ($r_S = 0.87$, $0.01 < p < 0.05$).

Discussion

There is a difference between a student's mental model - what they hold inside their head - and their expressed model - which is revealed to the world. However, the only way for a researcher to understand a student's mental model of a particular phenomenon is by eliciting one or more of their expressed models of that phenomenon. In this study, we elicited only one expressed model per student - for example, we did not also interview students about their drawings - and we did not probe students in any way - for example, by asking them to check whether they knew anything else and, if they did, to add it to their drawing, nor did we require them both to draw and to label their drawings.

The fact that Yr. 3 pupils attained only slightly lower levels than Yr. 9 or English undergraduates suggests to us that, despite (or even as a result of) the school biology they have received, most of these students lack much understanding of organ systems. For instance, they may know that they have bones but their drawings typically fail to show a skeletal system (simply defined as skull, spine, ribs and limbs). In other words, students fail to see what is inside themselves as a functioning whole. Their 'insides' rather consist of a scattered assemblage of isolated organs and incomplete organ systems. For us, as biologists, we would ideally like students not only to have a good knowledge of their various organ systems (i.e. a level 7 on our rating) but to appreciate the interconnections and interrelations of these various organ systems. After all, the skeletal system requires muscles; the muscular system requires nerves, etc..

As shown by Table 2, it is not the case that students are equally likely to draw the various organ systems. Both Gellert (1962) and we found that even very young children typically know about bones and hearts. Interestingly, such hearts are often shaped as on Valentine's cards. We do not know for certain whether students think that this is what the heart literally looks like or whether, in some cases, they are representing the heart 'symbolically' or as a shorthand derived from cards, cartoons or advertisements. However, the fact that several of the Yr. 9 and English undergraduates drew hearts, thus, suggests strongly that such hearts were intentionally being depicted in a non-anatomically 'correct' fashion.

References

Buckley, B., Boulter, C., & Gilbert, J. (1997). Towards a typology of models for science education. In J. Gilbert, Ed, *Exploring models and modelling in science and technology education* (pp. 90-105). Reading: University of Reading New Bulmershe Papers.

Caravita, S., & Tonucci, F. (1987, July). How children know biological structure-function relationships. Paper presented at the Second International Seminar: Misconceptions and Educational Strategies in Science and Mathematics. Ithaca, New York: Cornell University.

Cox, M. (1989). Children's drawings. In D. J. Hargreaves, Ed, *Children and the arts* (pp. 43-58). Buckingham: Open University Press.

Cox, M. (1997). *Drawings of people by the under-5s*. London: Falmer Press.

Gellert, E. (1962). Children's conceptions of the content and functions of the human body. *Genetic Psychology Monographs 65*, 293-405.

Guichard, J. (1995). Designing tools to develop the conception of learners. *International Journal of Science Education 17*, 243-253.

Leach, J., Driver, R., Scott, P., & Wood-Robinson, C. (1995). Children's ideas about ecology 1: Theoretical background. design and methodology. *International Journal of Science Education 17*, 721-732.

Novak, J. D., & Musonda, D. (1991). A twelve-year, longitudinal study of science concept learning. *American Educational Research Journal 28*, 117-153.

Osborne, J., Wadsworth, P., & Black, P. (1992). *Processes of life: Primary space project research report*. Liverpool: Liverpool University Press.

Osborne, R. J., & Gilbert, J. K. (1980). A technique for exploring students' views of the world. *Physics Education 15*, 376-379.

Tunnicliffe, S. D., & Reiss, M. J. (1999a). Building a model of the environment: how do children see animals? *Journal of Biological Education 33*, 142-148.

Tunnicliffe, S. D., & Reiss, M. J. (1999b). Students' understandings about animal skeletons. *International Journal of Science Education* 21, 1187-1200

White, R. T., & Gunstone, R. F. (1992). *Probing understanding*. London: Falmer Press.

Acknowledgements

We are especially grateful to all those who kindly allowed us to analyse and use their drawings and to the head teachers of the two schools and the principal of the college in which we worked.

Personal Context and Continuity of Human Thought; Recurrent Themes in a Longitudinal Study of Pupils' Understanding of Scientific Phenomena

Gustav Helldén
Kristianstad University, Sweden

Abstract

This study of personal context and continuity in twenty-three pupils' thinking builds upon data from a longitudinal study of pupils' conceptualisations of conditions for life, decomposition and the role of the flower. Each pupil was interviewed eleven times between the age of 9-15. At age 15 and 19, each pupil listened to what they had said four years earlier and described how they thought their understanding had developed. The occurrence of characteristic, individual elements of a content or structural nature can be followed through the interviews, year by year. As 15 and 19-year-olds, the pupils could recognise statements in the interviews as results of experiences from an early age. It is possible to follow a characteristic, individual theme in most of the interviews. Conceptions developed at an early age appeared to be important to future conceptual development. Early introduction of some scientific concepts would help pupils to develop a deeper understanding.

Subject

Many studies regarding pupils' conceptions in science have been carried out around the world. However, very few have been longitudinal in nature (White, at press). Such studies can provide researchers with the opportunity to study the nature of the learning process, learning pathways and the influence of everyday experiences on pupils' long-term learning. The present study will report on findings that emerged from my continuing analysis of data from a longitudinal study of 23 pupils' understanding of ecological processes (Helldén, 1995; Helldén, 1998; Helldén, 1999). Currently, an intensive debate is taking place between education researchers, who emphasise different perspectives on learning. One group argues that the situative perspective can provide a broader framework for understanding and improving educational practice (Greeno, 1997). Another group argues for a cognitive perspective (Anderson, Reder, & Simon, 1997). On the basis of empirical data from my longitudinal study, I will discuss situative and cognitive aspects of learning in relation to my analysis of long-term interviews of students.

The objectives of the present research project are:

- to analyse the interviews with pupils in order to identify personal themes;
- to study how personal context and continuity can influence pupils' conceptual development;
- to discuss the results of the analysis from a situative and cognitive perspective.

H. Behrendt et al. (eds.), Research in Science Education – Past, Present, and Future, 107–112.

Design and procedure

In order to make knowledge claims regarding long-term conceptual development of ecological processes, I conducted a six year, longitudinal study of 23 pupils' understanding of conditions for life, decomposition, and the role of the flower in plant reproduction. I, like many other researchers in science education, have found that clinical interviews can give in-depth information on pupils' thinking regarding natural phenomena (Duit, Treagust & Mansfield, 1996). Therefore, I interviewed the same pupils on 11 different occasions from grade 2 (9 years) to grade 8 (15 years) at Swedish comprehensive school. To challenge pupils' ideas about conditions for life, I grew plants in sealed transparent boxes. For my interviews on decomposition, I had soil and brown leaves on a table in front of the pupils. During my interviews about the flowers, I showed the students different kinds of flowers. During interviews with the same pupils at 15 years of age, they also listened to audio-tapes of my interviews with them four years earlier. I asked them to comment on their ideas at 11 years of age and explain why they had said what they had. I also asked them to describe how they thought their ideas had developed their after age 11 and what had been most important in the development of their ideas. At age 19, I interviewed the students again. I asked them these same questions, after they had listened to my interviews with them at 11 and 15 years of age. In this paper, I will report on findings from my continuing analysis of the interviews.

Analysis

Concept maps were constructed from the interview transcripts according to a design that has been developed at Cornell University (Novak, 1998). Ausubel's theory of meaningful learning had important implications for the analysis of the interview data and for the description of the pupils' differential conceptual development (Ausubel, Novak & Hanesian, 1978). By comparing the concept maps and the transcripts from the interviews with each pupil through the years, I have identified developmental patterns in the structure and content of the pupils' conceptions. The focus of the current research has been to identify personal contexts and continuity of thought in the pupils' conceptual development. I have also sought to explore examples of learning as part of a social practice that had then been integrated into the learners' ideas and how such examples of situated learning could develop through the years.

Findings

During analysis of the interviews, over the years, it has been possible to recognise personal themes in the pupils' conceptions about ecological processes. Such themes can have a structural nature – a way of explaining a phenomenon through the years. Other themes may concern the content.

At the beginning of the study, the pupils thought that the plants in the sealed transparent boxes would die because they lacked access to life supporting resources like water, air or oxygen. The pupils thought that the plants must take in matter of different kinds from the environment but did not consider the passing of matter flowing through the plants. In order to understand what was going on, the students described a 'use up model' that meant that the plant was the 'end point' for the necessary resources. Following the first interview, the teacher introduced the concept of the water cycle. Many pupils picked up this 'cycle model' and started to discuss the plants' possibility to survive from a new perspective. A new structure in the interviews can be identified. The pupils described individual 'cycle models' during the subsequent interviews to explain how the plants in the sealed box could maintain water, air, oxygen and carbon dioxide. They transferred the water cycle model to be a valid description of the cycle of other substances.

In most cases the twenty-three pupils' descriptions became more developed through the years. However, it is possible to recognise personal features in pupils' conceptions. Following the teacher's introduction of a cycle model, Eric used a 'cycle model' to explain why the plants could survive in the sealed box. *'It is sort of vacuum in there. The air evaporates but then it goes down again. And there will be air again. Up and down, up and down. The air rises, evaporates and becomes water. Then it falls down again and there is water there on the ground. Then the air comes up again.'* His 'up and down description' was replaced by a description in fewer words at 13 and 15 years of age, but he did not differentiate between air and water until he was 15 years old.

Several pupils seemed to have also a core idea, a personal theme that can be followed through the years, in the interviews about decomposition. In the interviews with Hanna, about decomposition of leaves, a characteristic feature appeared in her descriptions from 9 to 15 years of age. There is one theme, throughout the years, that concerns raining and drying that still exists far even if other perspectives are connected to this core idea. After Hanna had listened to earlier at 15 and 19 years old, she explained why she had described the process in that way. Near the house where she lived with her family were two big birch trees, from which lots of leaves fell down to the ground every autumn. At 19 years of age, Hanna also said that she remembered how she liked to go out after the rain playing mixing soil and water to mud even though she did not like to be awfully dirty.

Some students continued to express anthropomorphic explanations as personal themes year after year, especially concerning the role of the flower in plant reproduction and concerning defoliation. In the following case, it is possible to identify continuity in the development of the descriptions of defoliation in several pupils' statements, from leaf-centred towards more tree-centred views, and from explanations expressed in terms of a leaf's physical efforts towards those expressed in terms of physiological needs.

Oscar at 9y

'It doesn't get any water. Or it has no muscles left to be able to stay on the branch'

Oscar at 11y

'They don't have the strength to remain sitting there. They must jump off.'

Oscar at 13y

'They fall in autumn and they need much sun. Well, perhaps the tree does not have the strength to carry them any longer. It has enough to do getting nourishment itself and it drops the leaves.'

Oscar at 15y

'Well, it is ...... during winter the tree cannot nourish to the leaves and itself, so it drops the leaves. It cuts off the supply of nourishment to the leaf, doesn't it. Then they die and drop.'

At 19 years of age, after Oscar had listened to his earlier interviews, he claimed that these anthropomorphic features in his descriptions could be a result of his mother's way of explaining the phenomena to him. He said that he knew as a boy that the leaves did not have muscles. It was just a way of explaining why the leaves fell down from the trees in autumn. Also in other cases, he referred to his parents as important to the development of his understanding of ecological processes in the early ages.

Personal themes can also concern the content. Right at the very first interview with Oscar about conditions for life, he argued that the plants needed creepy crawlies in the soil. He expressed the same argument in the interviews at 11, 13 and 15 years of age. In fact this idea helped him to understand how the plants in the sealed box could get carbon dioxide from organisms in the soil. When Sofia explained why the plants could survive in the sealed box, she always mentioned the concept of dew as part of the cycle. At age 12y she said: *'In the morning there is dew on the leaves. And the dew rises up here to the lid and then during the night it drops down to the soil.'* She was the only pupil who used that concept. At 19 years of age, she expressed the following comment after having listened to the earlier interviews with her: *'It's from childhood. The dew has always fascinated me. It is unbehevably beautiful.'*

When Anders was interviewed about the decomposition of the leaves on the ground, he always referred to composting and described this process in a rather detailed way from 11 and 15 years of age. For example, from 9 to 13 years of age he always mentioned that you can put eggshells on the compost. He described the decomposition process as merely a fragmentation and mixing process without any organisms involved. When Anders heard this at 19 years of age he smiled and said: *'We had a neighbour who carried out composting in a special way. I liked to be there with him. The man even put eggshells and coffee grounds on the compost. I remember the first time I was there and he asked me to empty a bucket with coffee grounds and some eggshells on the compost heap. I was confused. I think I was 7 years old.'* This powerful experience seemed to have influenced Anders' thinking about decomposition but it did not help him to develop his idea. He did not develop a conception where organism activity was an important part of the process. Perhaps the powerful experience of the 'mixing process' in the bottom of the compost became an obstacle to the development of a deeper understanding.

Discussion

The longitudinal design of the present study has made it possible to identify individual themes in pupils' conceptual development through the years. These themes could be of a structural nature, a way of explaining a phenomenon. The teacher introduced a water cycle model early at school. Many pupils applied this model to describe cycles of different kinds of matter in an attempt to understand why the plants could survive in the sealed boxes. Personal themes have also been identified concerning the pupils' descriptions of the decomposition of leaves on the ground. Another personal structure that may be followed in the pupils' descriptions is of an anthropomorphic nature that pupils willingly used in order to understand and describe the phenomena.

It has also been possible to identify individual themes concerning the content of the pupils' statements. When the pupils, as 15 and 19 year-olds, listened to what they had said in earlier interviews, they could do more identify than just the themes. In some cases, they could explain how such a theme had built upon a concrete experience in childhood; what was learnt then became an integrated part of the learner's thinking. As in Hanna's case concerning her ideas about decomposition. Many pupils did not replace one understanding with another. Instead, they widened their range of ideas or increased their repertoire of ideas (Marton, 1998). However, core ideas developed at an early age, seem to be an important unit in many pupils' repertoires of ideas. There seems to have been powerful experiences in the early years that made it possible for an individual feature to exist in spite of all other influences while growing up. Even if there was substantial conceptual development, there was also a very strong element of personal context and continuity in the pupils' thinking about the ecological processes.

From a situative perspective it is argued that learning is participation in social practice and dependent on a contextual variation (Greeno, 1997). I do not deny the existence of such a dependence, but I have found, in my research, that there is another context, a personal context and continuity with important implications for the learning process. When pupils listened to what they had said about biological phenomena in earlier interviews, they could reveal particular events that they had experienced together with parents or playmates. The experiences were traced back to social situations but had become a part of the pupils' personal context through knowledge acquisition. Learning can not only be described as participation in a social practice. It is also important to pay attention to what goes on in an individual's mind.

I think there are great possibilities for improving science education by creating an atmosphere that provides pupils with opportunities to recognise and discuss their personal conceptions of scientific phenomena and compare them with other alternative conceptions. My study has shown that early experiences of different phenomena seem to play an important role in the development of many children's conceptual understanding. Would it not be possible to introduce some scientific concepts at an earlier age than we do today?

References

Anderson, J. R., Reder, L.M. & Simon,H. A. (1997).Situative versus cognitive perspectives: Form versus substance. *Educational Researcher* 26, (1), 18-21.

Ausubel, D. P., Novak, J.D. & Hanesian, H. (1978). *Educational psychology: A cognitive view,* 2nd ed. New York: Holt, Rinehart and Winston.

Duit, R., Treagust, D. & Mansfield, H. (1996). Investigating student understanding as prerequisite to improving teaching and learning in science and mathematics. In R. Duit & B. Fraser (Eds.), *Improving teaching and learning in science and mathematics* (pp. 17-31). New York: Teachers College Press.

Greeno, J. G. (1997) On claims that answer the wrong questions. *Educational Researcher* 26 (1), 5-17.

Helldén, G. (1995) Environmental education and students' conceptions of matter. *Environmental Education Research* 3, 267-277.

Helldén, G. (1998, April). A longitudinal study of pupils' conceptualisation of ecological processes. Paper presented at the Annual Meeting of the National Association for Research in Science Teaching in San Diego.

Helldén, G. (1999) A longitudinal study of pupils' understanding of conditions for life, growth and decomposition. In M. Bandiera, S. Caravita, E. Torracca & M. Vicentini. *Research in science education in Europe.* Dordrecht: Kluwer Academic Publishers.

Marton, F. (1998). Towards a theory of quality in higher education. In B. Dart & G. Boulton-Lewis (Eds.). *Teaching and Learning in Higher Education.* Melbourne: ACER.

Novak, J. D. (1998). *Learning, creating, and using knowledge: Concept maps as facilitative tools in schools and corporations.* Mahwah, NJ: Lawrence Erlbaum Associates.

White, R. (In press). The revolution in research on science teaching. In V. Richardsson (Ed.) *Handbook of Research on Teaching.* New York: Macmillan.

Entities of the World and Causality in Children's Thinking

Vassiliki Spiliotopoulou
Technical and Vocational Teacher Training Institute of Patras (SELETE), Greece
Philippos Alevizos
Department of Mathematics, University of Patras, Greece

Abstract

This paper examines the relationship between students' conceptual and experiential worlds. The research method combines qualitative and quantitative approaches in the analysis of data. It is based on the frequency with which categories of description concerning the causality of entities' movement and the origins of the entities are met across a wide age range of students for each entity. Through statistical methods, clusters of entities, for which students' answers fall into particular categories of description are investigated. The results show that a number of entities' features affect students' ways of perceiving the world around us and, more specifically, affect their sense of causality.

Introduction

Students' thinking of natural phenomena and scientific concepts has been the focus of a great deal of research on science education. Most of the studies are content-dependent and concern school science concepts, like mass, charge, force and momentum, that are characterized by Harré (1985) as the group of 'material concepts' and can be used in the description of things, materials and processes. There is also, according to Harré, another group of concepts, that he calls 'formal concepts' or alternatively 'structural' or 'organizational concepts. These concepts include 'causation', 'existence', 'identity' and the spatial and temporal concepts and are considered to bring structure and organization into our observations and descriptions. Causality is related to a number of scientific laws and is an important feature of both scientific and children's thought. The concept of causality has been recognized by Sabursky (1956) as one of the four basic features of the "scientific approach" to the physical world, during the ancient Greek period, which opened a new field in the history of systematic thinking. The causal principle, according to Bunge (1979), is not the only means of understanding the world, but it is one of them and "we are still in the process of characterizing our basic concepts and principles concerning causes and effects with the help of exact tools."

Causality is a neglected concept in school science curriculum; it is rather considered as naturally developed one and has not been extensively studied in the research field of science education. Awareness of the nature and status of children's causality can help us to think of ways to include it in the curriculum, to invent teaching approaches which will develop students' causal patterns and to create a deeper understanding of school science.

Piaget's first studies on the construction of reality show the child as constructing a basic ontology of things that are conserved, together with the construction of

H. Behrendt et al. (eds.), Research in Science Education – Past, Present, and Future, 113–118.

causes. Piaget (1930) dealt with causality, in its broadest sense, including every explanation of a material phenomenon, both the physical aspects of actions and their relationships to objects. Concerning physical causality, he found seventeen types of causal relation in child thought. He claimed that all early forms of causality are inexplicable if we do not allow that, between environment and consciousness, interposed schemas of internal origin occur, i.e. psycho-physiological schemas. The starting-point of causality is a non-differentiation between inner and outer experience: the world is explained in terms of the self. In his later studies of causality, Piaget (1971) stated that "to talk of causality is to presume that objects exist outside of us and that they act independently of us. If the causal model adopted includes an inferential part, the explanation of the phenomenon has the sole purpose of identifying the properties of the object." He admitted, "the causal explanation depends more on the objects than on the subjects."

The constitution of knowledge and the roles of the individual and the external reality in this constitution are still unresolved issues. The idea that the constraints presented by features of the physical world may act to shape personal knowledge construction is still under consideration (Driver, 1993). This matter is a basic concern of the domain of phenomenography, where it is suggested that "the perceived world, rather than perceiving child, would become thematized" (Marton, 1981). So, we focus here on apprehended contents of thought or experience" as a base for integrating findings. This study uses findings concerning students' general ways of thinking about causality and studies how these are related to characteristics of the physical world. More specifically it focuses on the dimensions of causality in students' thought and its relationship with the entities of the outside world. This is explored through children's responses to open questions about the movements and the origins of a number of physical entities. The choice is to study causality in the light of a more general 'knowledge interest' and not as content of science from a teaching point of view. Qualitative and quantitative research approaches have been adopted in order to make sense of the large number of data and to integrate findings.

The research

An essentially phenomenographic perspective has been adopted for the study reported in this paper (Marton, 1988). This study is part of broader research into children's whole system of thinking, which is considered, metaphorically speaking, as a cosmology (Spiliotopoulou, 1994). Cosmology involves ways of thinking that apply in particular areas only and others that apply to a wide range of aspects of the world. Thus, subsystems are explored through students' responses to a number of elementary questions in differing areas about a number of entities. Entities were selected according to the following criteria: 1) to be a familiar part of children's everyday or school life; 2) to belong to three groups of different scale in size; 3) to have physical existence; 4) to meet the constrains of the research.

Data collection

A questionnaire-grid, in which children had to answer 12 simple and basic questions about 22 entities of the physical world has been developed and used with 280 children aged six to sixteen. The research was carried out in a total of 8 schools (4 primary and 4 secondary) in the Merseyside area of England. Children worked, outside the classroom, in groups of four. Discussions were allowed and were audio taped (Spiliotopoulou, 1997). More specifically, this paper focusses on causal schemes that have been identified in children's explanations about entities' movements and entities' origins. In particular, children's answers to the questions: "Does it move?", "Why does it move?", "Who or what made it?" (the reference is to the entity) have been analyzed.

The qualitative data analysis

Analysis was carried out on the overall collection of answers to each and every particular question. Students' answers have been explored in terms of systemic networks, that have been specially constructed by abstracting on the triplets' 'entity-question-aspect of response'. Initial categories, in network form, were tested through our data and were changed or structured in a different way. This procedure has been repeated until we have obtained a certain degree of confidence that the network was a good description of all existing triplets for each question. Regarding the final networks, these can be considered as decontextualized descriptions of students' ways of answering the above questions. Three researchers participated in this phase of analysis and tested, whether the networks can be valuable descriptions of the data. Networks' characteristics, like terminals, delicacy, instantiation (Bliss, Monk & Ogborn, 1983) seem to be suitable for the phenomenographic analysis, as they allow decontextualized categories to be represented and the essential flavour of the data to come through.

Categories of description

The meaning of causality is inevitably related to movement. From an ontological point of view, movement is one of the main characteristics of reality; it is the most important aspect, in a dynamic world, as it has a determining value (Andriopoulos, 1991). Causality is also related to the origins of the entities. These meanings are related to the Aristotelian 'material' and 'final' causes. The relation of these two aspects of causality in children's thinking to the characteristics of the external world is explored in this study.

The main categories and further subcategories of the systemic network for the causality of movement have been discussed analytically elsewhere (Spiliotopoulou & Ioannidis, 1997). We consider here only the main exclusive categories of the network (Fig.1), that have been further analyzed. Concerning entities' movement, a small number of answers can be considered as *tautological* statements and show that students consider movement or non-movement to be so apparent that reasoning is not necessary. *Purposes*, *laws* and *causes* dominate their explanations and could be related to schools of thought known respectively as finality, legality and causality. Answers, like *'the earth moves to keep the world warm'* and *'a butterfly moves to get food'* show children's tendency to see *purposes* as explanations of movements and can be categorized as *general*, *specific* and *human*. Children's responses, like *'space moves because of a natural law'* and *'whatever has life moves-whatever is inanimate doesn't'*, refer to laws, that can be characterized as either *universal* or *local*. Children's explanations, that do not see a purpose or do not refer to laws, fall into the category *causes*. In this case, movement is attributed either to the *entity's ontology*, as in *'the table doesn't move, because it's heavy'*, *'a sea moves because of its waves'* or to *the powers of nature,* as in *'the wind makes trees move'* and *'a house moves, because of earthquakes'. Other factors* like other entities or other entities' actions can also cause motion for students: *'sugar moves, when man stirs it in the tea'* and *'the sky moves, because of the pressure of the atmosphere'.*

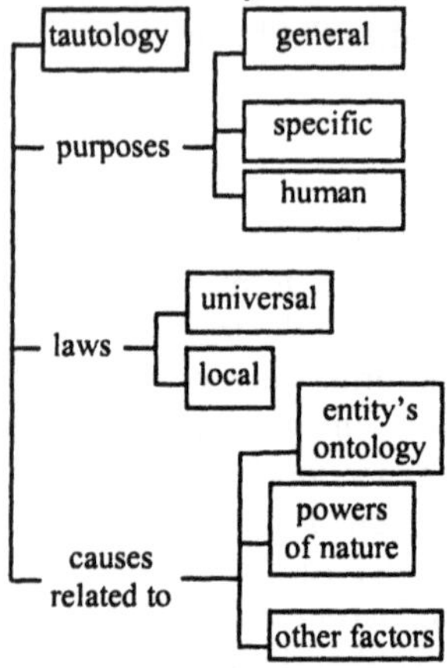

Figure 1 Part of the network for the movement and its causality

Concerning entities' origins, the main categories in children's descriptions express their view that either entities *don't have a creator,* or that the creator can be a *superior power* or *another cause.* When there is no reference to a creator students believe that entities have *always existed,* or that they are *self-created.* Superior power is usually god, while other causes can be *another entity,* an *event,* or a *process.*

The quantitative analysis

A computer program has been specially constructed, that reflects the shape of the particular systemic networks. Through this, data about the children's responses to the questions posed has been entered and analyzed. Frequencies of children's answers falling into the different categories for each entity were obtained. Further analysis was performed on the overall results of the programme, while differences between the age groups have not been considered here. The aim is to look at the results of the qualitative analysis, the decontextualized categories of description about the causality of entities' movement and origins and through the statistical analysis, based on tables of frequencies, to identify patterns in the distribution of categories over the entities.

The analysis proceeded in two stages and methods like the Principal Component Analysis (P.C.A.) and the Cluster Analysis (Lebart et al, 1982; Johnson & Wichern, 1992) have been adopted. The statistical programme SPAD.N has been used. The method of P.C.A. can work on tables with numbers that describe the answers (frequencies) for n statistical units (entities in our case) across a number of variables (categories), which are quantified. Thus, similarities between entities and relationships between variables can be revealed. The use of P.C.A. leads to the creation of axes, where the initial variables are arranged according to coordinates obtained through the analysis. Through the Cluster Analysis, entities can be grouped according to common characteristics, that are described now from the "synthetic variables", that have been produced from the P.C.A. Entities are clustered by Ward's criterion, that is based on the least possible reduction of variance. Thus, the set of entities breaks down into five clusters of entities with common characteristics, described with the help of the initial variables. How these clusters are formed in axes 1,2 and 1,3, is presented in Figure 2, where O1, O2, O3, O4, O5 are the centres of each group. These could be named as: entities of utility, entities of megacosm, entities of nature, alive-useful entities and entities of microcosm. The tendency of the entities to form the first cluster is due to students' reference to 'laws' when they justify entities' movement and to 'another entity', mainly man, when they think about the origins of entities like sugar, table and spoon. Students never refer to a 'superior power', when they discuss these entities' origins. The second cluster is positively characterized by 'tautological' explanations or by 'universal laws' and

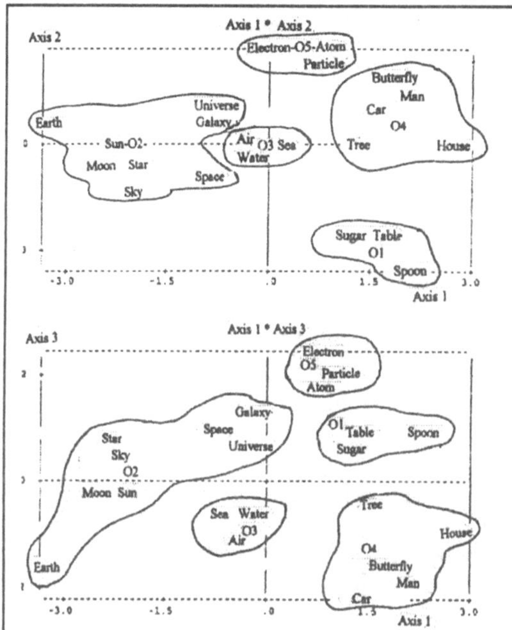

Figure 2 Clusters of entities on axes 1,2 and 1,3

negatively by 'causes due to entity's ontology'. For students, entities, like star, sky, moon, earth, sky, space, galaxy and universe, are created by 'an event', by 'a superior force', or they have 'always existed', or can be self-created, while the reference to another entity, as their creator, is rare. The cluster of 'entities of nature', like air, water and sea, is characterized mainly by students' views that their movement is caused by the 'powers of nature'. The cluster of 'alive-useful entities' is formed by their positive correlation of the entities' movement to causes related to their own ontology. 'Other factors' are rarely used to justify these entities' movement, and 'events' are rarely considered to create them. Entities of the microcosm form the last cluster mainly because they are not included in students' cosmologies and students often express their ignorance about them.

Reflections

It seems that the nature and the qualities of entities shape children's thinking about entities' movement and origins. As children try to make sense of the world and organize their experience, the world and the entities' characteristics in-versely organize and have a role to play in the formation of causal patterns in children's thought. This agrees with the phenomenological claims that ways of thinking are formed and knowledge is constituted from the relationship between the individual and the world. The status of the category of causation does not have an educational interest only, but poses a more general question. Bunge (1979) argued that causality is not a category of relation among ideas, but a category of connection and determination corresponding to an actual trait of the factual (external and internal) world, so that it has an ontological status – although like every other ontological category, it raises epistemological points. So the study of our exploration should be the world-as-experienced by students (Marton & Neuman, 1990) and the concern of our teaching plans could be the interaction of the ontological features of the world and the students' conceptions and the organization of it. Connecting educational research to educational practice we could follow a tradition, which is both content-oriented and makes use of descriptions from the learners' perspectives. For example, when we teach the laws of motion in physics, we never connect to the causes of the entities' movements or non-movement, neither do we explore students' experience; nor do we distinguish the different kind of motions between groups of entities (e.g. cosmic, macroscopic or microscopic entities). In biology, we present theories of creatures' appearance on earth, without considering the real difficulties that this field presents, or using experience from the history of science. We do not take into account students' experiences from everyday life, or systems of beliefs, that already exist in their broader environment and are proposed by social groups that have power in society (e.g. systems of religious beliefs). It is suggested that we would rather deal with both the conceptual and the experiential, as well as with what is thought of as that which is living. We would also deal with what is culturally learned and with what are individually developed ways of relating ourselves to the world around us (Marton, 1981). The consideration of such aspects in our educational plans and their integration in classroom practices could promote students' interaction with the world around them, broaden their thinking and deepen their understanding.

References

Andriopoulos, D.Z. (1988). *The concept of causality in presocratic philosophy. Philosophical Enquiry.* Thessaloniki: Vanias.

Bliss, J., Monk, M. & Ogborn, J. (1983). *Qualitative data analysis for educational research.* London: Croom Helm.

Bunge, M. (1979). *Causality and modern science.* New York: Dover Publications.

Driver, R. (1993). Constructivist perspectives on learning science. In P.L. Lijnse (Ed.), *European Research in Science Education. Proceedings of the first Ph.D.* Summerschool. Utrecht: CDß Press.

Harré, R. (1985). *The philosophies of science.* Oxford: Oxford University Press.

Johnson, R.A. & Wichern, D.W. (1992). *Applied multivariate statistical analysis.* New Jersey: Prentice Hall Inc., Englewood Cliffs.

Lebart, L., Morineau, A. & Fenelon, J.P. (1982). *Traitement des Donées Statistiques, Méthodes et Programmes.* Paris: Dunod.

Marton, F. (1981). Phenomenography – Describing conceptions of the world around us. *Instructional Science* Vol. 10.

Marton, F. (1988). Describing and improving learning. In R. Schmeck (Ed.), *Learning strategies and learning styles.* New York: Plenum Press.

Marton, F. & Neuman, D. (1990). Constructivism, phenomenology, and the origin of arithmetic skills. In L. Steffe & T. Wood (Eds), *Transforming children's mathematics education.* Hillsdale, New Jersey: Lawrence Erlbaum Associates.

Piaget, J. (1930). *The child's conception of physical causality.* London: Routledge & Kegan Paul.

Piaget, J. (1974). *Understanding causality.* New York: W.W.Norton & Company.

SPAD.N (1991). *Système Portable pour l' Analyse des Données Numeriques, Version P.C.. Saint-Mande.* France: CISIA (Centre International de Statistique et d' Informatique Appliquées).

Sabursky, S. (1956). *The physical world of the Greeks.* London: Routledge & Kegan Paul.

Spiliotopoulou, V. (1994). Children's cosmologies. In D. Psillos (Ed.), *European research in science education: Proceedings of the second Ph.D. Summerschool.* Thessaloniki: Art of Text S.A..

Spiliotopoulou, V. (1997). *The cosmologies of children, 6-16 years old.* Unpublished Ph.D. Thesis, University of Patras, Greece.

Spiliotopoulou, V. & Ioannidis, G. (1997). Causality in children's cosmologies and analogies with the historical studies' results. Paper presented at the First Conference of the European Science Education Research Association, Instituto di Psicologia, CNR, 2-6 September, Rome.

Using Media Reports of Science Research in Pupils' Evaluation of Evidence

Mary Ratcliffe and Patrick Fullick
Research & Graduate School of Education, University of Southampton, UK

Abstract

With development of 'scientific literacy' in mind, this classroom-based study explored skills of evidence evaluation of 15-16 year olds, through their interpretation of media reports of science research. Results indicate that most pupils can recognise research claims and evidence but are less able to recognise limitations of evidence and interactions with theory. They can offer sensible suggestions for generation of additional evidence. There appears to be limited pupil development across the one year of the study. More emphasis on evidence-theory interaction may aid pupil understanding of the scientific enterprise.

Background

This study arose from a desire to allow pupils to relate processes of science that they encounter in the classroom to contemporary science research reported in the media.

Many have argued that pupils should develop an appreciation of the conduct of science as a social enterprise with peer reviewed outcomes; pupils should recognise that generation of scientific evidence has a theoretical basis and may be inconclusive (Jenkins, 1997; Millar & Osborne, 1998). Giere (1991, p32) presents a simplified framework of scientific reasoning, along with an algorithmic approach to evaluation of scientific reports. He argues that pupils should consider the fit between the four elements of 'real world', 'model', 'data' and 'prediction'. There is conflicting and limited evidence of pupils' abilities in understanding the interrelationship between aspects of scientific reasoning. Kuhn *et al.* (1988), from presenting different age groups with problems in reasoning, argue that skills of distinguishing between evidence and theory are underdeveloped even in adults. Samarapungavan (1992), however, presents evidence that even young children are 'capable of co-ordinating theories with evidence and reasoning rationally in both theory selection and theory production contexts'. Driver *et al.* (1996) found that ability to be consistent in relating evidence to explanation increased with age, whilst there was a corresponding decrease in phenomenon-based reasoning. Other studies have highlighted that high school graduates have limited understanding of the scientific enterprise through evaluation of media reports of science research (Korpan *et al.*, 1997; Norris and Phillips, 1994; Phillips and Norris, 1999).

The intention of this study was to explore pupils' abilities and teachers' actions in evaluating evidence in media reports. The study was undertaken in classrooms rather than in clinical settings, in order to understand the factors that influence pupil development and classroom practice. This paper reports on one aspect only - the pupils' abilities.

H. Behrendt et al. (eds.), Research in Science Education – Past, Present, and Future, 119–124.

In order to allow discussions of media reports, on a regular basis, in science classes, teachers needed to relate the activity to the statutory science curriculum. In evaluating evidence as part of their science investigations, pupils are expected to: consider the sufficiency of evidence; reasons for anomalous data; reliability of results and to propose improvements and further investigations. This National Curriculum representation of the nature of scientific enquiry is not without its critics (e.g. Donnelly *et al.*, 1996) as it presents a narrow view of 'scientific method' with underdeveloped links between theoretical modelling, evidence generation and evaluation.

Research questions, explored in this paper, were:
i) What evaluation skills of media reports of research do pupils show?
ii) Do pupils improve evaluation skills of media reports with regular use?

Methodology

A new comprehensive school, X, had its first intake of pupils in year 10 (age 15) at the time of the research (1998/9). An evaluation of media reports of science research was incorporated into each of the six science modules for the year. (Earth Materials M1, Energy M2, Metals M3, Humans as Organisms M4, Maintenance of Life M5, Environment M6).

For each module, a report of up to 550 words from the popular scientific journal *New Scientist* was chosen to fit the curriculum content. Although *New Scientist* is aimed at adults, it contains summaries of recent research in a consistent format which coincide with Nwogu's (1991) analysis of such media reports.
The nature and structure of each report should allow pupils with mature reasoning to recognise that there are limitations in the evidence presented.

Although different reports were used across the modules, pupils attempted the same questions which linked with expectations of evidence evaluation in the National Curriculum:
1. What do the researchers claim? (i.e. what is the conclusion?)
2. What evidence is there to support this conclusion?
3. Is this evidence sufficient to support their claims? Explain your answer.
4. What further work, if any, would you suggest?
5. What scientific knowledge have the researchers used in <u>explaining</u> their results and claims?

Six science teachers each taught a class of 20 pupils. Apart from M6, Environment, each class was taught modules in a different order. Although the teachers may have different teaching styles and priorities, the format of the lesson, involving the report, was similar for each teacher, viz: the teacher presented the report in the context of the module's science content; made general, but not specific, links to evidence evaluation in investigations; explained ideas but did not provide answers to questions; summarised pupil responses after completion of the questions. The pupils recorded their responses to the questions and had the opportunity of discussion with their immediate neighbours. In theory, comparisons could be made between responses to the same report for pupils with limited and more extensive experience of such activities. As it was anticipated that this sole method of comparison of possible pupil development might prove problematic, a 'control' class was also used. The science department of another school, of similar pupil background, was also following the same syllabus. One class, Y, from this school answered questions on two reports, *without any teacher introduction or input during the pupil completion of questions or any emphasis on evidence evaluation during the year.*

Results

In practice, the six teachers engaged with the suggested use of *New Scientist* reports to different extents. Table 1 shows the pattern of use. Classes D,E and F experienced one report at most, hence discussion of any pupil development is restricted to classes A,B and C.

Table 1: Pattern of use of New Scientist reports by different classes

		Module				
Class	Sept	Nov	Jan	Mar	May	July
A	M1	M2	M3	M4a		M6
B	M3			M2		M6
C	M4b	M5		M3		M6
D	M4a			(interested but detailed scheme of work not used)		
E		M3		(limited engagement with research intentions)		
F				(no reports used despite interest expressed)		
'Control'						
Y		M5				M6

Classes A and B contained pupils expected to gain higher GCSE grades (estimated B-D across each class). The other classes, including class Y, had expected grades across the range C-G. Reports coded M1-M6. (M4a, M4b alternative reports for module 4)

Analysis of responses

Responses to questions 1 and 2 were categorised according to whether pupils reflected the key research outcome and a summary of the evidence. The specific categories for each report, full and partial, were identified by the first of us and validated separately by the second of us to allow an agreed protocol for marking.

Summarising across all responses in school X (n=188):
1. What do the researchers claim?
Full response 63%, partial response 32%, 'wrong' response 5%
2. What evidence is there to support this conclusion?
Full response 51%, partial response 28%, 'wrong' 13%, no response 7%
3. Is this evidence sufficient to support their claims? Explain your answer
No 47%; Yes 23%; Ambiguous 7%; not answered 23%
4. What further work, if any, would you suggest?

Suggestions for further work varied from a general exhortation for more testing to consideration of a range of different variables to alter. Many pupils made more than one suggestion. Of the 191 suggestions made, the largest proportion (53%) was to repeat the research for different named variables. Other suggestions, that were report dependent, included suggesting a larger sample size (17%) or a larger scale or longer time (17%). Only 3% suggested repeating the research as given, while 6%

were vague in suggesting 'further tests'. Only 4% referred to any underlying scientific model, which was explicitly discussed in two reports (M1 and M6).

5. What scientific knowledge have the researchers used in explaining their results and claims? On only five out of the fourteen occasions was this question attempted to any degree. Responses fell into four categories (total responses=78).

i) Citations of terminology and vocabulary used in the report (69%).

ii) Comments on the properties of the research subject (24%). These responses seemed to show a more general understanding of the content area.

iii) Comments reflecting scientists' use of experimental procedures (15%).

iv) Comments indicating no scientific knowledge was used (4%).

In order to make comparisons between individual pupils' responses, a reasoning level or 'score' was assigned to responses for the first 4 questions based loosely on the SOLO taxonomy, that has 5 levels based on use and integration of information (Biggs & Collis, 1982). (The 'extended abstract' level, 5, of the SOLO taxonomy was not relevant to responses seen):

1. Claim identified fully or partially and/or evidence identified fully or partially (response incomplete) *(pre-structural)*
2. As 1, plus evidence sufficient; no suggestions for further work *(uni-structural)*
3. As 1, plus evidence insufficient, no suggestions for further work or evidence as sufficient with suggestions for further work *(multi-structural)*
4. As 1, plus - evidence insufficient; suggestions for further work *(relational)*

Table 2 shows the responses for each class in the order in which they completed the tasks. Comparisons of responses on different occasions could only be attemted for classes A, B, C and Y. Two-tailed t-tests, for significance at the 0.05 level, were carried out in comparing the responses of different classes to the same report and the same class to different reports. A paired two-tailed t-test was carried out in comparing responses to M5 and M6 for both class C (11 pupils) and class Y (22 pupils), where sufficient matched samples could be clearly identified. It is recognised that this analysis has limitations, in particular the limited number of pupils with complete records for the reports attempted by their class and the limited validity of unpaired t-tests in this analysis. Nonetheless, the statistical analysis supports visual inspection of any differences in response patterns. These statistical comparisons showed that only in four cases were there significant differences at 0.05 level (Class B M6 vs M2; Class C M3 vs M5; M3 class B vs C; M6 class C vs Y). These are discussed below in the context of each report or class.

Individual Reports

It was anticipated that pupil performance is to some extent contextualised by the nature of the report. Thus comparison of performance by different groups on the same report undertaken at different times might show whether any progress in evaluation was made during the year. As table 1 illustrates, this comparison was necessarily limited to M3, where the one significant difference was between class B and class C. The better performance by class C indicates that there may be some developmental effect here. There was no significant difference in performance on M6, the final task, for classes in school X.

Table 2: Responses to the reports from each class

Class	Report	N	Reasoning 'score			
			4	3	2	1
A	M1	18	10	6	0	2
*A	M2	12	1	1	0	10
A	M3	16	8	3	1	5
A	M4a	15	8	2	3	2
A	M6	12	6	4	1	1
B	M3	17	5	6	1	4
B	M2	7	2	0	2	3
B	M6	6	5	0	1	0
C	M4b	9	0	0	0	9
C	M5	16	8	3	0	5
C	M3	15	13	0	0	2
C	M6	11	10	0	0	1
D	M4a	15	1	7	2	5
E	M3	19	12	3	4	0
total X		188	89	35	15	49
%			47	19	8	26
Y	M5	22	14	4	0	4
Y	M6	22	6	7	5	4

* excluded from statistical analysis - insufficient time given for completion

Individual Classes

Class A performed similarly throughout the year. A comparatively high proportion of pupils were able to reason effectively. Teacher A was confident in handling the material. She had a variety of ways of collecting outcomes from the activity to match the relevant content and skills in the module. Class B performed significantly better on M6, the last task, than on M2. Class C performed significantly better on M3, the later task, than on M5 and maintained this progress. This class seemed to be the only one that showed steady progress during the year with more experience of the tasks. The apparent progress of class C in comparison to other classes in school X can also be examined with respect to the 'control' class in school Y. Class C and class Y performed similarly on M5, despite class Y having no teacher input. Class Y's performance on M6 later in the year is poorer than on M5. There was a significant difference between class C and class Y's performance on M6, supporting the other comparisons showing that class C had made some possible improvement during the year.

Conclusion

There was an expectation that pupils showing mature reasoning would consider that additional evidence was needed to validate the main research outcome presented in the media report. Across the whole year, 47% of pupils in school X completed the first four questions, regarded the evidence as insufficient and argued for extensions or variations of the research. A larger number (66%) could indicate possible refinements, regardless of the extent to which they fully evaluated the evidence presented. The results are optimistic in showing the potential for developing skills of evidence evaluation, a similar outcome to a pilot study with younger pupils (Ratcliffe, 1999). It is difficult to attribute reasons for development, or lack of it, in different classes. Natural maturation and repeating the activity do not, of themselves, seem to improve evaluation skills. Extensive and explicit teaching seems needed to develop skills of evidence evaluation fully. This is also supported by high school graduates' limitations of evidence evaluation without prior teaching (Norris & Phillips, 1994; Ratcliffe, 1999). We need a greater understanding of *how* pupils develop in their skills of evidence evaluation. To be effective in supporting evidence evaluation, teachers need to be skilled in leading such activities and regard it as important to develop this aspect of scientific literacy. Some teachers in this study had limited engagement with the activities, despite declaring interest and receiving support in undertaking novel activities. Ongoing changes in the National Curriculum to teach pupils the nature of scientific enquiry require parallels in development and uptake of pedagogical strategies.

References

Biggs, J.B.& Collis K.F. (1982). *Evaluating the quality of learning.* London: Academic Press.

Donnelly, J.F., Buchan, A.S., Jenkins, E.W., Laws, P.M., & Welford, A.G. (1996). *Investigations by Order. Policy, curriculum and science teachers' work under the Education Reform Act.* Driffield UK: Studies in Education.

Driver, R., Leach, J., Millar, R., & Scott, P (1996). *Young people's images of science.* Buckingham: Open University Press.

Giere, R.N. (1991). *Understanding Scientific Reasoning* Fort Worth TX: Holt, Reinhart & Winston.

Jenkins, E (1997). Towards a functional public understanding of science. In R. Levinson & J. Thomas, Eds.; *Science Today* (pp. 137-150). London: Routledge.

Korpan, C.A., Bisanz, G.L. Bisanz, J. & Henderson, J.M. (1997). Assessing literacy in science: evaluation of scientific news briefs. *Science Education* 81, 515-532.

Kuhn, D., Amsel, E. & O'Loughlin, M. (1988). *The development of scientific thinking skills* London: Academic Press.

Millar, R. & Osborne, J. (1998). *Beyond 2000: Science education for the future.* London: King's College London, School of Education.

Norris, S.P. & Phillips, L.M. (1994). Interpreting pragmatic meaning when reading popular reports of science. *Journal of Research in Science Teaching* 31, 9, 947-967.

Nwogu, K.N. (1991). Structure of scientific popularizations. A genre-analysis approach to the schema of popularized medial texts. *English for Specific Purposes* 10, 111-123.

Phillips, L.M. & Norris, S.P. (1999). Interpreting popular reports of science: what happens when the reader's world meets the world on paper. *International Journal of Science Education* 21, 3, 317-327.

Ratcliffe, M. (1999). Evaluation of abilities in interpreting media reports of scientific research. *International Journal of Science Education* 21, 10, 1085-1099.

Samarapungavan, A. (1992). Children's judgement in theory choice tasks: Scientific rationality in childhood *Cognition* 45, 1-3.

Pupils' Perceptions of Science Education at Primary and Secondary School

Bob Campbell
Department of Educational Studies, University of York, UK

Abstract

This paper analyses the views of pupils before and after transfer to secondary school. While pupils reported an enjoyment of science the study raises concerns about the image, status and academic challenge of school science.

Introduction

Secondary schools in England are accused of ignoring pupils' primary schooling (Ponchard, 1997). There is also evidence that commitment and enthusiasm for science built up in primary school decreases on transfer to secondary school (Ofsted, 1994; SCAA, 1997). It is postulated that a factor in regression is that pupils' expectations of secondary science are not being met. While the literature on primary-secondary transfer is expanding, there is little on pupils' perspectives (Jarman, 1993, 1997). Against this background, a case study of a class of primary school pupils was undertaken. Three research questions were posed: (i) What are pupils' perceptions of learning science in their primary school? (ii) What are their expectations of learning science at secondary school? (iii) To what extent are these expectations realised in their secondary school?

Research Methods

Data were collected in June and July (Phase 1) as pupils came to the end of their primary schooling and again in December (Phase 2) at the end of their first term in secondary school. Phase 1 data were collected by questionnaire and semi-structured interviews. The questionnaire asked what pupils liked and disliked about primary school science and why; how they considered secondary school science would be the same as and different from primary school science and why; and what they were looking forward to and not looking forward to and why. Analysis of the responses generated categories of answers and reasons for each. Three sets of cards, each carrying a common category of response, were prepared. These were used in semi-structured group interviews. Pupils discussed the statements and sorted them into an order to reflect the majority viewpoint. Five, tape-recorded, group interviews of 15 to 20 minutes with 6 to 8 pupils were undertaken. Phase 2 data were collected by a questionnaire. This probed pupils' views of primary and secondary school science, differences between primary and secondary science and the reality of their expectations of secondary school science. Twenty-six pupils contributed data to both Phase 1 and Phase 2. This group had experienced National Curriculum Science at both schools but in different ways. At primary school they were taught in a classroom by a their class teacher but in secondary school they were taught in a

H. Behrendt et al. (eds.), Research in Science Education – Past, Present, and Future, 125–130.

laboratory by a specialist teacher. In both situations teachers used a wide variety of learning resources.

Findings

What pupils enjoyed most about learning science

Only aspects of content were reported by pupils as enjoyable at primary school whereas at secondary school, though again dominated by areas of content, pupils also mentioned equipment, experimentation and teaching methods. This suggests that at secondary school, pupils gained a broader perception of science. They see science as more than content and have opinions on how as well as what they are taught.

Aspect	Number of mentions by Primary School Pupils. n=26	Number of mentions by Secondary School Pupils. n=26
Content	74	29
Equipment	0	21
Experimentation	0	17
Teaching	0	5

Table 1: What pupils enjoyed most about learning science

Why pupils enjoyed learning science

The reasons given by pupils as to why they enjoyed learning science fell into 6 major categories as shown in Table 2.

Reason	Number of mentions by Primary School Pupils . n=26	Number of mentions by Secondary School Pupils. n=26
Discovery	21	23
Experimentation	15	
Teaching	12	13
Novelty	6	20
Fun	6	20
Interest	5	6

Table 2: Pupils' reasons for enjoying learning science

Regardless of what they stated they enjoyed, the responses from primary and secondary pupils explained this in terms of how science is learned (discovery, experimentation and the variety of teacher activities) rather than content. This interpretation is supported from the interviews in which 'doing experiments' was the main reason for enjoying learning science. 'Finding things out' and 'seeing things' were given as subsidiary reasons. Typical comments from primary pupils were such as: *It's exciting to do the experiments: to see what's happening* (I/P17); and *I like doing experiments. It is fun* (Q/P9). The underlying reason for content being viewed as enjoyable is that it is associated with a variety of activities; experimentation; the personal focus of control; the exciting uncertainty of the end result of practical investigation; and the uniqueness of science in relation to other areas of study. While the opportunity for discovery and the exposure to a variety of teaching approaches remained as explanations as to why secondary pupils enjoyed learning science, it is surprising that none made explicit mention of experimentation.

What pupils enjoyed least about learning science

Few aspects of learning science were listed as enjoyed least (Table 3). The majority related to areas of content but, unlike their earlier responses, the primary pupils also listed aspects of process.

Aspect	Number of mentions by Primary School Pupils. n=26	Number of mentions by Secondary School Pupils. n=26
Content	22	10
Teaching	13	4
Writing	3	10

Table 3: What pupils enjoyed least about learning science

There were some common underlying reasons why these aspects of learning science were least enjoyed. Writing was cited as something that was least enjoyed and also as a reason why other aspects were least enjoyable (Table 4).

Reason	Number of mentions by Primary School Pupils. n=26	Number of mentions by Secondary School Pupils. n=26
Teaching	11	1
Writing	10	5
Content	9	2
Difficulty	6	4
Boredom	5	9
Prior Knowledge	5	0

Table 4: Pupils' reasons why aspects of learning science were least enjoyable

Writing was the most mentioned negative factor in learning science. Writing was seen as boring, uninteresting and unnecessary. Primary and secondary pupils expressed similar views. Some primary pupils were quite explicit: *We hate writing* (I/P34); *Writing things up: that's what we do not like. That's definitely what we don't like* (I/P19). Some secondary pupils explained their dislike further in stating: *Writing up experiments: it gives you handache and it's boring* (Q/S1); *You have to write a lot and it gets boring* (Q/S16).

Boredom was the only reason mentioned more often than writing by secondary pupils. Boredom was linked not only with writing but also with a teaching style that required more passive than active participation and conflicted with pupils' expectations of doing science. Primary pupils saw themselves as active experimental investigators. What they did not enjoy was a teaching approach that required them to plan and record their investigations, so taking time away from experimentation. This was described in various ways such as: *We wasted time planning* (Q/P16); *You had to write and explain what you were doing* (Q/P25).

While experimentation was enjoyed in secondary school it was not stated as an explanation for enjoying other aspects of science, as was the case in primary school. This may be because it was an infrequent activity or because of a different teaching approach that located control with the teacher and not the pupil. For example, one girl characterised her experience of experimentation at secondary school thus: *We get told what to do and then go off in pairs (Q/S2).* Perhaps the reason experimentation is cited as enjoyable by secondary pupils is because, even directed experimentation, was a welcome break from less enjoyable activities.

Primary school pupils' expectations of learning science in the secondary school

Pupils listed more differences (26) than similarities (19) between primary and secondary school science. The majority thought that the greatest continuity between their primary and secondary science would be in experimentation and content. They expected the content to be the same but that it would be more difficult and more extensive. For example, pupils wrote: *We will do the same sort of topic but in more detail (Q/P12); We will learn the same things but in harder, different ways (Q/P3).* Most had an expectation of continuing personal experimentation. One pupil wrote: *You will still do experiments (Q/P25).* Another primary pupil considered that because experimentation is essential to science this would continue at secondary school: *We do experiments at both schools because it is science (Q/P30).*

There was also the notion that primary science was not genuine and that once at secondary school pupils would be *doing proper science (Q/P8).* This view links to the perception that the main differences related to learning environments. Pupils expected to be in laboratories with specialised facilities and sophisticated equipment. They contrasted this with the classroom science of the primary school and with simple apparatus. One pupil wrote: *They have proper science rooms and can do more serious experiments because they have tools and proper things to use for experiments* (Q/P26). While pupils expected learning to be more difficult they also expected to be given more responsibility. One pupil expressed this well: *They treat you more as adults and give you more time and equipment (Q/P7).*

When asked what they were most looking forward to, all the interview groups ranked using chemicals as either their first or second choice. All but one group had as their other top choice 'using equipment', 'doing experiments' or 'working in labs'. Primary pupils perceived learning science in secondary school as experimenting in specialist rooms, using chemicals, materials and apparatus not found in their primary school.

Expectations and reality of learning science in the secondary school

After a term of secondary science pupils, listed a similar number of differences (25) as at primary school but fewer similarities (7). Responses confirmed expectations on facilities had been met. They were taught in laboratories and used unfamiliar apparatus. This was evidenced in comments such as: *Using equipment and the equipment was more scientific* (Q/S1); *We used Bunsen Burners which I thought that we would do (Q/S7).* The frequent mention of the Bunsen burner suggests its symbolic significance as apparatus that characterises doing secondary school science. Comments on content indicated both elaboration of previously studied topics and new areas of science. With regard to difficulty, 11 of 74 primary comments indicated an anticipated increase in difficulty of science at secondary school but only 4 of 55 Phase 2 comments reported more difficult work. Two of these were linked to the amount of work: *You learn more at high school* (Q/S2); *You do a lot more science because you have 3 lessons of science a week* (Q/S9).

A number of pupils expressed an expectation that at secondary school they would be given more responsibility for their learning but this was not in evidence. This contrasts with corroboration that expectations about equipment had been met. However, no other anticipated aspect was mentioned as experienced by a significant number of pupils. The expectation of performing personal experiments with new and exciting equipment was unrealised. More than half the responses stated an unfulfilled expectation in this area. The statements: *I expected to do experiments everyday but we haven't* (Q/S3) and *I thought that we would do a lot more experiments than we have done so far* (Q/S20) signal the disappointment.

Three aspects dominated the 19 comments on what had surprised pupils. One related to new content (e.g. forms of energy) and another to 'dangerous apparatus' (a radioactive clock was given as one example). These are seen as positive aspects of pupils' new experience. The other aspect was writing. Comments such as: *I did not think that we would do big write-ups* (Q/S8); *We have to do a lot of writing* (Q/S27) are seen as negative comments and characterise a perception of science as a practical subject with little need for writing. In describing what she expected at secondary one school pupil stated: *We don't have to do more work (=writing) because you learn things in experiments and you don't have to do much work on them* (I/P8).

Discussion

While a case study, aspects of the primary-secondary interface identified here may be of value in informing the debate on conceptualising cross-phase curriculum continuity as advocated by Jarman (1999). While it needs to be recognised that, overall, pupils reported that they enjoyed learning science, the study raises three areas of concern.

Firstly, pupils left the primary school with a perception that the most important learning activity was personal, practical experimentation. There was a naive view that experimenting consists of just seeing what happens when, for example, chemicals are mixed. Pupils did not communicate an understanding of science as planned and systematic enquiry. Activities other than hands on experimentation were seen as barriers to 'real learning' and boring. This perception influenced their reaction to secondary school science. A low status was given to writing in both phases of schooling. Pupils did not convey either purpose or ownership of writing. From this analysis it would seem important that primary teachers present experimentation as just one way of learning science. Equally important is the need to present experimentation as planned investigation rather than random testing. Also, we need to give greater emphasis to the purpose of pupils' writing in science.

Secondly, pupils reported that an expected increase in the challenge of learning science in secondary school did not materialise. There is other evidence (e. g. Gunnell, 1999) that secondary science teachers underestimate the capabilities of entering pupils. SCAA (1996) advocate a transfer protocol for the move from primary to secondary school to ensure that information about pupils is passed on.

Kaur (1998) describes practice to make better use of national test scores provided by primary schools to monitor pupil progress. What may also be needed is more bridging projects such as that reported by Edwards (1999) in which outcomes of activities done in primary school science are used as starting points for secondary school science. In such ways primary and secondary teachers can reach common understandings of processes, standards and achievements and plan appropriate learning activities to achieve curriculum continuity and progression.

The third concern is with the perception of the status of science education in the primary school. While enjoyed by pupils, it was not seen as 'real science' when compared to that of the secondary school, where laboratories and specialised equipment made it seem exotic and superior. The danger is that pupils undervalue their primary school curriculum.

Pupils' perceptions of learning science are important to their future interest and attainment. Secondary school teachers have to deal with the perceptions developed before transfer. Thus, in primary schools we must work to ensure that these perceptions are balanced and realistic so that they form an appropriate basis for further learning. In secondary schools we must provide learning that recognises and builds on experiences and achievements of the primary school curriculum.

References

Edwards, R. (1999). Bridging Key Stages 2 and 3. *Education in Science* 181, 12-13.

Gunnell, B. (1999). Meeting the children. *Education in Science* 181, 10-11.

Jarman, R. (1993). Real experiments with Bunsen Burners: pupils' perceptions of the similarities and differences between primary and secondary science. *School Science Review* 74, 268, 19-29.

Jarman, R. (1997). Fine in theory: a study of primary-secondary continuity in science, prior and subsequent to the introduction of the Northern Ireland curriculum. *Educational Research* 39, 3, 291-310.

Jarman, R. (1999). Primary-secondary continuity in science. *Education in Science* 181, 4.

Kaur, B. (1998). Primary/Secondary liaison in science and value added form Key Stage 2 to 3. *Education in Science* 179, 9-11.

Ofsted (1994). *Science and mathematics in schools: a review*. London: HMSO.

Ponchard, B. (1997). Your Ofsted inspection is over - what next? *Primary Science Review* 49, 9-11.

SCAA (1996). *Promoting continuity between Key Stage 2 and Key Stage 3*. London: SCAA.

SCAA (1997). *Making effective use of Key Stage 2 assessment*. London: SCAA.

Acknowledgements

This paper draws on elements of the studies of two graduate students, Sally Ashworth and Gloria Beecroft. I thank them for access to their data.

Part 4: Teachers' Conceptions

Teacher Professionalism and Change: Developing a Professional Self Through Reflective Assessment

Manfred Lang
Institute for Science Education (IPN), Kiel, Germany

Abstract

Teachers and schools are, increasingly, subject to policies of educational change. However, as documented in different analyses, little will be accomplished by imposed change if teachers do not understand these reforms and become involved. One alternative is: professional "self-development", supported by a systemic framework. As a prototype of this process a continuous workshop of the German PING project for the development of integrated science teaching is described and analysed. Results are presented about the effect of concept maps as part of teachers' self-development, supported by a collaborative system. Teachers professional development through reflective self-assessment and feedback from researchers was found to be effective on components for improvement of planning science learning and development of integrated goals.

The role of professional development in educational change

Educational reforms are a result of rapid changes and developments in society and increasing political pressure (Black & Atkin, 1997; Beaton et al., 1996). Policymakers sometimes assume that organisational conditions are sufficient to be able to fit new ideas and strategies to unique, on-the-job conditions. However, as Sikes (1992) and Guskey (2000) note, little will be accomplished if teachers do not understand these reforms and get involved in a collaborative and time consuming process. They are required to change themselves in order to meet the specifications laid down by policy makers. A basic prerequisite for change is teachers' "self-development" or "self-cultivation" (Terhart, 1999). Teachers' professional self is the source of "self-cultivation" that guides professional development in a specific culture with its socialising forces. The "Self" cannot be produced from outside, but is formed by an individual's wisdom and ethical and moral conviction on a cultural and social background.

"Self-development" can be supported in a systemic framework (Yinger & Hedriks-Lee, 1998; Sikes, 1992) by creating conditions in which teachers can teach well and are challenged to improve their practice. Professional development and curriculum innovation are interlinked just as professionalism is part of curricular reflection in practice and innovation takes account of teacher professionalism. Teachers are part of an interacting network, that maintains a balance between their personal perception of educational needs and influences from outside. According to

H. Behrendt et al. (eds.), Research in Science Education – Past, Present, and Future, 131–136.

Bauer (1999), the professional self is a result of a professional effort to develop an action repertoire in practice, a system of values and goals, occupational and subject matter knowledge and a professional language in a professional culture. This "Self" develops in the context of an educational system, that gives teachers more or less "professional" freedom and supports partnership with openness, trust, mutual help and understanding. Sources for the development of the self are scientific and occupational expertise, feedback and social interaction through colleagues and teams and reflective decisions and actions in educational practice.

A central issue of professional development in various studies and analyses in research relates to an integrative knowledge structure of subjects and reforms that require teachers to move away from the twin securities of subject expertise and control of classroom agendas (Black & Atkin, 1996; Robertson, Cowley & Olson, 1998). Subject matter provides a highly complex frame for the work that teachers do (Roulet, 1999). It is difficult for teachers to alter that frame without support. Scientific knowledge suitable for the construction of a professional self understanding does not directly allow for mastery of everyday professional practice. Integration of science knowledge is, therefore, a demanding and time consuming task for teacher professionalism in a process of change as demonstrated in the following project.

Developing a professional self for integrated science teaching

Developing components of the professional self was studied in a collaborative workshop as part of a project for integrated science teaching (PING) in Germany, conducted between June 1996 and June 1998. Thinking in terms of integration is generally difficult for teachers, due to the lack of pre-service courses and material for integrated science teaching, disciplinary routines and a teacher-centered style of teaching and the exclusion of a variety of socially-oriented topics.

The national PING project is based on collaborative activities of teachers, researchers and teacher educators. Three principles comprise the approach to teacher education: (i) knowledge of the disciplines and teaching methods need to fit; (ii) development of materials is a collaborative process involving different groups of professional expertise; (iii) continuous reflection is central to professional development. Teaching is an open process of planning and learning for teachers and students. Experiences with this teaching and learning process are part of a reflective collaboration on development and revision of teaching materials and concepts, teacher training, research and administrative changes. Professional development in the project is maintained by agreement of teachers to participate in evaluation activities for material development and workshops for exchange of experiences. A core group of teachers, teacher trainers and researchers in the project are linked to a network for exchange with regional project groups and representatives from different federal states. The IPN – the research centre at the University of Kiel – provides a co-ordination centre for planning, information exchange, material development and revision to support co-operation.

The analysis of the PING project in the international OECD report (Black et al. 1997) on changes in mathematics, science and technology education suggests that change in science teaching is mainly due to the central role teachers play in a system of collaboration with university researchers and teacher educators. The report points out that, in some cases, practice is not sufficiently linked to elements of a curricular system allowing reflection in practice, which is a prerequisite for professional development. As a consequence, a collaborative approach for the PING workshop was adopted that fits the teachers' central role in the project and the aim of professional self-development in a curricular process. Self-assessment through concept mapping is assumed to support teachers' professional change (Mason, 1992). In the context of integrated science teaching, it is a powerful tool to design and analyse structural knowledge and to communicate complex ideas.

Results from research done during the continuous workshop were used in order to develop a more detailed picture of changes affecting teachers' professional self. What is the professional core of a teachers' self? How does self-assessment and collaboration influence aspects of teachers professional self? What educational changes may be expected?

Self-assessment of knowledge structure through concept-maps

During a period of two years a group of 22 teachers from 11 schools met in 9 sessions to prepare, plan and analyse lessons on 7 integrated topics. In two of these sessions, teachers developed concept maps for reflective self-assessment on planning and realisation of lessons on the topics "water" and "soil". These concept maps were discussed and changed during the workshop sessions. Final versions were copied and given to the teachers for their lesson planning. On the basis of these concept maps, they tried to conduct their lessons, followed by two other sessions reflecting on their work and revising their original planning.

Teachers appear to have adapted concepts for lesson planning based on the practical problems they encountered in teaching. They learned to use concept maps as flexible instruments for making choices in teaching. As one teacher commented in his post-map: "I choose concepts more carefully. I have these concepts at first, choose work-sheets on the basis of these concepts and have to judge their usefulness. This is the issue, whether they fit the goal." In addition, they also try to integrate proposals offered during the course into their maps: "The suggestion to use the aquarium was a shining example. I was looking for something objective that I could use in class or sometimes the class can look at and something that remains."

Teachers also mentioned their concerns about student interests or goals when choosing concepts for planning a lesson. One teacher said: "I was inclined to try to do everything that includes in my concept map. This was a reason to look for students' achievement. But it was by far too much. ... Now I would reduce it to the [topic] pond. I would do more with students and would emphasize pollution."

Concept maps are primarily perceived as instruments for analysis of subject matter but may not adequately reflect the realities of teaching. As one teacher confirmed: "For me concept-maps are important only for preparation of teaching on the basis of subject matter analysis. A teacher should keep this limitation in mind when planning his course for a particular grade. In addition, he has to think about goals and central activities in advance. It is not possible to transfer concept-maps completely into practice." The challenge for planning is to relate the integrative structure of a concept-map to actual practical activities as some teachers were able to do: "The new concept-map is less comprehensive but more related to practical activities.The Black River – our river in our village. We are now looking for the different interests children have when they work on different parts of the river. We didn't consider this in our first draft" .

Figure 1 and 2 show an example of a pre-map and modified post-map constructed by one group of teachers. They illustrate the reorganisation of central concepts and links and the reduction in the number of concepts and links that were typical among other groups in the course.

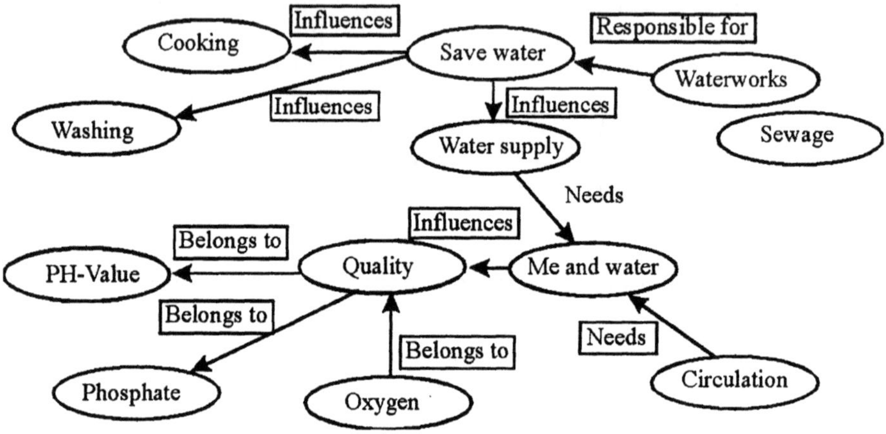

Fig. 1: Pre-map for planning to teach the topic "Me and water"

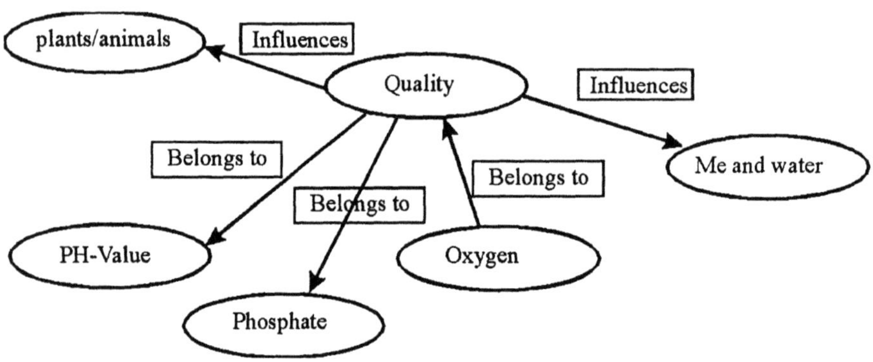

Fig. 2: Revised Post-map for the topic "Me and water"

Analysis of concept-maps and interviews indicates, that, during the workshop teachers' lesson planning changed from disciplinary structures related to the topics "water" and "soil" to conceptual structures that are more specific for integrated science teaching. Concept maps, constructed and discussed during the workshop were judged to be useful for teaching. They were limited in use through lack of time, different student interests and organisational barriers in school.

At the end of the workshop in June 1998, a questionnaire was used for inquiries on professional development as a prerequisite for integrated science teaching in the PING project. In order to assess participants' professional use of new knowledge and skills in integrated science teaching, indicators need to be identified (Guskey, 2000). These indicators were developed from empirical findings about teachers' professional self by Bauer (1999). In the context of integrated science teaching they are determined as abilities to develop goals for integration of science matter in education and its relation to everyday experience, knowledge about science learning, action repertoires and occupational knowledge.

One section of the questionnaire contained items asking about these indicators and reasons for professional development: Did the course help teachers to develop integrative goals, to integrate science and everyday knowledge, to plan actions for lessons, to improve science learning and to improve their occupational knowledge? Were possible reasons for this: the exchange of experiences among colleagues, feedback from research or self-assessment through concept maps?

Teachers rated the items on a four point scale ranging from 1 (very) to 4 (not at all). They gave high ratings for the usefulness of the PING workshop especially for relating science content and everyday experiences (M=1.50), the development of integrative goals (M=1.62) and the development of actions guiding lesson interventions (M=1.72) as aspects of professional development.

For each teacher a mean value of the items about aspects of professional development in integrated science teaching during the course was calculated. Hypotheses were tested about differences between mean professional development and its relation to usefulness of concept maps for reflective self-assessment, feedback from research and exchange among colleagues. Results from analysis of variance are summarised in Table 1.

	M (N=22)	SD	F (df=1,3)	Sign.
1. usefulness of concept maps	2.05	.90	10.06	.001
2. usefulness of feedback from research	3.05	1.13	9.51	.002
3. usefulness of exchange among colleagues	1.55	.74	8.58	.005

Table 1: ANOVA about mean ratings of professional development

F-ratios are statistically significant for differences on all three dependent variables. This means that differences in self-assessment with concept maps and collaborative feedback from researchers during the training course are related to differences on a general measure of professional development. Similarly, significant effects are found for exchange among colleagues. If we look at different aspects of professional development, significant correlations are found between the variables "usefulness of concept maps" and "planning for science learning" (r=.65) and between "feedback from research" and "development of integrative goals" (r=.60). In a subset on functions of concept maps development of ideas in groups is highly related to the aspect of creating integrated goals (r=.54) while planning of lesson content with concept maps is more related to improvement of science learning (r=.62) and occupational knowledge (r=.60). Exchange among colleagues is not specifically related to any aspect of professional development. Professional development as self-development in a supporting collaborative environment is very specifically related to different indicators. From the viewpoint of the teachers, they can control changes related to their science background for teaching. New demands regarding integrating goals need some external support from research. Teachers' action repertoire and occupational knowledge are not substantially changed within this framework.

Conclusions

Changes of the professional self are difficult and time-consuming because of a stable system of knowledge and routines, developed over many years. Concept maps, questionnaires and open interviews support self-development during a continuous workshop for the improvement of integrated science teaching. Collaboration with colleagues and researchers at the workshop were seen as helpful for improvement of professional knowledge. As a consequence from these findings, the continuous workshop for integrated science teaching can be specified as an effective model for teachers' professional development in a collaboration system. Results are encouraging enough to proceed with more differentiated questionnaire systems to analyse the complex structure for development of a professional self.

References

Bauer, K.-O. (1999). On teachers' professional self. In M. Lang, J. Olson, K.-H. Hansen, & W. Bünder (Eds.), *Changing practices/changing schools: Recent research on teachers' professionalism* (pp. 193-201). Leuven: Garant.

Beaton, A., Martin, M., Mullis, I., Gonzales, E., Smith T. & Kelly, D. (1996). *Science achievement in the middle school years.* Chestnut Hill: TIMSS/ IEA.

Black, P. & Atkin, J. (1996). *Changing the subject. Innovations in science, mathematics and technology education.* London: Routledge/OECD

Guskey, R. (2000) *Evaluating professional development.* Thousand Oaks: Corwin.

Mason, C. (1992). Concept mapping: A tool to develop reflective instruction. *Science Education* 76(1), 51-63.

Robertson, C., Cowell, B & Olson, J. (1998). A case study of integration and destreaming: Teachers and students in an Ontario secondary school respond. *Journal of Curriculum Studies* 30(6) 691-717.

Roulet, G. (1999) How curriculum reform affects teacher conceptions of practice. In M. Lang, J. Olson, K.-H. Hansen, W. Bünder (Eds.), *Changing schools/changing practices: perspectives on educational reform and teacher professionalism.* Louvrain: Garant.

Sikes, P (1992). Imposed change and the experienced teacher. In M. Fullen & A. Hargreave. (Eds.), *Teacher development and educational change.* London: Falmer.

Terhart, E. (1999). Developing a professional culture. In M. Lang, J. Olson, K.-H. Hansen & W. Bünder (Eds.), *Changing practices/changing schools: Recent research on teachers' professionalism* (pp. 193-201). Leuven: Garant.

Yinger, R. & Hendriks-Lee, M. (1998). Professional development standards as a new context for professional development in the U.S. *Teachers and Teaching: Theory and Practice* 4, 273-298.

Formative Assessment Using Concept Cartoons: Initial Teacher Training in the UK

Brenda Keogh and Stuart Naylor
Manchester Metropolitan University, UK
Max de Boo
University of Hertfordshire, UK
Rosemary Feasey
University of Durham, UK

Abstract

The research explores the use of the innovative concept, cartoon strategy, for probing understanding of science, focusing on student teachers in the UK. It examines the impact of the strategy on their attitudes to assessment, whether it helps them to begin to restructure their understanding and whether it might provide a possible strategy for them to use in their own teaching. The data suggest that the strategy is potentially valuable as a means of assessment of student teachers.

Introduction

Concept cartoons extend the range of pedagogical strategies available to teachers. They present learners with a set of alternative ideas about a scientific concept in visual form (see Figure 1). They are used mainly in the classroom to support teaching and learning in science by generating discussion, stimulating investigation and promoting learner involvement and motivation. Previous research (Keogh & Naylor, 1999) indicates that concept cartoons are useful for elicitation of understanding and might be used for the assessment of learning.

The research took place in the context of changes to the requirements to be met by student teachers in England. These requirements are defined in Circular 4/98 (DfEE, 1998), that includes the requirement to audit student teachers' scientific knowledge and understanding and identify gaps in their knowledge. No specific mechanism is required for this auditing process. The use of concept cartoons as an auditing mechanism formed the focus for the research in four University Schools of Education.

Numerous authors point to the significance of subject background knowledge for primary teachers (Osborne & Simon, 1996; Shulman, 1986). Mant and Summers (1995) claim that subject background knowledge enables teachers to diagnose pupils' learning, plan for progression and provide activities which help them to acquire the scientific view. Lack of a sound grasp of the scientific ideas which they are expected to develop in their pupils is a problem for many primary teachers and can lead to a lack of confidence in teaching science (Harlen & Holroyd, 1995). Research into student teachers' scientific knowledge and understanding shows that many of the alternative conceptions described elsewhere for children (eg Driver et al., 1994) have persisted into adulthood (Carré, 1993).

H. Behrendt et al. (eds.), Research in Science Education – Past, Present, and Future, 137–142.

Fig. 1: What do you think? – Why do you think this?

Black (1998) notes the impact of assessment on motivation and self-esteem. Many student teachers enter initial teacher training (ITT) courses with negative attitudes to science, often perceiving themselves as "failed learners" in this subject (Parker & Spink, 1997). Poor performance on an initial audit which simply tests their knowledge could reinforce their sense of failure, strengthen negative attitudes to science and provide poor conditions for learning about science.

Concept cartoons (Keogh & Naylor, 1997) were selected as a mechanism for auditing the student teachers' scientific knowledge and understanding. Potential advantages of using the concept cartoons included the fact that they are very different from most of the forms of assessment which student teachers would have experienced in the past, so that they would be less inclined to make premature judgements about this approach. The concept cartoons were viewed as having the potential for promoting metacognition in relation to the assessment process, which is an important aspect of student teacher development (Bell & Gilbert, 1996). The strategy was consistent with Bishop & Denley's (1997) claim that an effective auditing strategy will both diagnose the students' level of understanding of science concepts and help them begin to reconstruct their understanding. The concept cartoons are easy to use in the classroom, so using them for assessment also provided the student teachers with potentially valuable classroom materials.

Research methodology

The research addressed three specific questions in order to make a judgement about the extent to which concept cartoons might provide a useful auditing tool in ITT. These were:

- To what extent might the use of concept cartoons avoid student teachers developing negative attitudes to being assessed?
- Does the use of concept cartoons in assessment help the student teachers to begin to restructure their scientific knowledge?
- Does the use of concept cartoons help to provide a model for assessment, teaching and learning in the classroom?

The research took place over two years, with some modification to the approach occurring during the second year in the light of the evidence obtained. The different circumstances within each university led to some variation in data collection methods, with each researcher using the approach that was most relevant to their own circumstances and this allowed a degree of triangulation of the data. The data sources were a mixture of undergraduate and postgraduate student teachers with a range of personal histories, experience and science backgrounds.

Thirteen concept cartoons were used with the groups of student teachers, covering a broad range of science concepts such as electricity, forces, thermal insulation and ecological interactions. The concept cartoons provided a rich source of data for student teacher understanding of science concepts and for the types of alternative conceptions which were common, but this data is not the subject of this paper. Other complementary approaches to data collection included:

- Questionnaires (n = 318)
- Semi-structured interviews with students (n = 15)
- Informal feedback from all groups of students
- Taking note of references to concept cartoons in written assignments

A total of 635 audits have been analysed to date. The audits were given to the student teachers at the commencement of their course. Most of the data were gathered shortly after the completion of the audit. Informal feedback was elicited throughout the course and as part of final evaluation of the taught courses.

The data from the interviews, the questionnaire and the informal feedback were analysed in relation to the research questions using a combination of qualitative and quantitative analysis. Areas of agreement from these data sources were noted, what appeared to be significant issues were identified and any major discrepancies between the data sources were identified. The use of a range of data collection methods, a variety of data sources and interviews to follow up issues raised elsewhere enabled the data to be triangulated (Denzin, 1970) and helped validate the analysis.

Data and data analysis

Student teacher attitudes to assessment

The data showed that 47% (n = 333) had negative feelings about the idea of being assessed at the beginning of their course. Their views ranged from "apprehensive" or "nervous" to "anxious" or "horrified". They felt that assessment would demonstrate how little they knew, thus undermining their confidence and reinforcing their feelings of inadequacy in science.

Their feelings about the use of the concept cartoons as an assessment method were more positive, with 89% having positive feelings about this approach. Typical comments were that the approach was "more user friendly", "enjoyable and light hearted", "easy to understand" and "less threatening and more stimulating". It appeared that the experience of being assessed using the concept cartoons was more positive than many of the students had anticipated. The use of the concept cartoons appeared to have avoided the development of negative attitudes for the majority of the student teachers.

Restructuring understanding

More than half of the student teachers (69%) thought that using the concept cartoons helped them to think differently about the situations and to begin the process of restructuring their understanding. Several indicated that they had continued thinking and talking about the questions after the audit and, in some cases, had gone home and set up a practical investigation to clarify their thinking and find out more about the situations. Some felt so positive about this aspect of the concept cartoons that they used them in schools for the same purpose, echoing Gunstone (1988) who noted the overlap between the methods used to probe understanding and those used to develop understanding. Comments made by the student teachers included;

the concept cartoons promoted my thinking;
they allowed me to visualise the concepts;
I'm looking at forces and weights in a different way;
they made me think more widely;
they made me question what I thought I knew.

Self-reporting may not, necessarily, be a reliable indicator of an individual's level of understanding. However self-reporting of a change in thinking is much more likely to be significant, particularly when the frequency of reported changes in thinking is relatively high. In this study the directed attention of the student teachers, their active involvement with the concept cartoon situations and the opportunity for metacognition make conceptual change more likely.

It is particularly noteworthy that any conceptual change reported by the student teachers has been brought about by assessment, not by direct teaching.

Modelling assessment, teaching and learning

Nearly all of the student teachers (97%) held positive views about the value of the concept cartoons as a teaching approach at this point in their course. None expressed any negative views. They could often see multiple uses for the concept cartoons in the classroom, including elicitation, providing a context for discussion and identifying starting points for investigation. Many of them described how, in their experience, elicitation led naturally into wanting to find out more about the situation and to more focused investigation. They described how the concept cartoons were not only an assessment mechanism but that they could challenge learners, encourage thinking and stimulate them to justify their thinking more fully. They were able to identify parallels between the auditing approach used in their course and their classroom practice. Many of them had gone on to use the strategy in

their own teaching and, in some cases, had transferred the approach to another curriculum area. Typical comments included
they give me ideas about how to set up lessons;
they create initial interest and gain attention;
(they) get children thinking openly and brainstorming ideas.

These positive views are consistent with previous research in which the use of concept cartoons led to some student teachers developing their views of how constructivist approaches to teaching and learning might be implemented in the classroom (Naylor & Keogh, 1999).

Conclusions and implications

The concept cartoons do appear to be potentially valuable as an assessment method in initial teacher training. Although the student teachers were not uniformly positive in their responses, their attitudes to this approach to assessment in science were generally favourable. The use the concept cartoons for assessment purposes may offer the possibility of promoting positive attitudes to science teaching and learning amongst student teachers.

Many of the student teachers perceived the initial audit as a starting point for their learning. In principle, the purpose of an initial audit should be to enhance professional development (Bishop & Denley, 1997). In at least some cases the audit based on concept cartoons helped the student teachers to understand their learning needs and to begin the process of restructuring their understanding.

The use of the concept cartoons appeared to have a strong link with professional practice. They enabled most of the student teachers to identify links between assessment, learning and teaching and to recognise how they could provide possible models for assessment, teaching and learning in the classroom. Anecdotal evidence indicates that many of them have built on this experience and gone on to use the concept cartoons successfully in the classroom.

Some limitations of the research are evident. No comparisons were made with other assessment mechanisms and this may have influenced the student teachers' views. The study does not attempt to provide any long-term view of student teachers' thinking or ideas, nor the extent to which student teachers' attitudes to assessment and their classroom practice have been modified. Minor variations in approach across the four University sites made it difficult to standardise the data analysis, though this proved helpful for triangulation purposes. Issues such as whether the student teachers are allowed to talk as they complete their audits and whether they can take the audits away overnight may also have a significant effect on their anxiety levels and on the nature of their responses to the concept cartoons.

The implications for future teaching appear to be that the concept cartoons have clear potential value as an auditing mechanism in science for student teachers in England. They appear to avoid the risk of reinforcing negative attitudes to science and, potentially, can have a positive impact on professional practice. Although the range of concept cartoons used did not cover all of the National Curriculum for ITT, they appear to provide valuable starting points for student teacher professional development. They provide a useful mechanism for probing student teachers' understanding of scientific ideas; data collected on these ideas will be written up for publication elsewhere.

References

Bell, B. & Gilbert, J. (1996). *Teacher development: A Model from Science Education.* London: Falmer.

Bishop, K. & Denley, P. (1997). The fundamental role of subject matter knowledge in the teaching of science. *School Science Review* 286, 65-71.

Black, P. (1998). Formative assessment: raising standards inside the classroom. *School Science Review* 80, 291, 39-46.

Carré, C. (1993). Performance in subject-matter knowledge in science. In N. Bennett & C. Carré (Eds.). *Learning to Teach.* London: Routledge.

Denzin, N.K. (1970). *The research act in sociology: A theoretical Introduction to sociological methods.* London: The Butterworth Group.

Department for Education and Employment (1998). *Circular 4/98. Teaching: High Status, High Standards.* London, DfEE.

Driver, R., Squires, A., Rushworth, P. & Wood-Robinson, V. (1994). *Making sense of secondary science.* London: Routledge.

Gunstone, R. (1988). Learners in science education. In P. Fensham (Ed.), *Development and dilemmas in science education* (pp. 73-95). Lewes: Falmer.

Harlen, W. & Holroyd, C. (1995). *Primary teachers' understanding of concepts in science and technology.* Scottish Council for Research in Education.

Keogh, B. & Naylor, S. (1997). *Starting points for science.* Sandbach, UK: Millgate House.

Keogh, B. & Naylor, S. (1999). Concept cartoons, teaching and learning in science: an evaluation. *International Journal of Science Education* 21 (4), 431-446.

Mant, J. & Summers, M. (1995). Some primary school teachers' understanding of the Earth's place in the universe. *Research Papers in Education* 8(1), 101-129.

Naylor, S & Keogh, B. (1999). Constructivism in the classroom: theory into practice. *Journal of Science Teacher Education* 10(2), 93-106.

Osborne, J. & Simon, S. (1996). Primary science: past and future directions. *Studies in Science Education* 27, 99-147.

Parker, J. & Spink, E. (1997). Becoming science teachers: an evaluation of the initial stages of primary teacher training. *Assessment and Evaluation in Higher Education* 22 (1), 17-31.

Shulman, L.S. (1986). Those who understand: knowledge growth in teaching. *Educational Researcher* 15, 4-14.

Teaching Chemical Equilibrium in Australian and German Senior High Schools

David F. Treagust
Science and Mathematics Education Centre, Curtin University of Technology, Perth, Australia
Wolfgang Gräber
Institute for Science Education at the University of Kiel, Germany

Abstract

This paper briefly examines the teaching approaches, for the topic of chemical equilibrium, of a small sample of Australian and German senior high school teachers. The data are analysed using a framework comprising three dimensions of pedagogical practice, where each dimension is characterised by two extremes on a scale: a teacher-centred vs. student- centred dimension a teaching facts vs. teaching processes dimension, and a discipline-oriented vs. daily life-oriented dimension (Gräber & Nentwig, 1999). The research was designed to answer two questions: "To what extent do the teachers incorporate different aspects of these three pedagogical dimensions?" and "To what extent do the teachers each incorporate a similar organisation of content and examples?"

Introduction

Research studies during the past two decades, based on learning difficulties in chemistry classrooms, in which chemical equilibrium has been taught, have identified common alternative conceptions of chemical equilibrium held by secondary and tertiary students (Hackling & Garnett, 1985; Wheeler & Kass, 1978). These studies have recommended a greater emphasis on the quantitative aspects of equilibrium, a greater differentiation in the range of examples presented to students when discussing Le Chatelier's Principle and a greater emphasis on a laboratory approach.

However, there are few detailed descriptions in the literature about how these recommendations for teaching the topic of chemical equilibrium have been implemented and the effects on student learning. Exceptions are research in the Netherlands by Van Driel, De Vos, Verloop and Dekkers (1998), who analysed student learning following a specific teaching program and a detailed study in one Australian classroom by Tyson, Treagust and Bucat (1999), that illustrated the complexity of learning this topic. The Australian study described three explanations - collision theory, Le Chatelier's Principle and equilibrium law - that can be used to predict what will occur when equilibrium mixtures are disturbed, with the last of these being considered the most sophisticated of the three explanations. There is no comprehensive, classroom-based research in Germany concerning approaches to teaching chemical equilibrium, though there is strong anecdotal evidence that German chemistry teachers prefer a teacher-guided, inquiry approach, using experiments (Schmidkunz & Lindemann, 1976).

H. Behrendt et al. (eds.), Research in Science Education – Past, Present, and Future, 143–148.

Chemistry, in the last two years of secondary school in Germany and Australia, is similar in content and is designed to provide students with the opportunity to advance to university studies in chemistry. As the authors had conducted research on chemical equilibrium with one teacher in an Australian classroom (Tyson, et al, 1999) and with two teachers in German classrooms (Gräber & Treagust, 1999), they were interested in examining the comparative findings of these studies. There are few studies in the literature that compare teaching and learning in chemistry across different countries and this is an attempt to examine and analyse any commonalities and differences.

Purpose

This study was to compare Australian and German senior high school teachers' ways of teaching the topic "chemical equilibrium". The data were analysed using a three-dimensional framework:

1. Teacher-centered vs. student-centered
2. Teaching facts vs. teaching processes
3. Discipline-oriented vs. everyday life (context) –oriented

The leading questions were: "To what extent do the three teachers incorporate different aspects of these three pedagogical dimensions?", and "To what extent do the three teachers each incorporate a similar organisation of content and examples?"

Design and procedures

To answer these two research questions, the three teachers were observed teaching the topic of chemical equilibrium in their normal science laboratories. The study sought to describe and analyse the teaching and learning using an interpretive research methodology (Erickson, 1986) within a three dimensional framework of pedagogical practice:

(a) *The teacher centred <--> student centred dimension* describes either the teacher governing classroom activities, steering the students' learning processes and dominating the communication process or the students taking responsibility for their own progress and initiating their own learning processes in an autonomous way.

(b) *The teaching facts <--> teaching processes dimension* shows that the teaching/learning activities aim either at students learning science facts, laws and formulae contrasted with the acquisition of problem solving strategies and skills of processing information and interpreting data.

(c) *The discipline-oriented <--> daily life-oriented dimension* is illustrative of the aim of lessons being to either delineate the structure of a scientific discipline and to reproduce research findings on a reduced level or to provide means to understand daily-life phenomena, including their social, technological and economic implications.

The Teachers. The Australian teacher, Mrs. May, has 20 years experience, held science education administrative responsibilities in and out of the school and openly acknowledged the significance of prior knowledge on learning that occurs in the classroom. Of the two German teachers, Mr. Stegman has taught for about 35 years and worked as a pre-service teacher educator for 25 years. In his seminars with prospective chemistry teachers, he discusses different teaching approaches, comments on his own teaching experiences and those from colleagues, as well as the results from educational research. Mr. Wilhelm has taught for 12 years and has had experience in both types of secondary schools in Germany, the Gymnasium (grammar school) and the Gesamtschule (comprehensive school), which enabled him to reflect on the optimum teaching approaches in both types of schools. In each school, the classes observed were in Year 11 (16-18 years old) and students had selected to take chemistry as part of their final year or years of secondary school.

Documentary Resources. The three teachers were informed about the researchers' interest in how they made the concepts easier for students to understand; they were encouraged to teach in their normal style, despite the presence of a researcher in their classroom, during all lessons on the topic of chemical equilibrium. Mr. Wilhelm and Mr. Stegman were observed for 15 and 13, 45 minute lessons, respectively, over a period of five weeks and 16 lessons. Mrs. May was observed for 12, 1-hour lessons over four weeks, though reversible reactions were introduced in previous lessons. This small difference in observation frequency was indicative of the length of time that each teacher took to complete the chemistry topic.

The researchers audio-taped all aspects of the lessons, obtaining a record of the teachers' presentations and discussions and produced full transcripts[1]. The transcripts and observation notes were classified on a consensus basis and, to assist in the triangulation of the data collection, copies of the students' textbooks, work-sheets, notes and any supplementary materials, including homework and tests, were also examined (Mathison, 1988). Throughout the study, each teacher was available to answer the researchers' questions and to discuss the activities from the day's teaching. At the end of the observation period, the teachers were interviewed separately about their teaching philosophy and class organisation. Interviews were held with all students in both German classes, usually as groups of three and with four targeted students in the Australian class. By means of questionnaires, information was sought about students' knowledge of chemical equilibrium and their reactions to the lessons.

Data Collection and Analysis. Data were analysed and interpreted at three levels. Initially, the teachers' teaching was examined, along the three dimensions described by Gräber and Nentwig. The second level of analysis involved an examination of transcripts and classroom observation notes in terms of the structure of the content of the chemical equilibrium being taught and learned in the lessons. For the third analysis, which is not reported in this paper, students' understanding of the chemical equilibrium phenomena and their reactions to their lessons were taken into account. The findings are reported as two assertions.

Results

Assertion 1: To varying degrees, each teacher illustrated student-centredness and an orientation towards teaching processes and placed emphasis on the structure of the discipline.

All three experienced teachers differ from the majority of German teachers (Stork, 1984) and particularly from beginners (Fischler, 1994) in being less teacher-centred and teaching processes rather than predominantly facts. In these two

[1] Acknowledgement: Louise Tyson collected these data as part of her doctoral studies.

dimensions, they also differed from each other on the positions on these dimensions but, according to the third dimension, all are discipline-oriented.

The teacher centred <--> student centred dimension

Mr. Wilhelm's philosophy is that students should find and define questions and problems by themselves, whereas Mr. Stegman, who also supports students' active process of finding and solving problems, prefers to provide more teacher assistance. Mr. Wilhelm supported autonomous learning through group work; Mr. Stegman has tried group work but prefers teacher-led class discussions with group experiments. Mrs. May used both whole class teaching with questions and also provided many opportunities for students to work independently and in groups.

The teaching facts <--> teaching processes dimension

Mr. Stegman and Mr. Wilhelm ensured that students learned facts and concepts but each taught with a stronger orientation towards processes. Mr. Wilhelm was also concerned that students developed cross-curricular competencies such as co-operatively solving problems while working in groups. It is likely that these differences in teaching approaches have originated from the different kinds of social atmospheres in the Gymnasium and Gesamtschule schools, the latter promoting teamwork and social competencies as parts of its explicitly stated goals. Mrs. May taught processes by having students solving problems using one of three explanations - collision theory, Le Chatelier's Principle and the equilibrium law - as a strategy to help predict changes in a chemical reaction that was at equilibrium.

The discipline-oriented <--> daily life-oriented dimension

All three teachers approached this topic based on the discipline of the underlying chemistry. Mr. Stegman expressed the view that scientific concepts can be taught by integrating a science-technology-society approach, with the teaching of chemical concepts of everyday life issues. However, he decided, explicitly, not to use this approach in chemical equilibrium, but rather to concentrate on the underlying chemistry, because he believed that everyday life issues are too complex. Mr. Wilhelm introduced the topic from the basis of a practical application of using a chemical equilibrium reaction with $Fe(SCN)_3$ in understanding how a photometer works, but he did not provide other daily-life examples during the lessons. For Mrs. May, the nature of this Year 11 chemistry course needed to be discipline-oriented, because a considerable number of students in the school went on to study chemistry at university. However, the textbook used by Mrs. May included several pages denoted to the applications of chemical equilibrium, primarily industrial applications - the production of ammonia and oxygen transport in the blood. However, these aspects were left for students to read on their own rather than being discussed in class.

Assertion 2: Despite different teaching approaches using different chemical exemplars, the organisation of the content for chemical equilibrium presented by the three teachers was similar.

In both countries, the syllabus for the topic of chemical equilibrium was similar and teachers taught the subject matter set out by the relevant educational authority, though for each teacher the content was arranged in a different order and involved dissimilar teaching approaches. Neither German teacher used a textbook, while Mrs. May followed the sequencing of concepts in the students' chemistry textbook 'Foundations of Chemistry' (Garnett, 1997). Both German teachers focussed on the conceptual development of the content through the chemistry concepts themselves as opposed to providing a range of analogical bridges, though Le Chatelier's Principle was introduced at the end of the teaching sequence. In the Australian classroom, students were introduced to the equilibrium law as a means of predicting the effect of changes to chemical equilibrium mixtures and were then introduced to Le Chatelier's Principle as another means of making these predictions. However, in the Australian classrooms, students were expected to solve textbook problems by deciding which explanation - collision theory, Le Chatelier's Principle or the equilibrium law – was the best predictor of changes to equilibrium reactions.

Despite dissimilarities in teaching approaches, analysis of the lessons and discussions showed a common structure, despite the content being covered in different ways. Basically, the content structure for the organisation of the lessons on chemical equilibrium for the three teachers was similar:

- Introduction of the idea of a reaction going both ways
- Introduction of the idea that not all reactions are complete
- Examination of the reaction, quantitatively, to show that there is a constant relationship between products and reactants
- Interpretation of the data to understand the meaning of the equilibrium constant
- Interpretation of the effects on the equilibrium system by changing parameters such as the concentration of reactants or products, pressure and temperature and introducing Le Chatelier's Principle.

In both German classes, the teaching had a common pedagogical structure of five phases -- students experience a problem; students reflect to solve the problem; a range of suggestions are considered for students to solve the problem; the results are interpreted and where possible generalised and the learning outcomes are consolidated. For example, in these classes, much emphasis was initially placed on the nature of chemical reactions being reversible. In Mr. Wilhelm's class, group practical work involved (a) reaction of copper sulfate with water, (b) heating hydrated copper sulfate, (c) reaction of carbon dioxide with water and (d) heating carbon dioxide solution. Mr.. Stegman's class experienced these same experiments, but lessons were more teacher-directed; in addition students conducted an experiment involving iron (II) ions with silver ions.

In contrast, students in the Australian classroom were less involved in developing an understanding of reversible reactions but carried out laboratory exercises that examined the effect of disturbing equilibrium mixtures of chromate/dichromate and cobalt chloride/cobalt hexaqua ions. The focus on the textbook chapter involved classic reactions between nitrogen and hydrogen to form ammonia and gaseous reactions between nitrogen dioxide and dinitrogen tetroxide. Emphasis was placed on determining the equilibrium constant expression and how to best predict the effect of changes in these systems using collision theory, Le Chatelier's Principle and the equilibrium law.

Conclusion

This paper has brought together data from two different studies with teachers who were acknowledged experts in the chemistry classroom. Based on our experience of working with many teachers in their classrooms, we claim that they are representative of effective teachers in each of the two countries. There are many commonalities about the teachers' approach to this topic, despite the fact that they are from different education systems and teach in different kinds of schools. One could argue that the nature of the chemistry subject matter, the strong knowledge base of the teachers and their epistemological commitments about optimum learning brought them to a similar decision about how to organise the concepts within the topic of chemical equilibrium.

In summary, in addition to ensuring that basic factual knowledge was learned, each teacher focused on teaching science processes that are different to the majority of chemistry teachers in Australia and Germany and to beginning teachers who stress the teaching of concepts and facts. Both German teachers used exemplars and experiments that were simpler and more accessible to the students than those used by the Australian teacher. Further, the German teachers used more experiments in directing students in problem solving about the nature of chemical equilibrium reactions; problem solving in the Australian classroom was largely based on textbook problems. Despite these different teaching approaches and using different chemical exemplars, each teacher held clear, similar, underlying expectations of what was deemed important for students to understand about the topic of chemical equilibrium.

References

Erickson, F. (1986). Qualitative methods in research on teaching. In M. C. Wittrock (Ed.), *Handbook of research on teaching. 3rd Ed.* (pp. 119-161). New York: Macmillan.

Fischler, H. (1994). Concerning the difference between intention and action: Teachers' conceptions and actions in physics teaching. In I. Carlgren, G. Handel, & S. Vaage (Eds.), *Teachers' minds and actions: Research on teachers' thinking and practice* (pp. 165-180). London, England: Falmer Press.

Garnett, P. J. (Ed.). (1997). *Foundations of chemistry.* Melbourne, Australia: Longman Cheshire.

Gräber, W., & Nentwig, P. (1999). *Scientific literacy: bridging the gap between theory and practice.* Paper presented at the meeting of the Association for Teacher Education in Europe, Spring University, Klaipeda, Lithuania.

Gräber, W., & Treagust, D. F. (1999, March). *Content and structures of chemical equilibrium: Commonalities in teachers' practice.* A paper presented at the annual meeting of the National Association for Research in Science Teaching, Boston,MA.

Hackling, M. W. & Garnett, P. J. (1985). Misconceptions of chemical equilibrium. *European Journal of Science Education* 7, 205-214.

Mathison, S. (1988). Why triangulate? *Educational Researcher* 17(2), 13-17.

Schmidkunz, H. & Lindemann, H. (1976). *Das forschend-entwickelnde Unterrichtsverfahren: Problemlösen im naturwissenschaftlichen Unterricht.* München: Paul List Verlag.

Stork, H. (1984). Zur Aufrechterhaltung von Motivation und Lebensnähe in einem fachlich anspruchsvollen Chemieunterricht. *Chimica* 38, 145-157.

Tyson, L. M., Treagust, D. F. & Bucat, R. B. (1999). The complexity of teaching the topic of chemical equilibrium. *Journal of Chemical Education* 76, 554-558.

van Driel, J. H., de Vos, W., Verloop, N. & Dekkers, H. (1998). Developing secondary stunts' connections of chemical reactions; the introduction of chemical equilibrium. *International Journal of Science Education* 20, 379-392

Wheeler, A. E. & Kass, H. (1978). Students' misconceptions in chemical equilibrium *Science Education* 62, 223-232.

The Ideas of Spanish Primary Teachers on how to Develop an Understanding of Processes in Science and their Support in Textbooks

Susana García-Barros, Christina Martínez-Losada,
Pedro Vega and Matilda Mondelo,
University of A Coruña, Spain

Abstract

Textbooks are one of the most widely used teaching aids by primary school teachers in our country. The activities suggested in some of the latest editions, include different types of procedures that are, however, often deficient in that they fail to devote the necessary amount of attention to enquiry and its associated procedures. Furthermore, teachers, who inherit to a specifically theoretical tradition in the teaching of science, do not appear to be aware of the educational significance of procedures.

Introduction

A wide consensus of opinion currently exists regarding the importance of the development of ideas in primary school children through the use of processing skills (Harlen, 1989). Due to this, enquiry activities are especially recommended. Nevertheless, research, even at higher levels has revealed that the enquiry activities that are in fact used (Hodson, 1994) and those, that are contained in textbooks (Tamir & García Rovira, 1992) do not always encourage the development of enquiry techniques. In Spain there exists a long tradition of theoretical and conceptual priorities in the teaching of science and this has been influential in the fact that many teachers fail to attach sufficient importance to procedures (De Pro, 1998). Procedures, however, such as manipulation, intellectual and communication skills, must be specifically taught through enquiry activities.

Educational innovation in primary school science classes and the adaptation of such innovation to the new trends reflected in the recent educational reforms that our country has undergone, depend to a great extent on the teaching staff. Their opinions about what children must learn and in what ways they must learn it, affect the selection of contents and activities. On the other hand, teachers' opinions are influenced by the materials selected, especially textbooks, coming from a limited range of publishers, and constituting the main source of classroom material used by primary school teachers within our framework (Martínez Losada et al., 1999). The innovation of teaching materials and their dissemination must be seen, therefore, as a key part of educational innovation. As a result of the above, this paper intends to a) discover the types of activities used by teachers and the criteria with which these are chosen, trying to detect the extent to which procedural activities are cited as specific objectives; b) analyse what type of activities are included in school textbooks and what procedures they allow to develop and c) compare the aspects as stated.

H. Behrendt et al. (eds.), Research in Science Education – Past, Present, and Future, 149–154.

Methodology

We have gathered the opinions of 554 teachers from 150 primary schools in Galicia (north-west Spain), all of whom habitually use textbooks. These teachers answered a questionnaire, in which they were asked to indicate the frequency (always, mostly, sometimes or never) with which they use so-called paper and pencil activities and how often they used practical activities They were also asked in an open-ended question, to specify the objectives of the activities they proposed. We chose primary school (6-12 years) textbooks published recently by three widely-used publishers in this field that we refer to as A, B and C. A total of 18 books were used, analysing the activities related to two areas of study - water and animals - that are particularly important at this educational level. Using a total of 482 activities - 128 dealing with water and 354 with animals - the potential objectives to developed and the procedures encouraged were analysed. The study of procedures was based on the classifications proposed by various authors and collected by De Pro (1998). The procedures that were analysed are included in table 1. The analysis of the texts and the teachers' replies to the open question were taken independently by two different people.

Results and discussion

What type of activities do Primary School teachers carry out?

We have observed that the primary teachers questioned use paper and pencil activities on a regular basis (46.9% in all subjects and 31.6% in some subjects), but they only use practical activities in some subjects (47.8%) or in none whatsoever (8.0%). With respect to textbooks, it must be noted that the proportion of practical work included in these, with respect to the total amount of activities, is excessively small. Only 25 of the 482 activities analysed (5.2%) can be considered as practical activities. In addition, we have noticed differences concerning the subject dealt with, given that all the practical activities, except for one, dealt with water (18.7%). No differences were noticed from one publisher to another.

The results obtained, if we consider both the teachers' replies and the analysis of texts, reveal that priority is still given to theoretical activities in our country. This is consistent with a deep-rooted methodological tradition, that is difficult to change, in spite of the fact that teachers "theoretically" place special importance on practice (Martínez Losada et al., 1993) and in spite of the fact that the recently established reform supports educational innovation in this respect. Furthermore, it must be stated that the absence of practical proposals, aimed at studying animals in the texts used is specially noticeable. Particularly as this subject is especially appropriate as a means of bringing children closer to the study of real specimens, both in the classroom and in the children's own environment. This deficiency, detected in primary school books, represents an important problem when promoting innovation in science education at this level; as the teacher mainly uses the publishers, analysed in this work, as a guide for classroom programming (Martínez Losada et al., 1999). The introduction, therefore, of practical work would require the teacher to possess sufficient scientific knowledge and concern for innovation to enable the preparation of specific activities not found in the normal textbook (Mellado et al., 1998). On the other hand, the integration of practical proposals in school textbooks is valuable for the children themselves, as they get used to realising that studying science is more efficient and gratifying when one gets the chance to directly observe the phenomena.

What objectives do the teachers attribute to the activities proposed for their students? What objectives permit the development of school textbook activities?

In figure 1 we see that more than 55% of the teachers consulted considered that practical activities and paper and pencil activities are used mainly to reinforce learning (*establishing concepts, assimilating knowledge, aiding understanding* and *applying theory*), while the use of these activities in the acquisition of new knowledge was less favoured (21.4% of the paper and pencil activities and 15% of the practical activities). References to the development of procedures (*gathering data, observation, handling, experimentation and research*) were especially scarce, although the highest percentage was found in practical activities (25.7%). No procedures from the oral and/or written communication field nor the development of enquiry were mentioned at any time. About 10% of the teachers questioned considered assessment to be an objective of the activities, although the detection of the students' prior ideas was not mentioned. This latter aspect is, nevertheless, the objective of 12.9% of the activities included in the textbooks. The remaining activities are aimed at applying theory (50%), or developing new knowledge (35%). On this occasion, we have detected differences from one publisher to another. Publishers A and B include a greater percentage of theory application activities (56.5%/publisher A; 59.4%/publisher B) than publisher C (31.1%) (X^2 = 22,2 and 23,9 respectively; p<0,001). With respect to the 25 practical activities analysed, it must be stated that their objectives were based on obtaining knowledge (52%) and on directly applying the theory (36%), with no noticeable difference between the three publishers.

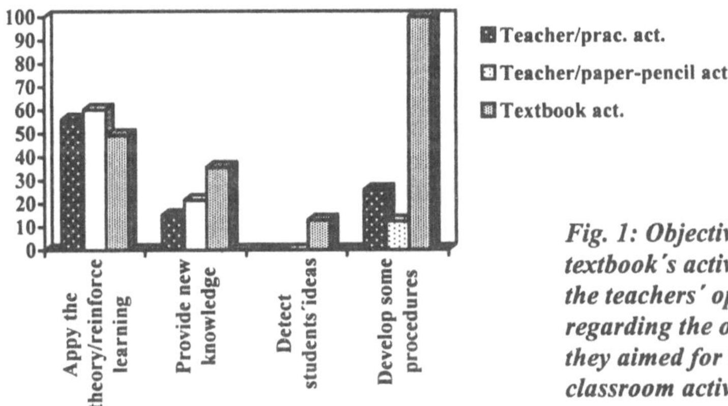

Fig. 1: Objectives of the textbook's activities and the teachers' opinions regarding the objectives they aimed for with classroom activities

The opinion of the primary school teachers about the objectives of the activities developed in the classroom tells us that, in general, they are especially interested in conceptual development, as opposed to other aspects such as procedures and attitudes, which although possessing educational interests in themselves, also contribute to conceptual development. In this respect, the importance of the procedures being taught specifically in the learning activities is highlighted (De Pro, 1998), above all at lower levels, where the children must develop their ideas by developing process skills (Harlen, 1989). On the other hand, the teachers favour the same aspects of practical work, defending the opinion that the main objective is to test the theory, with very few teachers recognising that the student can develop new knowledge by such a means. This idea of teaching is restrictive and incoherent with the latest trends in science teaching, which defend the educational possibilities of practical work (Duggan & Gott, 1995; Gott & Duggan, 1996) and the need to offer, through this, a suitable vision of scientific endeavour (Hodson, 1991; Hodson, 1994). Another aspect related to the teachers' ideas that is worth stressing, is the practical lack of reference made by this group to objectives relating to the detection of prior ideas or appraisal. Both of these features being fundamental to the learning process, which demands the introduction of specific activities (Kempa, 1986; Driver, 1988; Jorba & Sanmartí, 1996).

What procedures are used in the activities of the primary school textbooks?

All the activities proposed by school textbooks demand the use of some procedures (see figure 1). In table 1 we specify the procedures used in activities analysed in this study. All these activities require the use of communication, predominantly written communication (69,4%). Less than 22% propose oral answering, whilst less than 2% of the activities explicitly promote discussions or debates. We have not found any activities that include drawing up conceptual maps. 38.2% of the activities suggest making observations, although these are mainly indirect by way of drawings and illustrations. Observation is a procedure used significantly more in water-related activities (64,8%) than in animal-related activities (28,5%) (X^2= 50,99; p<0,001). Practically 60% of the activities promote procedures related to organising information. Although classification, a procedure traditionally associated with the study of living beings, is not found so often in this area. On the other hand, the interpretation of facts and situations is a procedure, that is found in only 16.8% of the activities. Moreover, both the development of manipulating skills and enquiry are rarely present in the activities analysed.

The analysis of the procedures used in the activities suggested by the larger publishers in our country show marked deficiencies. Firstly, it is important to note that, in spite of the fact that communication is always contemplated, this is usually limited to the writing of short sentences or definitions and to oral replies to certain questions, generally aimed at detecting prior ideas. In addition and despite the fact that the teacher may encourage discussion in small or large groups, it would be advisable for the publishers to specifically contemplate this aspect, given the great educational value that exchanging ideas has in the learning process (Solomon, 1987). Even though observation is a procedure which is regularly used, it is usually on an indirect level, through illustrations, which seems particularly surprising when dealing with science. It should finally be stressed that the absence of procedures related to solving problems - important in the first education levels (Harlen 1989) - is an important deficiency in recently published school textbooks.

Comparing the procedures indicated by the teaching staff and those fostered by the activities analysed, it must be stressed that, although all the activities propose procedures, these procedures were hardly ever suggested by the teachers (see figure 1), whose suggestions were excessively generic. This tendency of the teaching staff, participating in this study, is consistent with the data obtained by other researchers in our country (De Pro et al., 1999). It is especially surprising to find that the primary teachers make no reference to communication that is, in fact, a widely-used procedure in the classroom. Teachers do not, however, seem to grant it the importance it really has in the science learning process (Tamir & Garcia Rovira, 1992; Lawson, 1994), considering it to be a mere mechanism which is necessary to discover what the students know, what they have observed, etc. Finally, considering the deficiencies we have discovered in this study, we believe that primary school science education requires the preparation of innovative materials consistent with current research and the development of quality teacher-training in both the scientific and psychological/pedagogic aspects (Osborne & Simon, 1996), that takes, as a starting point, the teachers' thinking (Gil et. al., 1991).

CONTENTS	ANIMALS				WATER				
	A n= 147	**B** n= 122	**C** n= 85	**Tot.** n= 354	**A** n= 46	**B** n= 21	**C** n= 61	**Tot.** n= 128	**Tot.** n= 482
PLANNING *Proposal of a hypothesis, design of experience*	-	-	-	-	4,3	-	3,3	3.1	1,2
OBSERVATION *Direct or indirect - drawings*	23,8	35,3	27,0	28,5	60,9	80,9	62,3	64,8	38,2
SEARCH FOR INFORMATION *In the text itself, the press, other texts*	22,4	10,6	5.9	14.4	54,3	9,5	14,7	28.1	18,0
O.F.: *Simple description*	18,4	12,3	10,6	14.4	34,7	33,3	50,8	42.2	21.8
O.F.: *Identification of characteristics*	29,2	47,5	48,2	40,1	17,4	28,5	19,7	20.3	34,8
O.F.: *Establishment of relationships*	6,8	9,2	14,4	9,3	28,3	19,1	21,3	23,4	13,1
O.F.: *Classification*	6,8	15,6	23,5	13,8	4,3	9,5	3,3	4.6	11,4
COMMUNICATION *Oral, written, graphs*	100	100	100	100	100	100	100	100	100
INTERPRETATION *Facts, objects, situations*	11,5	23,7	17,6	17,2	15,2	14,3	16,4	15.6	16,8
MANUAL SKILLS *Use of material, gathering samples, numerical operations*	2,4	-	1,2	1,1	19,5	23,8	16,4	18,8	5.2
CONCLUSIONS ARGUED	-	-	-	-	-	-	-	-	-

O.F.: ORGANIZING THE INFORMATION

Table 1: Procedural content developed in activities proposed in primary education texts from three publishers: A, B & C (percentages)

Conclusion

1. In our country, science education still bears the influence of traditional methodological approaches, where priority is given to the development of conceptual content, as opposed to other types of content, both procedural and moral, which contribute decisively to the scientific education of the ordinary citizen.

2. The primary education school textbooks from large publishers have shown deficiencies with respect to the importance granted to practical work and to the type of procedures that are developed through their activities. Procedures that are of such great importance at the lower levels of education, such as those associated with

enquiry are still insufficient.

3. The teachers' opinions regarding the objectives of the activities that they choose to develop in the classroom are, if possible, more traditional than those included in the texts analysed. That suggests to us that the teachers are specially influenced by an educational tradition, characterised by emphasis on memorising and reproducing facts and concepts. Both the lack of importance granted to science in primary education (Harlen, 1989) and the little importance that this educational level has in the investigation in science education, may contribute to maintaining this tradition.

References

De Pro, A. (1998). ¿Se pueden enseñar contenidos procedimentales en las clases de ciencias? *Enseñanza de las Ciencias* 16(1), 21-41.

De Pro, A., Saura, O. & Sánchez Blanco, G. (1999). ¿Qué contenidos procedimentales seleccionan los profesores de ciencias cuando planifican unidades didácticas. En Martínez Losada, C. & García Barros, S. (Eds.), *Didáctica de las Ciencias tendencias actuales* (pp. 115-127). A Coruña: Universidade da Coruña.

Driver, R. (1988). Un enfoque constructivista para el desarrollo del currículum de Ciencias. *Enseñanza de las Ciencias* 6(2), 109-120.

Duggan, S. & Gott, R. (1995). The place of investigations in practical work in the UK National Curriculum for Science. *International Journal of Science Education* 17(2), 137-147.

Gil, D., Carrascosa, J., Furió, C. & Martínez Torregrosa, J. (1991). *La Enseñanza de las Ciencias en la Educación Secundaria*. Barcelona: ICE Universidat de Barcelona. Horsori.

Gott, R. & Duggan, S. (1996). Practical work: its role in the understanding of evidence in science. *International Journal of Science Education* 18(7), 791-806.

Harlen, W. (1989). *Enseñanza y aprendizaje de las ciencias*. Madrid: Morata-MEC.

Hodson, D. (1991). Practical working science; time for a reappraisal. *Studies in Science Education* 19, 175-184.

Hodson, D. (1994). Hacia un enfoque más crítico del trabajo de laboratorio. *Enseñanza de las Ciencias* 12(3), 299-313.

Jorba, J. & Sanmartí, N. (1996). *Enseñar, aprender y evaluar: un proceso de regulación continua. Propuestas didácticas par las áreas de Ciencias de la Naturaleza y Matemáticas*. Madrid: MEC.

Kempa, R.F. (1986). *Assessment in Science*. Cambridge: Cambridge University Press.

Lawson, D.E. (1994). Uso de los ciclos de aprendizaje para la enseñanza de destrezas de razonamiento científico y de sistemas conceptuales. *Enseñanza de las Ciencias* 12(2), 165-187.

Martínez Losada, C., García Barros, S. & Mondelo, M. (1993). Las ideas de los profesores de ciencias sobre la formación docente. *Enseñanza de las Ciencias* 11(1), 26-32.

Martínez Losada, C., García Barros, S., Vega, P. & Mondelo, M. (1999). Enseñar Ciencias en educación primaria: ¿Qué tipos de actividades realizan los profesores? En Martínez Losada, C. & García Barros, S. (Eds.), *La Didáctica de las Ciencias. Tendencias actuales* (pp. 199-210). A Coruña: Universidade da Coruña.

Mellado, V., Blanco, L. & Ruíz, C. (1998). A framework for learning to teach science in initial primary teacher education. *Journal of Science Teacher Education* 9(3), 195-219.

Osborne, J. & Simon, S. (1996). Primary Science: past and future directions. *Studies in Science Education* 26, 99-147.

Solomon, J. (1987). Social influences on the construction of pupil's understanding of science. *Studies in Science Education* 14, 63-82.

Tamir, P. & García Rovira, M.P. (1992). Características de los ejercicios prácticos de laboratorio incluidos en los libros de texto de ciencias utilizados en Cataluña. *Enseñanza de las Ciencias* 10(1), 3-12.

Pre-Service Elementary Teachers Constructing the Nature and Language of Science

John A. Craven III
Queens College City University of New York, USA
Brian Hand
Iowa State University, USA
Vaughan Prain
La Trobe University, Bendigo, Australia

Abstract

This paper reports on a two-year effort to understand and affect the conceptions of the nature of science held by pre-service elementary teachers. The first year of the study, examined the change in the ways twenty seven students defined and described science, following a series of tasks designed to have them 1) explore explicit and tacit conceptions of science and 2) negotiate a definition for science. Findings from the first year, include notable shifts in complexity and sophistication of the language used to describe science. Data from the second year of the study, suggests a strong influence of conceptual views of science of the pre-service teachers upon the selection of curriculum-related children's literature. The findings suggest: 1) that the limited language and simplistic structure typically used to describe science may belie a deeper and richer understanding of the subject and 2) opportunities to explicitly construct their conceptions regarding the nature of science positively, influences their selection of children's literature within science.

Problem

Calls for the inclusion of the Nature of Science (NOS) in the science curriculum have gone too long without providing distinction between the knowledge needed by elementary and secondary level teachers. Further, we need to know 1) what is reasonable and justifiable for elementary teachers to understand about the nature of science? and 2) what is the fruitfulness of particular pedagogical approaches used for teaching the NOS to those with limited backgrounds in the disciplines? Thus, there is a need for studies that provide insights into methods and outcomes of including NOS in the preparation of elementary teachers.

Background

In recent years, much attention has been given to the role of the nature of science in science education (Lederman, 1998; 1992; Matthews, 1998; Norris, 1997; Cleminson, 1990). Indeed, the American Association for the Advancement of Science and the National Research Council champion the inclusion of the NOS in the science curriculum (NRC, 1996; AAAS, 1990). By including the NOS in the curriculum, science educators enable students to develop more complex understandings of the subject. Some link an understanding of the nature of science to the ability of teachers to develop deeper understanding, interest and engagement in the subject (e.g., Lederman, 1998; Matthews; 1994; King, 1991). Specifically,

H. Behrendt et al. (eds.), Research in Science Education – Past, Present, and Future, 155–160.

Matthews (1994) argues that, when teachers are equipped with understandings of the nature of science, they can make more informed decisions on ways of teaching science. Hewson and colleagues (Hewson et al., 1992) state that the conceptions teachers hold about the content they teach, strongly influence the way they teach. Also, evidence is revealing the connections between teacher beliefs about the subject of science as well as knowledge and classroom practices (e.g., Kennedy, 1998; Hashweh, 1996; Stodolsky & Grossman, 1995; Kagan, 1992; Cleminson, 1990; Prawat & Anderson, 1989). While few would argue against the inclusion of the NOS in the science curriculum, the exact nature of science remains contested (e.g., Lederman, 1998, 1995; Norris, 1997; Slezak, 1994a, 1994b; Cleminson, 1990). Thus, a unified conception of the NOS remains elusive. Matthews (1998) maintains that it is unrealistic to expect students to develop the understandings and insights of experts in the history and philosophy of science. Rather than overwhelming students with highly complex questions, teachers should set modest goals when teaching the NOS (Matthews 1998; 1994). To resolve the issues, it has been recently proposed that an approach toward teaching the NOS based on broader characteristics may be less contentious to experts in science (Lederman, 1998). In their review of eight international science standards documents, McComas and colleagues (1998) offer some tenets of science that, in part, include:

1. Scientific knowledge is durable yet has a tentative character.
2. Scientific knowledge relies heavily on, but not entirely on, observation, experimental evidence, rational arguments and skepticism.
3. There is no single way to do science (therefore, there is no universal step-by-step scientific method).
4. New knowledge must be reported clearly and openly.
5. Scientists require accurate record keeping, peer review and replicability.
6. Scientists are creative.
7. The history of science reveals evolutionary and revolutionary characteristics (pp. 6, 7).

Design

This longitudinal, two-year study examined the pre-existing and developing conceptions of science of one cohort of students in year one and one cohort of students in year two (N = 27 per cohort). The students were pre-service elementary teachers from a large, urban teacher preparation institution within the eastern US. Data collected included: 1) students' written responses to open-ended questions, 2) student projects (individually constructed projects, lesson plans and whole-class, negotiated projects) and 3) instructor journal notes of observations of small group and whole-class student-to-student interactions.

Procedure

In the first year, students were given a sequence of tasks that included individual writing assignments, collaborative writing assignments and small group as well as whole-class discussions. These tasks were designed to elicit their explicit and tacit knowledge about the subject science. The goal of the first task, Task One, was for students to individually select examples from the public, printed media (i.e., newspapers and journals) to illustrate and articulate the qualities and/or characteristics that made writing either "scientific" or "pseudo-scientific". The goal of the second cycle, Task Two, was for students to construct a negotiated rubric that articulated the qualities and/or characteristics across three types of writing including "scientific", "semi-scientific", and "pseudo-scientific". Over the duration of the course, the question "What is Science?" was posed to the students. In year two, 27 new students engaged in the same series of tasks as the first year students. However, an additional task of the students in the second year was to find a children's book within science that best exemplifies what science is about. The students were twice asked to select a book and to provide (each time) a rationale for their selections.

Data Analysis

The students' conceptions of science were organized according to major themes expressed in the written responses to open-ended questions The data were analyzed using the interpretative-descriptive approach described by Strauss and Corbin (1990) and the method of constant comparative analysis (Glaser & Strauss, 1967). In this approach, the researchers used an iterative process of describing and interpreting the data and transforming it into something meaningful. Analysis of the second year's cohort followed the same procedures as the first.

Findings

In the first year of the study, a notable shift was shown to have occurred, in the pre-service elementary teachers' description of science, following the series of tasks. Over time, in written student responses to the question, "What is science?" and through small group and large group discussions, the pre-service teachers articulated more complex conceptions of science. As students were required to call upon their more deeply seated understandings (i.e., their tacit understandings) of science, several other changes took place. The initial and concluding perceptions of science as reported by first year students are reflected in Table 1 (Initial) and Table 2 (concluding).

First, the descriptions of science became more complex. For example, the students stated that there are several kinds of scientific research. Additionally, the scientific method reported at the beginning of the class came to be described as scientific methods that followed some general rules and procedures. These methods could be used to either confirm or rebut ideas and explanations, whereas earlier in the semester the students described the scientific method solely as a means of producing knowledge.

Second, the students' use of language expanded to include more expert terminology related to science. Terms that were frequently used in both written and verbal communications of the students towards the latter part of the course included validity, reliability, repeatability, evidence-based conclusions, tentative, biases and interpretation.

Third and last, their views on the nature of science and knowledge became more "scientific". For example, the students, collectively, described knowledge as the interpretation of data that is often subject to alternative explanations. Furthermore, the students were more inclined to talk about knowledge as something "proven to date" and in terms more tenuous than those used earlier in the semester.

Year 2 Study

In year two, the structure of the learning experiences with regard to NOS were almost identical to that described earlier. Learning outcomes were similar across cohorts. The outcomes of the additional task (selecting children's literature) are presented in Table 3 (Initial Categories) and Table 4 (Concluding Categories). Results from the second year of the study suggest that the conceptual and epistemological orientations of the pre-service teachers strongly influence the type and nature of children's literature in science used within the curriculum.

Table 1: Categories and Descriptions For Initial Definitions of Science (Year One)

Category N (total)=27	Description
1 n =13	Science is defined by listing discrete topics or facts in science with little or no elaboration. catch-all expressions characterizing science as "everything".
2 n=9	Science is defined as a method that includes experiments or as a universally applicable, lock-step procedure.
3 n=5	Science is defined as questioning and pursuing knowledge.

Table 2: Categories and Descriptions For Concluding Definitions of Science (Year One)

Category N (total)=27	Description
1 n=10	Science is described as a method but not as a single "method". Science operates on "rules" and "principles" established by the community. These methods can confirm or rebut ideas.
2 n=17	Science is described as inquiry with a focus on questioning. Ideas are confirmed or disproved through the inquiry and data is subject to multiple interpretations. Science is seen to have significant social components.

Conclusions

The authors of this study do not suggest that the students held highly sophisticated understandings of the NOS. However, there is considerable support to argue that the language used by the students to describe science could certainly serve as a solid foundation for advancement towards more expert understandings within the subject. The outcomes reported in this paper are consistent with the modest goals suggested by Matthews (1998). Further, the limited language and simplistic structure typically used to describe science may actually belie a deeper, richer understanding of the subject and, with opportunities to examine their explicit and tacit conceptions regarding the nature of science, significant gains can be made. The fruitfulness of this study in exploring the conceptions of the nature of science held by pre-service elementary teachers, when specific attention is given to the more deeply seated tacit knowledge of the students, generates implications for further research in (as well as curriculum design for) the preparation.

Table 3: Initial Categories and Descriptions For Children's Books within Science (Year Two)

Category N (total)=27	Description
1 n=10	Encyclopedic. These books include visual dictionaries and books primarily used to look up terms and/or vocabulary.
2 n=9	Topic book. These books describe, in detail, one or several related topics. Emphasis of the book is on introduction of vocabulary and explaining concepts.
3 n=8	Activity Book. The activities are nearly always "cook book" with step-by-step instructions for getting a particular outcome.

Table 4: Concluding Categories and Descriptions For Children's Books within Science (Year Two)

Category N (total)=27	Description
1 n=2	Encyclopedic. These books include visual dictionaries and books primarily used to look up terms and/or vocabulary.
2 n=4	Topic Book I. These books describe, in detail, one or several related topics. Emphasis of the book is on introduction of vocabulary and explaining concepts.
3 n=7	Topic Book II. These books describe in detail one or several related topics. The reader is provided with explanations by the authors of "how we know".
4 n=1	Activity Book. The activities are nearly always "cookbook" with step-by-step instructions or they treat science "experiments" as "magic".
5 n=8	Investigation Books. These books provide descriptions of experiments and investigations that children can conduct, but the emphasis is on teaching children the processes of science.
6 n=5	Scientists. The life and times of scientists are explored in these books. Often, these books provide the reader with insights into the historical development of ideas.

References

American Association for the Advancement of Science (1990). *Science for all Americans.* New York. Oxford University Press.

Cleminson, A. (1990). Establishing an epistemological base for science teaching in light of contemporary notions of the nature of science and of how children learn science. *Journal of Research in Science Teaching* 27(5), 429-445.

Glaser, B.G., & Strauss, A.L. (1967). *The discovery of grounded theory.* Chicago, IL: Aldine.

Hashweh, M. Z. (1996). Effects of science teachers' epistemological beliefs in teaching. *Journal of Research in Science Teaching* 33(1), 47 - 63.

Hewson, P. W., Zeichner, K. M., Tabachnick, B. R., Blomker, K. B., & Toolin, R. (1992). A conceptual change approach to science teacher education at the University of Wisconsin-Madison. Paper presented at the Annual Meeting of the American Education Research Association, San Francisco, CA.

Kagan, D. M. (1992). Implications of research on teacher beliefs. *Educational Psychologist* 27(1), 65-90.

Kennedy, M. M. (1998). Education reform and subject matter knowledge. *Journal of Research in Science Teaching* 35(3), 249-263.

King, B. B. (1991). Beginning teachers' knowledge of and attitudes toward history and philosophy of science. *Science Education* 75(1), 135-141.

Lederman, N. G. (1998). The state of science education: Subject matter without context. *Electronic Journal of Science Education* 3(1) Sept. 1998. ISSN 1087-3430 Available: http://unr.edu/homepage/jcannon/ejse

Lederman, N. G. (1992). Students' and teachers' conceptions of the nature of science: A review of the research. *Journal of Research in Science Teaching, 29*(4), 331-359.

Matthews, M. R. (1998). In defence of modest goals when teaching about the nature of science. *Journal of Science Education* 35(2), 161-174.

Matthews, M. (1994). *Science Teaching: The Role of History and Philosophy of Science.* Routledge, New York.

McComas, W., Clough, M., and Almazroa, H. (1998). The role and character of the nature of science in science education. In W. McComas (Ed.), *The nature of science in science education rationale and strategies* (pp. 3-39). Norwell, MA: Kluwer Academic Publishers.

National Research Council (1996) *National science education standards* Washington D.C.: National Academy Press.

Norris, C. (1997). *Against relativism: Philosophy of science, deconstruction and critical theory.* Malden, MA: Blackwell Publishers.

Prawat, R. S., & Anderson, A. L. (1989). Eight teachers' control orientations and their students' problem-solving ability. *Elementary School Journal* 89, 99-111.

Slezak, P. (1994a). Sociology of scientific knowledge and science education Part 1. *Science and Education* 3, 265 - 294.

Slezak, P. (1994b). Sociology of scientific knowledge and science education Part 2: Laboratory life under the microscope. *Science and Education* 3, 329-355.

Stodolsky, S., & Grossman, P. (1995). The impact of subject matter on curricular activity: An analysis of five academic subjects. *American Educational Research Journal* 32, 227-249.

Strauss, A., & Corbin, J. (1990). Basics of qualitative research: *Grounded theory procedure and techniques.* Newbury Park: Sage.

Combining Knowledge of Physics and Chemistry in Teaching: The Behaviour of a Narrow Jet of Water in the Presence of Charged Insulators

Liisa Kyyrönen
Department of Teacher Education, University of Helsinki, Finland
Maija Ahtee
Department of Teacher Education, University of Jyväskylä, Finland

Abstract

In this study the aim was to find out how student teachers applied their knowledge, in physics and chemistry, in an upper secondary science topic. The first stage consisted of an analysis of the upper secondary students' interpretations of an experiment in which a charged rod is brought near another charged rod and a narrow jet of water. The second stage of the study was based on these results. The main research question was: How do the student teachers apply their knowledge to their explanations? The main finding was that the student teachers' answers borę a strong resemblance to those of the upper secondary students. The earlier learning situation influenced the student teachers' interpretations and they did not make use of the theoretical knowledge obtained during their further studies.

Introduction

The main idea of this study was to raise questions or challenges, in which student teachers need support to develop into novice teachers. Demonstration forms the core of the research situation. The behaviour of charged rods, a familiar school experiment, formed the first part of the demonstration. In another experiment, the behaviour of a narrow jet of water in the presence of charged insulators was examined.

This study is based on the conception that the teachers' content knowledge plays a central role when he/she guides the upper secondary students' way of constructing knowledge. In science teaching, the central demand is that the teacher has to have a clear view of the theory to which the observations of the phenomenon can be matched and which will also reveal the relations between the concepts needed to describe and explain the observations. (Shulman, 1987; Villani & Pacca, 1994; Howe, 1996; Ogborn *et al.*, 1996; Hodson & Hodson, 1998; Mintzes & Wandersee, 1998). In this study, in particular, the teacher has to understand macroscopic-level knowledge as a knowledge that is transmitted through observations and that informs about the properties of matter and microscopic-level knowledge as an explanatory knowledge.

Design and problems

In refining the research task, it was decided to concentrate on considering the way student teachers apply scientific knowledge in explaining an experiment

H. Behrendt et al. (eds.), Research in Science Education – Past, Present, and Future, 161–166.

familiar to them from secondary school. Therefore, it was decided to analyse, first, how the upper secondary students interpret the demonstration and to base the further research design on these results. Thus the study consists of two stages with different subject groups but the same demonstration. In the first stage, A, in a chemistry lesson at the upper secondary school, the polarity of the water molecule was taught as new knowledge. The upper secondary students were studying the compulsory course in chemistry in the first year. In the second stage, B, the chemistry student teachers should have had the scientific knowledge needed to explain the phenomena in question. The student teachers had all studied physics and chemistry at the corresponding departments and chemistry, at least the minimum number of courses required for chemistry teacher competence. Now they were carrying out their studies in teacher education.

The chosen demonstration extended over a long time scale in both the upper secondary students' and student teachers' experiences. All of them had earlier seen or carried out the first experiment. All of them were able to describe the familiar electrostatics experiments. The deflection of a jet of water was new to the upper secondary students but familiar to the majority (15/18) of the student teachers. The effect of the material of the charged rod on the deflection was new also to the student teachers.

The first stage A

The research question in this stage was: What kind of interpretations do the upper secondary students connect to an experiment in which the behaviour of a jet of water is studied in the presence of either a positively or negatively charged rod?

The video tapes of the teaching situation and the notes written by the upper secondary students during the chemistry lesson formed the basic documents. During the lesson of the first teaching group (21 students, age 16-17), the teacher demonstrated the interactions of two charged rods. After this, the negatively charged rod was brought near a jet of water. The students predicted how the jet of water would deflect when a positively charged rod was brought near it. The second teaching group (19 students divided into five study groups) studied the behaviour of the jet of water in the proximity of the charged rods without the teacher's introduction about the electrostatics experiments.

The following three findings are based on the data from both groups: (i) The upper secondary students remembered the electrostatics experiments well, but rather as experiential events than from the conceptual point of view. For example, a student remarked spontaneously: "*It's this cat fur thing again!*" (ii) The deflection of the water jet in the same direction in the proximity of the oppositely charged rods was against the upper secondary students' predictions and it seemed hard for them to believe what they saw: "*The water behaves strangely*" or "*It (the water) goes towards it (the ebonite rod) and not away from it!*" (iii) Only one upper secondary student even tried to analyse the conflict between the prediction and the observations: "*Why do you rub the glass rod with different material – with plastic and not with cat fur (like the ebonite rod)?*"

The second stage B

The upper secondary students do not naturally connect the experiment on the deflection of the water jet to their everyday experiences, and the scientific knowledge needed for the explanation is high in the theoretical hierarchy. The student teachers do not specifically encounter this experiment during their subject studies but they should have acquired the necessary knowledge to be able to give a theoretically justified explanation

The upper secondary student will acknowledge the existence of his expectations when he has to write down his prediction. His theoretical framework will then notably influence the resolution of the conflict between observations and expectations The student's way of relating an event is more often a description of what happened rather than an explanation of how and why it happened. A science teacher's main task is, however, to guide the students toward the explanations; that is, to guide them to ask how and why questions (Ogborn *et al.*, 1996).

The research questions in the second stage were the following: (1) What kind of memories does the experimental design awake in the student teachers' mind? (2) How do the student teachers differentiate the two experimental designs? (The interactions of the charged rods and the interaction of the jet of water with a charged rod.) (3) How do the student teachers apply and integrate their knowledge in physics and chemistry in their explanations?

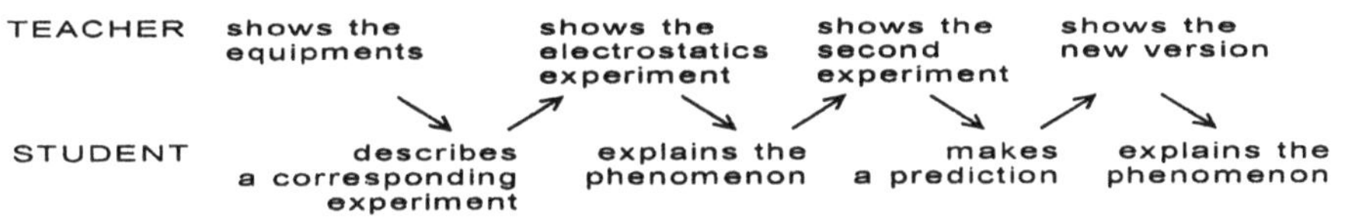

Fig. 1: The experimental design

Data collection and analysis

The student teachers' responses to the questions introduced to them separately in the different stages of the experimental design, (Fig. 1) formed the raw data.

In all student teachers' minds, the sight of the equipment awakened almost similar memories of how they were used at school during the physics lessons. The lack of any competing descriptions can be explained by the fact that the experiments, as such, have no exact substitute in student teachers' everyday experiences. The same thing was found with the upper secondary students in stage A.

The student teachers' explanations were divided into three groups according to the view of the nature of scientific knowledge: (i) The concepts have been learnt as facts and the scientific knowledge is thought to be verified through the observations. The account gives an answer only to the question *What*? (ii) In the explanations they also considered the question what kind of things could lead to the charging of the rod during rubbing? The account gives an answer to both questions *What and How*? (iii) A scientific way, at least to some extent, of treating the phenomenon.

Some of the respondents restricted their description to just telling what they had been taught at school, even though they had been asked to explain the phenomenon as exactly and deeply as possible and, therefore, the effect of their university studies on the quality of the explanation was not revealed. The difficulties in giving an explanation were considered to be due to forgetfulness. For example, a student teacher claimed that he had once known, in detail, how to explain the phenomenon but that now: "*I do not quite remember the reason for the opposite signs of the ebonite and glass rod*". Some of the respondents pondered also the adequacy of their own knowledge: "*The physics argumentation is not clear to me. Why is it the ebonite rod that gives out the electrons?*"

Two thirds of the student teachers were convinced that they were able to give at least, a satisfactory explanation. In many cases, the written accounts conflicted with this view. For example, a student teacher wrote that, in rubbing, a charge will be formed. Otherwise, the answer contained only concrete, observable details. However, at the end of his answer he had written "*the formation of charges?*"

The answers to the new version of the second experiment (replacement of the negatively charged glass rod with the positively charged ebonite rod) were classified into three groups: (i) Answers in which the jet of water was predicted to behave in the same way in both cases and the reasoning was quite correctly based on the polarity of the water molecule. (ii) Answers in which it was stated that the ebonite rod would repel the jet of water; in other words, the reasoning was based on the interaction of the opposite charges. However, in some of these answers the polarity of the water molecule was mentioned. (iii) Answers in which the student teachers were muddled in their observations or in the related concepts and in which, therefore, the predictions had no meaning even though it may have been superficially correct.

In their answers assigned to the first group (i) the student teachers did not problematize, the interaction between the dipoles, even though they had drawn beautiful pictures, in which the water molecules were in a regular formation with the same charges side by side. The student teachers whose answers were assigned to the second group (ii) noticed the conflict between their prediction and the real event. However, their ways of overcoming the conflict varied. For example, one student teacher was very surprised and did not even try to find an explanation: "*Something quite unexpected happened in this phenomenon... It strikes me as crazy*". She gave an explanation of her own and wrote about the jet of water as "*a neutral partner to both the charged glass and ebonite rods*". Another student teacher strongly believed that she now understood the whole thing. However, she did not think over what might happen between the dipoles. The answers within the third group (iii) showed that the student teachers' difficulties in analysing their observations prevented them, also, from noticing the conflict in their own presentation.

Findings and conclusions

The results of this study are provisional. The student teachers' answers support the research strategy. The assumption, based on the results of the first stage, A, that the student teachers would also remember the experiments familiar to them from their school time turned out to be true in almost all cases. Half of the respondents mentioned that cat fur would be suitable for rubbing the ebonite rod. There were, however, no traces of the use of theoretical knowledge in the explanation of this experientially remembered experiment.

Some of the student teachers (6/18) only said what the observations were and what concepts were related to these. Their belief about the nature of science, thus, appears as factual and unambiguous knowledge (route I, Fig. 2). Others of the student teachers (12/18), tried to explain what, during rubbing, causes the rods to be charged. These answers were classified into two groups. Some of the student teachers (8/12) answered the question how?; that is, they described the relations between the observations and the corresponding concepts (route II, Fig. 2). These responses still contained a factual view of scientific knowledge. Some of the student teachers (4/12) proceeded in their explanations to the level described by the question "why" and, thus, brought forward the question of how to interpret scientific knowledge (route III, Fig. 2).

The same basic model of how to describe the electrostatics experiments was present both in the upper secondary students' and in the student teachers' answers:

The taught fact "the rods will be charged in rubbing" can be verified by observing the interactions between the rubbed rods. The experimental design did not question the fact that only two kinds of electricity exist. In this way, the empirical situation also supported the presentation of the taught knowledge as factual knowledge. This conclusion is also supported by the fact that all student teachers said *what* observations and concepts belong to the examination of this phenomenon.

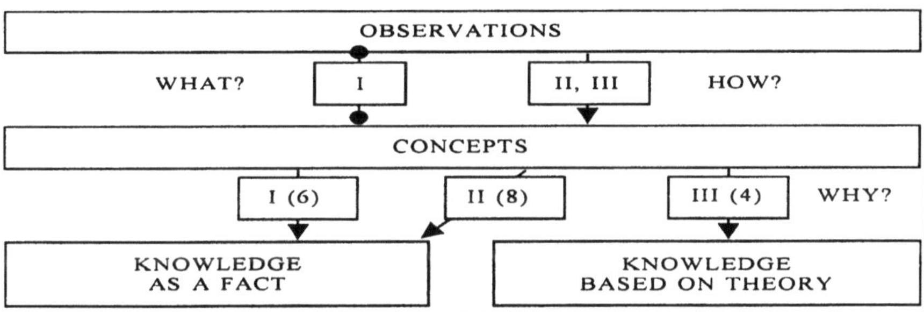

Fig. 2: Student teachers' conceptions of the nature of scientific knowledge

Student teachers' difficulties in explaining the phenomenon scientifically came out when they tried to think over *how* the charging of the rods could happen (see Fig. 2). The student teachers' answers (12/18), in which they looked for a microscopic level explanation for the charging of the rods, could be divided into two groups (Arons, 1990, 144). In half of the answers, the rubbing was said to create charges and in the other half, charges were said to change place. In one answer, belonging to the latter group, the charges were always supposed to move from the rubbing material to the rod; in the other answers, the charging was determined according to whether electrons moved from the rubbing material to the rod or vice versa. In spite of the efforts to give an explanation, the observations and the concepts remained separated, the construction of the explanation was supported with facts learnt by heart, and the theory did not guide the formation of an integrated whole. Also the meaning of the concepts was confused so that the presentation became internally conflicting. For example, the charging of the rod was seen as equivalent to the polarization in an electric field.

As to the second experiment (see Fig. 1), the analysis revealed the student teachers' way of simplifying water with a water molecule and using it to interpret the behaviour of the jet of water. One student teacher described how "*the oppositely charged end of the dipole will always be attracted by the rod and then the whole molecule will follow after*". The interaction between the water molecules was clearly not taken into account in any of the answers. In some answers the interpretation of the structure of the jet of water corresponded to the conception of the continuous structure of matter.

The explanation of the behaviour of the jet of water near a charged insulator would have needed, in addition to the analysis of the information obtained from the two types of experiments also a synthesis of this information (Arons, 1990; Taber, 1998). However, no student teacher took into account the electric interaction between the water molecules and, thus, the behaviour of the matter was identified as the behaviour of a single molecule. The importance of theory in scientific explanation and in the evaluation of the validity of the explanation did not come up in any answer in such a way that the object of the explanation would have pointed to the question *Why*? The way four student teachers approached the task was classified as more profound than just a repetition of taught facts. One of the student teachers showed a critical attitude when she considered the possibility of different explanations. The second asked meaningful questions: "*Why do the rods become charged in rubbing as they do? Why is it the ebonite rod which will give up electrons?*" The third analysed the difference in the experimental designs (Roth *et al,*. 1997): "*In the second phenomenon only the rod was charged and the water molecules are polar ...?*" The fourth student teacher wondered: "*Does the orientation of the water molecules affect the size of the angle of the water jet?*"

From this study, the following challenges were found for physics and chemistry teacher education: The student teachers should clearly state on which theoretical framework they base their explanation of the phenomenon they have studied. They should be guided to evaluate their own thinking model used to explain different phenomena, by asking about both the relevance of the theory they have used and the validity of the theory in their interpretation. The student teachers should be required to state their own explanation in a form in which the meaning of the concepts defined by the theory is expressed in a precise scientific way. If there is no explicit guidance as to what a scientific explanation means in linking the observations, the concepts and the theory together, the result will be like those found in this study. The student teacher will return to his school days and look at the school information without applying the possibilities given by his own university studies "to see more and understand more profoundly".

References

Arons, A.B. (1990). *A guide to introductory physics teaching*. New York: John Wiley & Sons.

Hodson, D. & Hodson, J. (1998). Science education as enculturation: some implications for practice. *School Science Review* 80, 17-24.

Howe, A.C. (1996). Development of science concepts within a Vygotskian framework. *Science Education* 80, 35-51.

Mintzes, J.J. & Wandersee, J.H. (1998). Reform and innovation in science teaching: A human constructivist view. In J.J. Mintzes, J. H. Wandersee, & J.D. Novak, *Teaching Science for Understanding. A Human Constructivist View*, San Diego: Academic Press.

Ogborn, J., Kress, G., Martins, I. & McGillicuddy. (1996). *Explaining science in the classroom*. Buckingham: Open University Press.

Roth, W.-M., McRobbie, C.J. & Boutonné, S. (1997). Why may students fail to learn from demonstrations? A social practice perspective on learning in physics. *Journal of Research in Science Teaching* 34, 509 - 533.

Shulman, L.S. (1987). Knowledge and teaching: Foundations of the new reform. *Harvard Educational Review* 57, 1-22.

Taber, K.S. (1998). The sharing-out of nuclear attraction: or "I can't think about physics in chemistry". *International Journal of Science Education* 20, 1001-1014.

Villani, A. & Pacca, J.L.A. (1994). *Scientific knowledge and teaching ability. Proceedings of an international conference on Thinking Science for Teaching: The Case of Physics*, September 1994, in Rome, Italy.

Intuitive Rules: A Theory and its Implications to Mathematics and Science Teacher Education

Pessia Tsamir, Dina Tirosh, Ruth Stavy and Ilana Ronen
School of Education, Tel-Aviv University, Israel

Abstract

This article presents the intuitive rules theory, relating to students' responses to different tasks in science and mathematics. We argue that many alternative conceptions apparently related to different mathematical and scientific domains originate in a small number of intuitive rules: "More A-More B", "Same A-Same B" and "everything can be divided". This theory has a strong predictive power. This paper demonstrates this power and discusses possible implications to science and mathematics teacher education.

Introduction

It is widely reported that students at different grade levels in different countries tend to give incorrect responses to tasks related to diverse topics in science and mathematics. Various theoretical frameworks were developed to explain this phenomenon, of which a dominant one has been the alternative conception paradigm. According to this paradigm the child takes an active, constructive role in the knowledge acquisition process and brings alternative, internally coherent, robust and persistent conceptions to the learning situation (e.g., Driver, 1994; Fischbein, 1987). Yet there is evidence that students tend to respond inconsistently to tasks related to the very same mathematical or scientific concept (e.g., Clough & Driver, 1986). This constitutes a challenge to the alternative conception paradigm.

Through our work in both science and mathematics education, we have observed that students react in similar ways to a wide variety of conceptually non-related tasks. These tasks differ with regard either to their content area and/or to the required reasoning but share some common, external features. Based on these observations, we have suggested an alternative theory that explains and predicts students' responses to mathematics and scientific tasks: Intuitive Rules Theory. We argue that many alternative conceptions, apparently related to different, specific mathematical and scientific domains, are actually specific instances of these rules. The rules have the characteristics of intuitive thinking: It was observed that responses based on these rules are taken as self-evident, they are used with great confidence and perseverance. Moreover, intuitive rules have attributes of globality and coerciveness.

So far, we have identified four intuitive rules: two (*More A-More B*, and *Same A-Same B*) relate to comparison tasks, and two (*Everything comes to an end*, and *Everything can be divided*) to successive division tasks. The two main strengths of this approach are: (1) it accounts for many of the observed, alternative conceptions in science and mathematics education, and (2) it has great predictive power. Thus, the theory of the intuitive rules can enable researchers, teachers and curriculum

H. Behrendt et al. (eds.), Research in Science Education – Past, Present, and Future, 167–175.

planners to foresee students' inappropriate reactions to specific problems, and this can then help them to plan appropriate sequences of instruction.

This paper consists of three main sections. The first section presents intuitive rules related to comparison tasks and the second describes intuitive rules related to successive division tasks. The last section presents some possible implications of the Intuitive Rules Theory to mathematics and science teacher education.

Intuitive rules and comparison tasks

Comparison tasks share some common features: In each task two objects (or systems) that either differ or resemble in a certain, salient quantity (A) are described. The student is then asked to compare these objects (systems) with respect to another quantity (B). An adequate response depends on the specific task. Students, however, tend to base their comparison of the quantity B on the salient quantity A.

The intuitive rule: *More A-More B* is reflected in students' responses to comparison tasks in which two objects which differ in a certain, salient quantity (A) are described ($A_1>A_2$). The student is then asked to compare the two objects with respect to another quantity B ($B_1=B_2$ or $B_1<B_2$). In such cases, many students responded inadequately that $B_1>B_2$, according to the rule More A (the salient quantity)- More B (the quantity in question).

The intuitive rule *Same A - Same B* is observed in students' responses to many comparison tasks in which two objects to be compared are equal in respect to one quantity A ($A_1=A_2$) but differ in another quantity B ($B_1\neq B_2$). A common incorrect response to such tasks, was that $B_1=B_2$ because $A_1=A_2$, according to the intuitive rule *Same A - Same B*.

In the following sections we discuss the intuitive rules *More A-More B* and *Same A–Same B*.

The intuitive rule: More A-More B

In previous papers we have shown that the intuitive rule *More A-More B* accounts for students' responses to many comparison tasks, including classical, Piagetian conservation tasks, tasks related to intensive quantities and other mathematical and scientific comparison tasks (e.g., Stavy & Tirosh, 1996; Tsamir, 1997; Tsamir, Tirosh & Stavy, 1997). In all these cases, a substantial number of students respond according to the rule More A (the salient quantity)- More B (the quantity in question) arguing, incorrectly, that $B_1>B_2$ because $A_1>A_2$. Here we describe several studies aimed at testing the predictive power of the intuitive rule *More A-More B* in different content domains (geometry, mechanics and biology).

I. <u>Larger area - larger perimeter</u>. A total of 100 students from grades 1,3,5,7, and 9 from an upper, middle-class background were individually interviewed, with reference to the following task:

> *Two identical, plastic rectangles are presented. Each of these rectangles consists of a small square at the upper right corner and a polygon. The small square is removed from the upper right corner of one of the rectangles. A polygon is obtained. Is the perimeter of the obtained polygon equal/not equal to the perimeter of the original rectangle? If equal, explain why? If not equal, which is bigger? Why?*

In this task, the areas of the two objects differ but the perimeters are equal. We predicted that due to the salient differences in the areas of the rectangle and the polygon, students would incorrectly argue that the perimeter of the rectangle is larger than that of the obtained polygon. It was indeed found that in all these grade levels, not less than 70% of the respondents claimed incorrectly that the perimeter of the rectangle was larger than that of the polygon. Common justifications were that the rectangle "is larger", "has more area", etc. These high percentages of *More A* (area, or size of rectangle) - *More B* (perimeter) confirm our prediction that responses to this comparison task are strongly affected by the intuitive rule *More A-More B*.

II. Longer distance-longer time. A questionnaire, including the following problem, was presented to a total of 360 students in grade levels 7-12:

A motorboat crosses a river, from point A to point B. Its engine speed is constant, and the direction of its wheel is perpendicular to the riverbank.

One morning the boat took off from point A and arrived at point B. On this day the river was quiet, without any current whatsoever. On another morning, the same boat, again, was meant to cross the river from point A to point B. This time, however, a strong current swept the boat away, and it reached the other side at point C.

Choose the correct statement, and explain your choice: 1. The crossing time on the first day was shorter than that on the second day. 2. The crossing time on the second day was shorter than that on the first day. 3. The crossing times on the first and second days were equal.

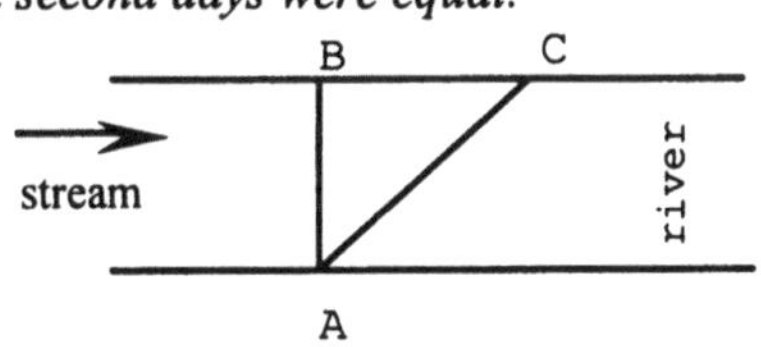

The correct answer to this question is that the crossing times of the boat on the first and second days are equal. We predicted that students, affected by the differences in distance, would argue, in line with the intuitive rule *More A-More B*, that the crossing time on the first day was shorter than that on the second day. Indeed, 53%, 61%, 83%, 58%, 67%, and 74% of the students in grades 7, 8, 9, 10, 11 and 12 respectively incorrectly claimed, in line with the rule *More A* (distance between endpoints) - *More B* (duration of time).

III. Larger animals- larger cells. The following problem was submitted to a total of 120 students in grade levels 7 and 8:

Is the size of a muscle cell of a mouse bigger than/equal to/ smaller than/ the size of a muscle cell of an elephant? Yes/No. Explain your choice.

The correct answer is "equal to". We predicted that students would tend to argue in line with the intuitive rule *More A - More B*, that the cells of the larger animal are larger. Indeed, the majority of the students (65% in grade 7 and 51% in grade 8) incorrectly claimed that larger animals have larger cells. Common justifications were: "The elephant is bigger than the mouse".

We have so far described several comparison tasks from different content domains, all aimed at testing the predictive power of the intuitive rule *More A -*

More B. The findings confirm our prediction that students tend to claim that $B_1>B_2$ because $A_1>A_2$.

We suggest that students' responses to such tasks are determined by the specific, external characteristics of the task which activate the intuitive rule and not necessarily by students' ideas about the task's specific content or concepts. This type of response, although valid in many situations (the more one eats the fatter one gets, the more money one has - the more one can spend), does not apply to others.

We have mentioned before that the rule is often used in situations in which it is not applicable. Yet, we have shown that, with regard to many such tasks, people at different ages and/or with different levels of instruction, at some point start using the rule selectively. We suggest that with age and/or instruction, schemes, rules, and bodies of knowledge related to specific tasks are developed or reinforced. Consequently, in respect to these tasks, the rule loses its power in favour of other, competing knowledge. For instance, in the case of the conservation of quantity of matter, "The higher - the more" is replaced by identity or compensation considerations. It is also possible that with age and/or instruction, children become aware of the need to examine their initial responses, to consider other factors which might be relevant to the task, and to avoid conflicting arguments. Thus, they gradually learn the limits within which *More A-More B* is applicable. Still, although children cease to use the rule in certain instances at certain ages, they never stop using it altogether and it continues to dominate in various other situations. In fact, in many of the previously described instances (e.g., comparing segments, comparing angles, free fall) older children and adults kept using the rule, even after formal, related instruction.

The intuitive rule: Same A-Same B

The mathematics and science tasks related to the rule *Same A - Same B* share some common features. In each of them, two objects (or systems) equal in a certain quantity A ($A_1=A_2$) but different in another quantity B ($B_1\neq B_2$) are described. Students are asked to compare B_1 and B_2. Our prediction, based on the theory of the intuitive rules, was that a substantial number of students would claim that $B_1=B_2$ because $A_1=A_2$. Here we shall describe several studies aimed at testing the explanatory and the predictive power of this intuitive rule in different content domains (numbers, geometry, probability, structure of matter and biology) (Tirosh & Stavy, 1996; Tsamir, Tirosh & Stavy, 1998).

I. Same percentage - same perimeter. Thirty-five 11th graders were presented with a questionnaire, including the following problem:

Consider a rectangle. Side a of this rectangle is increased by 20% and side b is reduced by 20%. Is the perimeter of the rectangle before the change bigger than/ equal to/ smaller than/ the perimeter of the rectangle after the change?

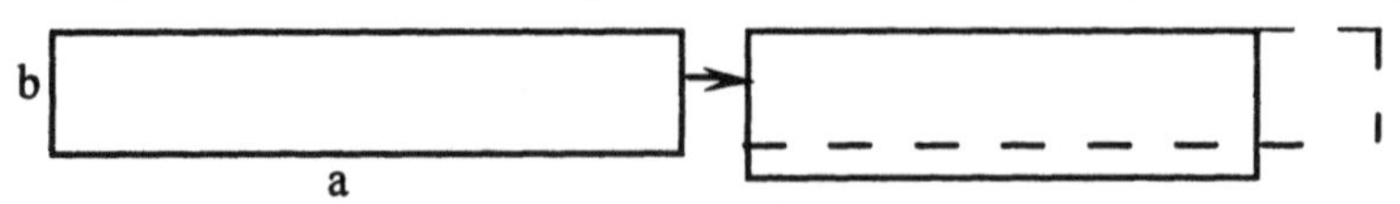

We predicted that students would incorrectly argue that the perimeter before and after the change is the same because "the addition and the reduction are the same". Indeed, the vast majority of the students (72%) claimed exactly this.

II. Same volume - same resistance. The following task, related to the resistance to dryness of bacteria, was included in a written questionnaire submitted to about 60 students each in grades 10,11 and 12:

Bacteria are usually shaped spherically (cocci), rod-like (bacilli) or like a spiral (spirillae). The cell volume of each of these bacteria is equal. Is the resistance to dryness of these three types of differently-shaped bacteria equal/ non-equal? Why?

If you think their resistance is different, which of these three types of bacteria is most resistant? Why?

The spherical bacteria are the most resistant to dryness. Yet, although all the participant students received formal instruction related to the ratio surface area and volume and its role in biological phenomena, about a third of the students claimed, in accordance with the intuitive rule *Same A - Same B* that: "the cells have the same volume and therefore their resistance to dryness is the same.

In the comparison tasks presented here, the two objects or systems to be compared were equal in respect to one quantity (A_1=A_2) but different in respect to another one ($B_1 \neq B_2$). A common incorrect response to all these tasks, regardless of the content domain, took the form of B_1=B_2 because A_1=A_2. We regard all these responses as specific instances of the use of the intuitive rule *Same A - Same B*. It seems that our cognitive system tacitly assumes that when two objects are equal in a certain quantity, they are equal in other quantities as well. This assumption could evolve from a more general tendency to extrapolate given information to new situations. This type of extrapolation, although valid in many situations, does not apply to others.

Intuitive rules and subdivision tasks

Successive division tasks share some common features: In each of them a process of successive division is described and the student is then asked to determine whether the process will come to an end. The adequate response is that the process is infinite when the object is a mathematical one and finite when the object is physical. It was found that many students in grade levels 7 to 12 gave the same responses to successive division tasks related to both material and mathematical objects. The younger ones claimed that *Everything comes to an end*. The older students argue that these processes will never end, as *Everything can be divided*. Since these two rules were identified due to the responses given by different aged students to the same tasks, the following section relates to both these intuitive rules.

In this paper we describe students' responses to successive division tasks in mathematics and science. In mathematics, successive division tasks were used to examine students' conceptions of infinity (e.g., Fischbein, Tirosh & Hess, 1979). In the physical sciences successive division of material objects tasks were used to investigate students' conceptions of matter as particulate (e.g., Pfundt, 1981).

We presented students of different ages with various tasks related to successive division of mathematical and physical objects (e.g., Cohen, 1997). In all these tasks, students were asked whether the described successive division of a specific object would come to a halt. The purpose of these studies was to explore the generality, the explanatory and the predictive power of two intuitive rules: *Everything comes to an end* and *Everything can be divided*. Here we present a sample of these tasks.

A. Repeated Division Tasks.

1. Consider a rectangle. Divide it into two equal rectangles. Divide one rectangle into two equal rectangles. Continue dividing in the same way.
   Will this process come to an end? Yes/ No. Explain your answer.
2. Consider a rectangular piece of aluminum foil. Divide it into two equal parts. Divide one half into two equal parts. Continue dividing in the same way. Will this process come to an end? Yes/ No. Explain your answer.

B. Seriation Tasks.

3. "Consider the following series:
   1, 1/2, 1/4, 1/8, 1/16, 1/32, 1/64, 1/128,...
   In this series each number is half the previous one.
   Will this process of dividing the numbers come to an end? Yes/ No.
   Explain your answer.
4. A teaspoon of sugar is put into a cup of water and stirred well into it. Half of the sugar water is poured out, half a cup of water is added to the cup and is mixed thoroughly with the remaining sugar water. This process is executed again: half of the sugar water is poured out, half a cup of water is added, etc. This process is repeated.
   Is it possible to reach a stage at which no sugar at all will be found in the cup? Yes/No. Explain your answer.

We presented these (and other) tasks to 282 students from grades 7 to 12. All had studied the particulate theory of matter and received some instruction in Euclidean geometry. Half the students in each grade level first received the problems related to mathematical objects (1, and 3) on one sheet of paper. The problems related to physical objects (problems 2, 4) were then provided on a different sheet of paper. The other half of the participants received the problems in the reverse order.

Two types of responses were given by students in all grade levels to all tasks: (a) the process is finite, and (b) the process is infinite. The percentages of finite responses to both the mathematical and material object tasks decreased with age, while the percentages of infinite responses increased. The majority of the younger students provided finite responses to all tasks. Many of them explained their judgement by claiming, in line with one intuitive rule, that everything comes to an end. The older students, who became aware that subdivision processes may continue endlessly, tended to provide infinite responses to both the mathematical and material objects tasks, claiming, in line with another intuitive rule, that everything can be divided.

In this section we have shown that students' responses to successive division of mathematical and material tasks do not necessarily accord with scientific concepts and conceptual frameworks. We claim that the intuitive rule *Everything comes to an end* is extrapolated from experience in daily life while the intuitive rule *Everything can be divided* is a direct consequence of the natural tendency of our cognitive system to extrapolate. In this case, the extrapolation is from a visible stage in the process of subdividing a given object, to the next stage, and then to the next stage, and so on. Starting from a certain age, subjects see no reason to argue that at a certain stage, this extrapolation will not be valid anymore.

Applying knowledge of the intuitive rules in teacher education

Extensive documentation exists regarding students' alternative conceptions in science and mathematics. Teachers are encouraged to consider students' ways of thinking as a springboard for teaching (e.g., Borasi, 1986; NCTM, 1991). Thus, it is essential that teachers are familiar with students' conceptions related to specific content areas and their possible sources.

The theory of the intuitive rules attempts to explain and predict students' common incorrect responses in science and mathematics. This theory specifies intuitive rules underlying students' reactions to mathematical and scientific tasks. It could also be used to analyse given tasks. For instance, a task for which the correct answer is in line with an intuitive rule is expected to be "easy" while a task for which the correct answer contradicts an intuitive rule would probably be "difficult". Hence, it seems important that teachers be aware of the theory of intuitive rules and consider them when designing teaching sequences.

In this spirit, courses based on the Intuitive Rules Theory (IRTC), were designed for mathematics and science teachers. The main aims of these courses were (a) to introduce the theory of the intuitive rules by demonstrating the strong effect of the rules on participants' own responses and on their (future) students' responses to science and mathematics tasks; (b) to train teachers to use the theory of intuitive rules as a means to predict students' possible responses to given tasks; (c) to identify given tasks as "easy" or "difficult" and (d) to plan learning sequences, taking account of all the above.

This section will present the outline and some initial observations of the course for prospective elementary and middle school teachers.

The IRTC *for Prospective Teachers*

The IRTC was first tried with a group of twenty-five prospective teachers in twelve lessons of an hour and a half each. In the first two meetings, the participants were asked to respond to two questionnaires: The first, *Prospective Teachers' Questionnaire* (PTQ) included forty comparison mathematics and science tasks known to elicit responses of the type More A - More B, and Same A - Same B. Application of these intuitive rules yielded correct responses to some tasks and incorrect responses to others. The second, *Students' Thinking Questionnaire* (STQ) included comparison tasks for which the prospective teachers were expected to provide correct answers. None of these tasks was included in the PTQ questionnaire. Part I of the STQ presented the subjects with various correct and incorrect students' responses to each task (correct answers were designated). Participants were asked to suggest possible reasons for both the correct and incorrect responses. In Part II of the STQ, subjects were asked to predict common responses to given, comparison tasks and to list possible sources for each of these responses. In Part III, subjects were presented with several, "easy" and "difficult" comparison tasks related to a specific concept or topic. They were asked to provide reasons for presenting/ not- presenting each task in their future classes, and to choose one or two tasks that would definitely be presented in classes.

Prospective teachers' responses to both these questionnaires were used as a springboard for defining the intuitive rules and discussing their possible impact on their own as well as their future students' responses to comparison tasks. In

Meetings 3-7 which were devoted to these purposes, participants were divided into small groups. Each group was responsible for 6-7 comparison tasks: they discussed the responses given by each member of the group, and had to reach a consensus, acceptable to all members in the group. The participants documented their discussions and handed in the transcripts along with their decisions. A representative of each group reported on the work of his/her group. He/she related the initial responses of each member, the arguments, and final conclusions. During these discussions the prospective teachers noticed that they tended to provide responses of the types *More A–More B, Same A–Same B*. They also noticed that sometimes the application of these rules yielded correct responses and, often, incorrect responses. They were then asked to read several related articles (e.g., Stavy & Tirosh, 1996), to prepare brief reports on the main issues, and to present them in class. Meeting 8 was devoted to this purpose.

Meetings 9-12 related to possible implications of the theory of intuitive rules to instruction (e.g., Tirosh, Stavy & Aboulafia, 1997; Tirosh, Stavy & Cohen, 1998; Tirosh & Tsamir, 1996). Participants were asked to find tasks that are most likely to be given a response of type *More A–More B,* or *Same A–Same B*. They were asked to classify each of these tasks as either "easy" or "difficult" and justify their choice. Each pair revised their work in the light of detailed comments made by another pair (non-emphasized comments had to be justified). The instructor then reviewed these assignments, highlighting interesting issues related to the original problems, the comments and the revised version. A summary of interesting findings was then anonymously presented to the class, raising a discussion.

The participants then tried the revised versions with 6-8 middle school students. They analysed students' responses with reference to correctness, students' dependence on the intuitive rules, and their predictions regarding the degree of difficulty of each task. Finally, each pair presented their findings in class.

Initial Observations

Prospective teachers' own responses to comparison tasks (as reflected in their response to the PTQ) were affected by the intuitive rules: they provided incorrect responses in line with either *More A-More B* or with the *Same A-Same B* rules to several tasks.

Participants tended to provide only one (if any) reason for each students' expected response to Part I of the STQ. They argued that correct responses resulted from formal knowledge (such responses were typical even in tasks when an intuitive rule was consistent with the formal answer). In part II of the STQ, when asked to predict students' responses to given comparison tasks, participants usually mentioned a single common response for each problem, prevalently an erroneous one. In their responses to Part III of the STQ, many participants expressed their belief that "difficult" tasks, which are expected to trigger incorrect, intuitive responses, should not be presented in class. The idea that "it is not fair to present students with 'difficult', counter-intuitive problems" and that "teachers should choose problems that trigger success and avoid failure" were repeatedly expressed both in writing and in the discussions.

Searching the curriculum for problems that are apt to trigger the application of the intuitive rules was an extremely difficult task for the participants, even for those who exhibited a satisfactory understanding of the intuitive rules. The sessions with the middle school students contributed a great deal to participants' attitudes towards these rules. They were impressed by the explanatory and predictive power of these rules. In the concluding discussion, the prospective teachers related to this learning experience, mentioning that they had studied the theory of the intuitive rules, and then in addition, studying this unit had given them an opportunity to restudy the various content domains included in the curriculum from a fresh, unusual angle. They stated that the Intuitive Rules Theory is "a must" for teachers.

These findings suggest that the Intuitive Rules Theory has an impact on prospective teachers. Clearly, additional, systematic research is needed to study the effects of prospective and in-service teachers' acquaintance with this theory on their instruction.

References

Borasi, R. (1986). Using errors as springboards for the learning of mathematics. *Focus on Learning Problems in Mathematics* 7, 1-132.

Clough, E.E. & Driver, R. (1986). A study of consistency in the use of students' conceptual frameworks across different task contexts. *Science Education* 70(4), 473-496.

Cohen, S. (1997). *Tacit intervention and student responses to mathematical and material object tasks.* Unpublished M.A. thesis. Tel-Aviv University, Israel.

Driver, R. (1994). *Making sense of secondary science.* London: Routledge.

Fischbein, E. (1987). *Intuition in science and mathematics: An educational approach.* Dordrecht, Netherlands: Reidel.

Fischbein, E., Tirosh, P. & Hess, P. (1979). The intuition of infinity. *Educational Studies in Mathematic* 12, 491-512.

National Council of Teachers of Mathematics [NCTM] (1991). *Professional standards for teaching mathematics.* Reston, VA: NCTM.

Pfundt, H. (1981). Pre-instructional conception about substances and transformation of substances. In W. Jung, H. Pfundt, & C. V. Rhoeneck (Eds.), *Problems concerning students' representation of physics and chemistry knowledge* (pp. 320-341). Ludwigsburg, Germany: Pädagogische Hochschule.

Stavy, R. & Tirosh, D. (1996). Intuitive rules in science and mathematics: The case of "more of A-more of B". *International Journal of Science Teaching* 18, 653-667.

Tirosh, D. & Stavy, R. (1996). Intuitive Rules in Science and Mathematics: The case of "Everything can be divided by two". *International Journal of Science Education* 18, 668-679.

Tirosh, D., Stavy, R. & Aboulafia, M. (1998). Is it possible to confine the application of the intuitive rule: "subdivision processes can always be repeated?". *International Journal of Mathematics Education in Science and Technology* 29, 813-815.

Tirosh, D., Stavy, R., & Cohen, S. (1998). Cognitive conflict and intuitive rules. *International Journal of Science Education* 20, 1257-1269.

Tirosh, D., & Tsamir, P. (1996). The role of representations in students' intuitive thinking about infinity. *Journal of Mathematical Education in Science and Technology* 27, 33-40.

Tsamir, P. Representations of points. (1997). In E. Pehkonen (Ed.), *Proceedings of the 21st Annual Meeting for the Psychology of Mathematics Education.* Vol IV (pp. 246-253). Lahti: Finland.

Tsamir, P., Tirosh, D., & Stavy, R. (1997). Is the length of the sum of three sides of a pentagon longer than the sum of the other two sides? In E. Pehkonen (Ed.), *Proceedings of the 21st Annual Meeting for the Psychology of Mathematics Education.* Vol IV (pp. 214-221). Lahti: Finland.

Tsamir, P., Tirosh, D., & Stavy, R. (1998). Do equilateral polygons have equal angles? In A. Olivier & K. Newstead (Eds.), *Proceedings of the 22nd Annual Meeting for the Psychology of Mathematics Education.* Vol IV (pp. 129-136). Stellenbosch: South Africa.

Part 5: Conceptual Change – Teaching and Learning Processes

Conceptual Change Research and the Teaching of Science[1]

Stella Vosniadou,
National and Capodistrian University of Athens, Greece

Abstract

In this paper, I will try to define the conceptual change approach to the learning of science, as it has been developed over the years, on the basis of cognitive/developmental research. An effort will be made to differentiate this approach from the naive empiricism adopted by many science educators as well as from the Piagetian framework. It will be argued that the conceptual change approach can explain the phenomena observed in the acquisition of science concepts better than other approaches. At the end of the paper implications of this approach for teaching of science will be drawn.

What is the Conceptual Change Approach?

Many science educators adopt an empiricist approach to describe the process of learning science. According to this approach, there is little or no predisposition for learning. Knowledge acquisition is based on experience and proceeds in a continuous manner, enriching existing conceptual structures. Science learning depends on increased experiences, that, at the beginning give rise to concrete ideas but that later, these ideas, become more abstract and more widely applicable. Instruction should provide children with more experiences and opportunities to understand the process of doing science.

The development of science concepts has been interpreted differently by Piaget (1970). Piaget has also given a great deal of attention to experience but he claimed that the process of developing more abstract conceptual structures depends on the constructive activity of the learner. He chose to provide a structural account of the intellect in terms of a mathematical model. According to this model, the process of intellectual development proceeds through a series of stages, each of which is characterized by a different psychological structure. In infancy, intellectual structures take the form of concrete action schemas. During the pre-school years these structures acquire representational status and later develop into concrete operational structures (described in terms of groupings based on the mathematical notion of sets and their combinations). The last stage of intellectual development,

[1] Plenary Address

H. Behrendt et al. (eds.), Research in Science Education – Past, Present, and Future, 177–188.

formal operational thought, is characterized by the ability to engage in propositional reasoning, to entertain and systematically evaluate hypotheses, etc.

The process of cognitive development described by Piaget has been characterized as "global restructuring" (Carey, 1985) and is considered to be the product of the natural, spontaneous process of intellectual development and not of explicit learning. The implications of this approach for instruction is that we should encourage the constructive ability of learners and provide them with experiences that may be interpreted differently at different stages but which, by the time students reach adolescence, will be transformed into scientific learning and understanding.

The conceptual change approach, which will be described here, differs significantly from the Piagetian and empiricist approaches. It focuses on knowledge acquisition in specific subject-matter areas and describes the learning of science concepts as a process that requires significant reorganization of existing knowledge structures and not just their enrichment[2].

The proposal that the learning of science involves "conceptual change" has its roots in the work of science educators like Novak (1977), Driver and Easley (1978), and Viennot (1979) who were among the first to pay attention to the fact that students bring task alternative frameworks, preconceptions, or misconceptions to the science learning, that are robust and difficult to extinguish through teaching. Posner, Strike, Hewson and Gertzog (1982) drew an analogy between Piaget's concepts of assimilation and accommodation and the concepts of "normal science" and "scientific revolution" offered by philosophers of science such as Kuhn (1970) and derived from this analogy an instructional theory to promote "accommodation" in students' learning of science. The work of Posner et al. (1982) became the leading paradigm that guided research and practice in science education for many years but also became subject to a number of criticisms that have not yet been answered (e.g. Caravita & Halden, 1994).

In my view, the instructional questions that Posner et al. (1982) tried to answer cannot be adequately approached until we have a better understanding of *how* students learn science. The conceptual change approach described here is based on cognitive/developmental research and attempts to provide a description of the process of learning science and the mechanisms that bring it about. The implications of this approach for instruction are described later. More specifically, the following arguments are being made:

(1) The human mind has developed, through evolution, specialized mechanisms to pick up information from the physical and social world. This results in very quick and efficient learning which starts immediately after birth. Some kinds of things are easy to learn, not because they are less complex, but because human beings are prepared through evolution for this kind of learning. This seems to apply to the learning of language and to intuitive physics. Intuitive physics is the knowledge about the physical world that develops early in infancy and allows children to function in the physical environment.

(2) Learning which is acquired early in life and which is not subject to conscious awareness and hypothesis testing can stand in the way of learning science. This happens because scientific explanations of physical phenomena often violate

2 This type of knowledge reorganization is also known in the literature as "domain-specific restructuring" as opposed to Piaget's "global restructuring" (Carey, 1985).

fundamental principles of intuitive physics, constantly confirmed by our everyday experience. After all, the currently accepted scientific explanations are the product of a long historical development of science characterized by revolutionary theory changes that have restructured our representations of the physical world.

(3) Conceptual change is required in the learning of many science concepts (and not only science). This is because the initial explanations of the physical world in intuitive physics are not unrelated and fragmented observations but form a coherent whole. I have argued in previous work (Vosniadou, 1994) that intuitive physics can be said to be organized in a framework theory which constrains the process of acquiring further knowledge about the physical world and can cause misconceptions[3]. Many so-called misconceptions can be explained as synthetic models formed by individuals in their effort to assimilate new information into the framework theory. The change of the framework theory is difficult because it represents a coherent explanatory system based on everyday experience and tied to years of confirmation.

What are the phenomena that the conceptual change approach explains?

A great deal of information has been accumulated on how students learn science. Most researchers would agree on the following three conclusions about this process:

(1) Science learning is difficult. Even after many years of science instruction, students still seem to have difficulty understanding science concepts. This applies even to the students who perform above average in terms of test scores and teacher evaluations.

(2) Science learning is characterised by misconceptions. Misconceptions have been noted in practically all subject areas of science. Hundreds of misconceptions, enough to fill out tens of volumes, have been reported in the literature. Research conducted in my laboratory has revealed several representations formed by elementary school children regarding the shape of the earth and the explanation of the day/night cycle (Vosniadou & Brewer, 1992, 1994) that can be seen as "misconceptions". Figure 1 shows the range of mental representations of the shape of the earth obtained by elementary school children in a study conducted in the United States. Some children believe that the earth is shaped like a flat rectangle or a disc, is supported by ground below and covered by the sky on top. Other children think that the earth is a hollow sphere with people living on flat ground deep inside it, or a flattened sphere with people living on its flat top and bottom. Some other children form the interesting model of the dual earth, according to which there are two earths: a flat one on which people live, and a spherical one that is a planet up in the sky. These representations of the earth are not rare. In fact, only 23 of the 60 children that participated in this study (mostly fifth graders) had formed the culturally accepted model of the spherical earth. This finding has been confirmed by a series of cross-cultural studies investigating concept of the earth in children from India, Greece, and Samoa (Vosniadou, 1994a).

---

[3] The term *theory* is used to denote a relational, explanatory structure and *not* an explicit, well formed, mathematical and socially shared *scientific theory*.

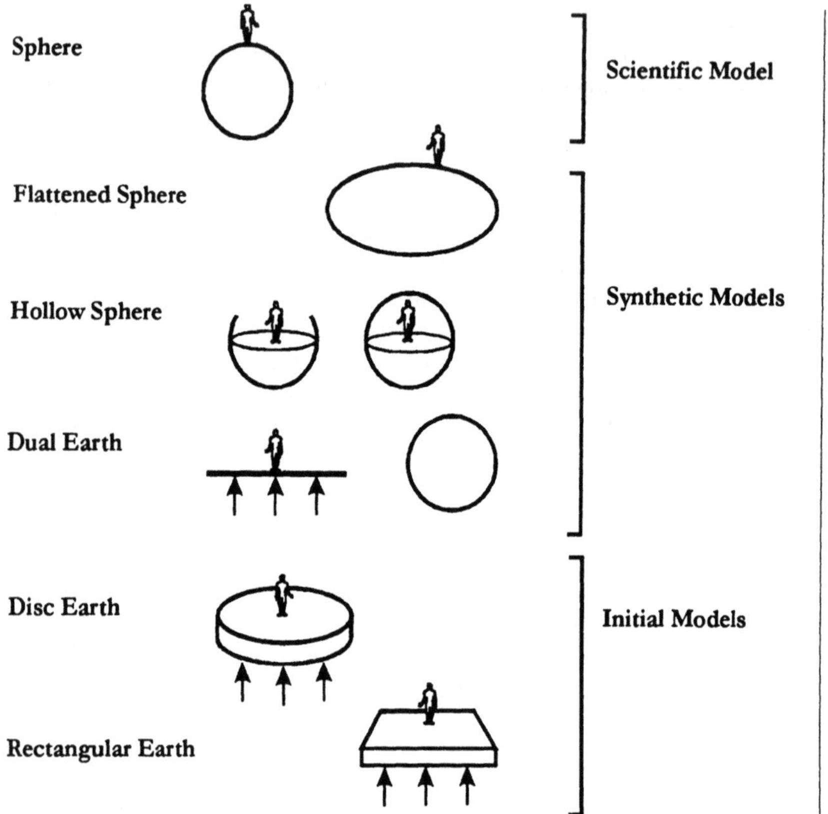

Figure 1: Mental models of the earth

(3)Science learning is inert. The term 'inert knowledge' has been used by Bereiter (1984) and Bransford et al. (1989) to describe problems of knowing something but failing to use it when it is relevant. Inert knowledge is considered to be knowledge accessible only in a restricted set of situations. Although, potentially, it could apply to many more. Science knowledge is often inert in the sense that students learn how to solve science problems at school but fail to apply this knowledge to explain physical phenomena outside of school (diSessa, 1982).

Most researchers would agree on the above-mentioned description of the process of learning science. Despite this agreement, the interpretations made of these difficulties differ. Some researchers think that science learning is difficult because students have limited experiences and/or because they do not know how to interpret the limited experiences they have. They claim that children do not know how to test hypotheses, accept explanations that should be rejected on the basis of the available evidence, base their explanations on what they perceive through their senses not on the logic of things, or do not even see the need to explain why things happen (Harlen, 1995).

Other researchers believe that novices' thinking is based on superficial interpretations of physical reality that may be able to explain a limited set of situations but that do not constitute a coherent and systematic theory. According to this view, learning science is basically a process of organising this 'knowledge in pieces' into more complex and systematic knowledge structures governed by the laws and principles of physics (diSessa, 1993).

I find a great deal of truth in the explanations mentioned above. There is no doubt that students base their explanations on everyday experiences that are by definition limited, that they need to develop better procedures for testing and evaluating hypotheses, and that the thinking of the expert is more coherent, more systematic and more closely linked to the laws and principles of physics. On the other hand, children's thinking does not appear to be quite as limited as suggested above. Vosniadou and Brewer (1994) found that 38 out of the 60 elementary school children they examined provided well defined explanations of the day/night cycle. These explanations were empirically accurate, in the sense that they were consistent with the empirical evidence expected within their range. In addition to being sensitive to issues of empirical accuracy, the children seemed to show sensitivity to issues of logical consistency and of simplicity in their explanations.

Limitations in experiences and in logical thinking cannot fully explain the phenomena of misconceptions and of inert knowledge which are observed not only in elementary school students but in high school and college students as well. In order to explain the above-mentioned phenomena, we need a theory of learning not only as a process of enriching existing knowledge but also as a process of conceptual change.

Let us consider, for example, the previously mentioned misconceptions found in elementary school children's representations of the shape of the earth. Even very young children are now exposed to considerable information regarding the spherical shape of the earth through children's books, TV programmes, discussions with parents, globes, etc. In our studies in the United States (e.g. Vosniadou, 1994b; Brewer, 1992) we had to go as far as testing three-year-olds to find children who had not been exposed to this information. Many four-year-old children already knew something about the spherical shape of the earth. It is therefore difficult to claim that children's misconceptions about the shape of the earth result from limited experiences or even from limitations in logical thinking. The explanation of misconceptions and of inert knowledge which I have provided in my work, is that they are caused by students' attempts to reconcile incompatible pieces of information, some of them stemming from everyday experience and some coming from the surrounding culture, often in the form of science instruction in schools.

We will start with misconceptions. If we look carefully at the misconceptions of the earth presented in Figure 1, we can see that they can be explained as students' attempts to incorporate the information received from culture, according to which the earth is a sphere, into their existing "theory" according to which the earth is a flat physical object and that people live on its top. For example, the children who form the model of the hollow sphere, seem to understand that the shape of the earth is spherical, but they believe that people live on flat ground inside the earth. On the other hand, the children who form the model of the flattened sphere think that the earth is spherical but also a little flat on the top and maybe the bottom where the people live. The children who form the dual earth model think that there are two earths: a round one which is up in the sky and which has all the characteristics of the adult model, and a flat one on which people live.

All misconceptions regarding the shape of the earth encountered in the American sample as well as the Indian, Greek and Samoan samples in our studies (Vosniadou, 1994a) can be explained as attempts on the part of the children to synthesize two inconsistent pieces of information: the information they receive from instruction according to which the earth is a sphere, and the information they receive from their everyday experiences, that the earth is flat.

Now, we can all understand how children may form an initial representation of the earth as a flat, physical object supported underneath, with people living on top and solar objects, such as the moon and the sun, located above it. Our studies of pre-school children's ideas about the earth do indeed confirm the hypothesis that children start with this simple mental representation. The question is: why children do not change their flat earth representation to the representation of a spherical earth when we tell them so and when we show them a globe?

My answer to this question is that the representation of the earth as a flat, physical object is not a simple belief but a complex construction supported by a whole system of observations, beliefs and pre-suppositions that form a relatively coherent and systematic explanatory system. Figure 2 shows a pictorial representation of some of the beliefs and pre-suppositions that underlie the representation of a flat, supported earth that I assume to be the first representation that children form.

I cannot go into detail here about this explanatory system, which is described in detail in previous work (see Vosniadou, 1994b). The important point to make for the purpose of this paper is that the representation of a flat earth is based on the assumption that the earth is a physical body and that, as a physical body, it is constrained by all the pre-suppositions that apply to physical bodies in general. Some of these pre-suppositions are the organisation in space terms of the directions of up and down and the pre-supposition that unsupported objects fall 'down'.

Such pre-suppositions, which stand in the way of understanding the spherical shape of the earth, are not addressed by science instruction. Examination of the curricula used to teach astronomy to elementary school students in the U.S.A. and in Greece have shown that students are not provided with explanations of how it is possible for the earth to be round and flat at the same time or how it is possible for people to live on the 'sides' and 'bottom' of this globe without falling 'down'. It seems particularly important to teach children something about gravity in order to understand how people can live on a spherical, rotating earth.

The mechanism of adding information into an existing knowledge base can produce a misconception if the two pieces of information belong to two incompatible explanatory frameworks, as is the case in the shape of the earth. In these situations the understanding of a scientific explanation requires a more fundamental restructuring of the knowledge base - the revision of fundamental pre-suppositions and beliefs - before the additive mechanisms can work. This is what we mean by conceptual change. Children must understand the earth as an astronomical rather than as a physical object. This change in ontological categories is really a kind of theory change and does not necessarily imply theory replacement.

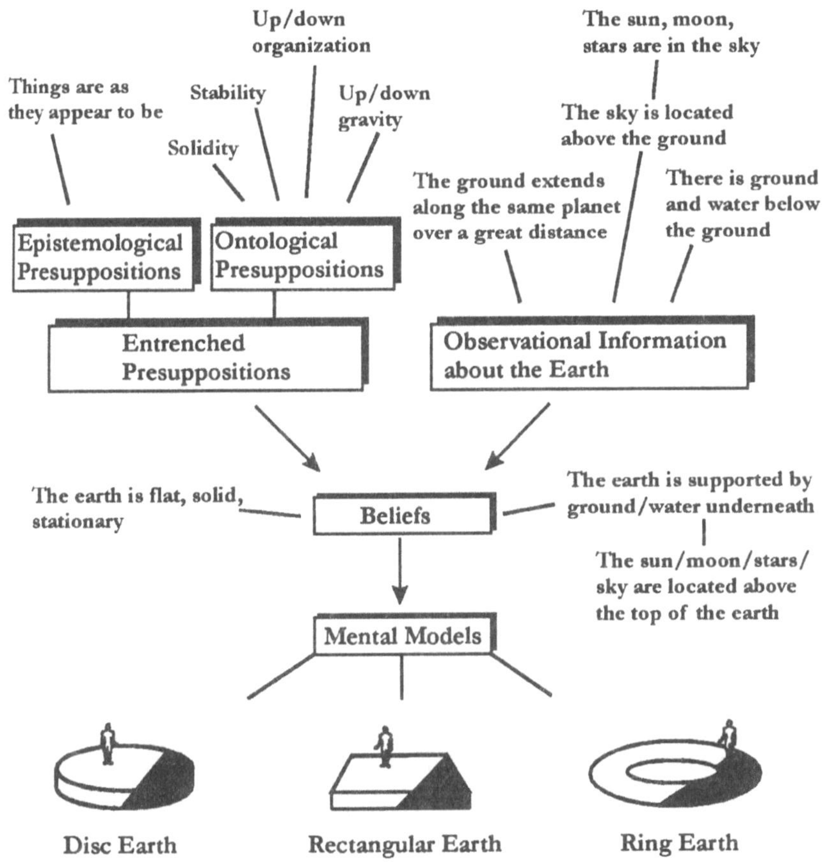

Figure 2: Domain of knowledge about the physical world

The conceptual change approach can also explain the phenomenon of inert knowledge. One of the reasons why scientific information acquired in school settings is inert and therefore not utilised in real life is because this information is so incompatible with existing conceptual structures that students do not realise that the two belong to the same category.

The suggested analysis is confirmed not only in the case of astronomy but in many other subject-matter areas of physics. Our studies of the process of conceptual change in mechanics and thermal physics (Ioannides & Vosniadou, 1991; Vosniadou, 1994b; Vosniadou & Kempner, 1993), show that the successive mental models of force and of heat constructed by elementary and high school students can be explained as students' attempts to assimilate the information they receive from instruction into a fundamentally inconsistent explanatory framework.

For example, in the area of mechanics children construct an initial concept of force according to which force is a property of objects that feel heavy. This 'internal' force appears to represent the potential these objects have to react to other objects with which they come into contact. It is also central in explaining the motion of inanimate objects. In the ontology of the young child, the natural state of inanimate objects is that of rest, while the motion of inanimate objects is a phenomenon that objects is that of rest, while the motion of inanimate objects is a phenomenon that needs to be explained, usually in terms of a causal agent. This causal agent is the force of another object.

The initial concept of force is very different from the way the linguistic term 'force' is currently interpreted by the scientific community. In Newtonian physics, force is not an internal property of objects but a process that explains changes in the kinetic state of physical objects. In the framework of the accepted view, motion is a natural state that does not need to be explained. What needs to be explained are changes in kinetic state.

The process of understanding the meaning implicit in the scientific concept of force is usually a slow and gradual affair, likely to give rise to misconceptions. It appears that students gradually differentiate the concept of weight from the concept of force and replace the notion of an internal force with the notion of an acquired force (otherwise known as impetus) that is a property of the objects that move. Despite important changes in the concept of force, which occur with development, certain entrenched pre-suppositions of the framework theory, such as that of force being a property of objects and that the motion of inanimate objects requires an explanation, continue to remain in place in the conceptual system of high school or even university students, despite the fact that these students have been exposed to systematic instruction in Newtonian mechanics (Ioannides & Vosniadou, submitted).

Designing instruction to facilitate the learning of science in the early years

There is no doubt that the first step in teaching science to young children is to provide them with an environment rich in new experiences and opportunities to observe interesting phenomena and to encourage them to try to make sense of these experiences. As children grow they need to be introduced to a deeper qualitative understanding of selected subject matter areas in science. This qualitative understanding can provide the necessary foundations for the more systematic approaches to science that follow later. The conceptual change approach outlined above has specific recommendations to make for science instruction in the elementary school years, about curricula and instruction, which go behind providing the necessary experiences.

Breadth of coverage of curricula

The finding that the understanding of science concepts and explanations is a difficult and time-consuming affair, likely to give rise to misconceptions, calls for a reconsideration of current decisions regarding the breadth of coverage of the curriculum in science education. It may be more profitable to design instruction that focuses on the deep exploration and understanding of a few, key concepts in one subject-matter area rather than cover a great deal of material in a superficial way. For example, the science curriculum for the fifth grade in Greece includes short units on mechanics, thermodynamics, energy, the particulate nature of matter, and the processes of life. This approach encourages the casual memorisation of facts, does not develop the qualitative understanding of science concepts and is very likely to lead to logical incoherence and misconceptions. It also makes teachers very anxious about covering all the material, with the result that not enough attention is paid to what students actually understand.

Sequence of acquisition of the concepts to be taught

Research in the learning of science has also shown that the concepts that comprise a subject-matter area have a relational structure that influences their order of acquisition. This structure needs to be taken into consideration when designing curricula and instruction. For example, in the subject-matter area of astronomy students understand the spherical shape of the earth only after they have acquired an elementary notion of gravity. Explanations of the day/night cycle on the basis of the earth's axis of rotation cannot be understood before students know not only that the earth is a rotating sphere but also that the moon revolves around the earth. Otherwise they form misconceptions such as that the sun and the moon are stationary at opposite sides of an up/down rotating earth (Vosniadou & Brewer, 1994). Similarly, a scientific explanation of the seasons only occurs in students who have formed a mental model of a heliocentric solar system, know the relative sizes of the earth, the sun, and the moon, and understand the scientific explanation of the day/night cycle. As Saddler's studies with Harvard undergraduates show, very few college students understand how the seasons occur, despite the fact that this is a piece of science included in the elementary school curriculum in the United States (Sadler, 1987).

At present, such findings are not taken into consideration in the design of science curricula. A detailed investigation of the astronomy units in four leading science series in the United States (Vosniadou, 1991), as well as an examination of the national curricula teaching astronomy to elementary school children in Greece has shown that many concepts are introduced in a sequence that does not provide students with all the information necessary for understanding them.

As mentioned earlier, the kind of instruction that elementary school students typically receive regarding the shape of the earth involves a simple statement that the earth is "round like a ball" sometimes accompanied by a class demonstration of a globe. In this type of instruction, teachers do not explain to students how it is possible for the earth to be spherical when it appears to be flat, or how it is possible for people to live on the 'sides' and bottom of this sphere without falling 'down'. I have not found reference to the notion of gravity associated with astronomy instruction in the elementary school grades. This is because gravity is considered to belong to the subject-matter area of mechanics and astronomy. It is obvious that this type of instruction does not address students' pre-suppositions, and, therefore, does not provide them with the information they need in order to construct an approved representation of the earth as a sphere.

Taking into consideration the students' point of view

The realisation that students do not come to school as empty vessels but have beliefs and pre-suppositions about the way the world operates that are difficult to change, has important implications for the design of science instruction. Teachers need to be informed about how students see the physical world and must take their students' points of view into consideration when they design lessons. Instructional interventions need to be designed so that students will become aware of their implicit beliefs and pre-suppositions need to be provided. Experiences so that students will understand the limitations of their explanations and be motivated to them. Finally science instruction in the school should be related with activities that take place outside the school.

Facilitating metaconceptual awareness

Although children seem to be relatively good interpreters of everyday experience, they do not seem to be aware of the explanatory frameworks they have constructed. Moreover, they are not aware that their explanations of physical phenomena are hypotheses that can be subjected to experimentation and falsification. Their explanations remain implicit and tacit. Lack of metaconceptual awareness of this sort prevents children from questioning their prior knowledge and encourages the assimilation of new information to existing conceptual structures. This type of assimilatory activity seems to form the basis for the creation of synthetic models and misconceptions, and lies at the root of the surface inconsistency so commonly observed in students' reasoning.

To help students to increase their metaconceptual awareness it is necessary to create learning environments that facilitate group discussion and the verbal expression of ideas. Recently technology-supported learning environments have been constructed that make it easier for students to express their internal representations of phenomena and compare them with those of others. Such activities may be time-consuming, but they are important for ensuring that students become aware of what they know and understand what they need to learn.

It is important to emphasise here that science learning does not only mean that a student acquires a different explanatory system from the layman's; it also means a more flexible conceptual system, a system that makes it easier to adopt different perspectives and different points of view. What brings about this cognitive flexibility (and this is an important area for future research) is, in my opinion, increased metaconceptual awareness. It is difficult, if not impossible, to understand other points of view if you do not even recognise your own point of view. Increased awareness of one's own beliefs and pre-suppositions is a necessary step in the process of understanding the presuppositions and beliefs of others and probably the first step in the process of conceptual change.

Addressing entrenched pre-suppositions

Students often do not see the need to change their beliefs and pre-suppositions because they provide good explanations of their everyday experiences, function adequately in the everyday world, and are tied to years of confirmation. In order to persuade students to invest the substantial effort required to become science-literate and to re-examine their initial explanations of physical phenomena, it is necessary to provide them with additional experiences (in the form of systematic observations or the results of hands-on experiments), that prove to them that the explanations they have constructed are in need of revision. If we want these experiences to be useful in the process of belief revision we need to select them carefully so that they are theoretically relevant. It is not the case that any new experiences will do the job.

The linguistics of science learning

To make this problem even worse, there is a serious communication problem associated with science learning. The semantics of terms such as heat, force, weight, etc., are completely different in everyday language than in scientific language. This becomes a source of errors and misunderstandings that could perhaps have been avoided if different linguistic terms were used in the science vocabulary. Educators need to be more sensitive to the linguistic difficulties associated with the learning of science concepts and need to discuss them with their students.

Cultural support for science learning

This brings us to a last but very important point that has to do with cultural support for science learning. Although scientific explanations are the ones our culture supports, they have not yet filtered down to everyday culture. Whatever science learning takes place in school, it is not really supported outside the school. Except in cases where children have scientifically literate parents who provide them with books, take them to science parks and museums, and talk to them about science. It is important that science becomes more a part of everyday reality through TV programmes, popularised books, science museums for children, and other cultural activities than is currently the case.

Concluding comments

I have argued that learning science is not a process that can be explained by assuming that new knowledge is simply added onto existing knowledge structures, but a process that often requires that basic pre-suppositions and beliefs about the physical world are revised. It appears that this restructuring process is not trivial but a rather difficult one, because the beliefs and pre-suppositions about the physical world that are based on everyday experience are robust and resistant to change. The realisation that the learning of science is a difficult and time consuming affair is the necessary first step in the direction of preparing the education community to put together the concentrated efforts needed for designing the kinds of curricula and learning environments that have the potential to make students both knowledgeable and enthusiastic about science.

References

Bereiter, C. (1984). How to keep thinking skills from going the way of all frills. *Educational Leadership* 42, 75-77.

Bransford, J.D., Franks, J.J., Vye, N.J., & Sherwood, R.D. (1989). New approaches to instruction: because wisdom can't be told. In S. Vosniadou & A. Ortony, Eds., *Similarity and Analogical Reasoning*. New York: Cambridge University Press.

Caravita, S. & Halden, O. (1994). Re-framing the problem of conceptual change. *Learning and Instruction* 4, 89-111.

Carey, S. (1985). *Conceptual change in childhood.* Cambridge MIT Press.

diSessa, A. (1982). Unlearning Aristotelian physics: a study of knowledge-based learning. *Cognitive Science* 6, 37-75.

diSessa, A. (1993). Towards an epistemology of physics. *Cognition and Instruction* 10, 105-225.

Driver, R. & Easley, J. (1978). Pupils and paradigms: A review of literature related to concept development in adolescent science students. *Studies in Science Education* 5, 61-84.

Ioannides, C. & Vosniadou, S. (1991, August). The *development of the concept of force in Greek children.* Paper presented at the biennial meeting of the European Society for Research on Learning and Instruction, Turku, Finland.

Ioannides, C. & Vosniadou, S. (Submitted). The changing meanings of force: A developmental study.

Kuhn, T. (1970). *The structure of scientific revolutions.* Chicago, IL: University of Chicago Press.

Piaget, J. (1970). *Genetic epistemology.* New York: Columbia University Press.

Posner, G.J., Strike, K.A., Hewson, P.N., & Gertzog, W.A. (1982). Accommodation of a scientific conception: Toward a theory of conceptual change. *Science Education* 66, 211-227.

Sadler, P.M. (1987). Misconceptions in astronomy. In J.D. Novak, Ed., *Proceeding of the Second International Seminar: Misconceptions and Educational Strategies in Science and Mathematics. Vol. 3* (pp.422-425). Ithaca, NY: Cornell University.

Viennot, L. (1979). Spontaneous reasoning in elementary dynamics. *European Journal of Science Education* 1, 205-221.

Vosniadou, S. (1991). Designing curricula for conceptual restructuring: Lesson from the study of knowledge cognition in astronomy. *Journal of Curriculum Studies* 3, 219-237.

Vosniadou, S. (1994a). Universal and culture specific properties of children' models of the earth. In L.A. Hirschfield & S.A. Gelman, Eds. *Mapping the Mind.* New York: Cambridge University Press.

Vosniadou, S. (1994b). Capturing and modelling the process of conceptual change. *Learning and Instruction* 4, 45-69.

Vosniadou, S. & Brewer, W.F. (1992). Mental models of the earth: A study of conceptual change in childhood. Cognitive *Psychology* 24, 535-585.

Vosniadou, S. & Brewer, W.F. (1994). Mental models of the day/night cycle. *Cognitive Science* 18, 123-183.

Vosniadou, S. & Kempner, L. (1993, April). Mental models of heat. Paper presented at the biennial meeting of the Society for Research in Child Development, New Orleans, LA

Rhetoric and Science Education

Isabel Martins, Eduardo Mortimer, Universidade Federal de Minas Gerais, Brazil
Jonathan Osborne, Kings' College London, UK
Charalampos Tsatsarelis, Institute of Education, University of London, UK
Maria Pilar Jiménez Aleixandre, Universidade de Santiago de Compostela, Spain

Abstract

The work described in this paper draws from a theoretical perspective that sees rhetoric, argument and discourse as an important feature of science teachers' practice and the learning of science. It aims to contribute to an on-going debate about a wide range of issues such as: challenges to current conceptions of practical activity in teaching; analyses of the role of metaphors in the conceptualisation of science entities; and discussions of the contribution of argumentative practices in the construction of knowledge. It draws from studies in the field of semiotics and the rhetoric of science to analyse strategies of argumentation in science communication and to consider the social process of validating scientific knowledge and the construction of scientific "authority" in the classroom, offering fresh insights on the practice of science teaching and learning.

Introduction: why rhetoric?

Research in science education has been traditionally concerned with eliciting and analysing both students' and teachers' conceptions about specific topics in school science, about the nature of science, or about aspects of classroom activity. Usually, results have been organised as systems of beliefs, attitudes, conceptions and epistemologies. More recently, a number of researchers have shifted their attention to analysing discursive interactions in the classroom. This shift in focus has led to a perspective that beliefs, epistemologies and conceptions are no longer things that an individual or a group of individuals possesses or shares. They become a property of discourse. That is, something which is realised through the discursive interactions themselves (Roth & Lucas, 1997). This new approach makes it possible to go beyond descriptions of ideas held by people and allows an investigation of the processes through which they are proposed, questioned and negotiated. This paper presents four studies which are representative of such novel perspectives in science education and proposes an interesting departure: to discuss the role of rhetoric in teaching and learning. Although coming from different theoretical perspectives, all four studies share a concern about the need to understand the nature of discursive interactions between students and teachers and the process through which meaning is made in school science.

Because of the strong anti-rhetorical tradition in science, it may at first seem strange to talk about rhetoric and science teaching. Traditionally scientific knowledge is regarded as resting on evidence. Understanding science is equated with rational conviction and does not depend on persuasion. The role of rhetoric in science education has, to some extent, been neglected. Nevertheless, there are a few key studies, in this area, which provide a context for the pieces of research described here, such as the previous work on how teachers explain science constructing

H. Behrendt et al. (eds.), Research in Science Education – Past, Present, and Future, 189–198.

entities (Ogborn, Kress, Martins & McGillicuddy 1996); or Millar's (1989) analysis of the function of rhetoric in experiments and demonstrations; or Solomon's (1989,1992) discussion of the role of analogy in scientific argument.

Nowadays, contemporary scholarship attempts to view rhetoric as a more general term. That is, as any type of instrumental expression calculated to influence an audience towards some end. However, for researchers who attempt to broaden the conception of what rhetoric includes, the analysis of verbal texts is still regarded as its main focus. For us, a new approach to rhetoric should aim at describing not just verbal texts but to include other choices of resources to realise meaning in different semiotic modes. The term rhetoric is, therefore, used to refer to the articulation of different modes of communication, such as language, images and gesture, to produce coherent texts which help shape a given view of the world.

In this symposium, four papers discussed different aspects of that which is involved in thinking about science teaching as rhetoric by focussing on analyses of theoretical perspectives, classroom interactions and textbook materials.

Promoting rhetoric and argument in the science classroom

This paper develops the case for any authentic science education to pay attention to the role of rhetoric and argument—an area to which this field gives only scant regard (Driver, Newton & Osborne, in press). For, if rhetoric and argument are a central feature of the practice of science (Fuller, 1997; Taylor, 1996), it is surprising that science teaching displays such little interest in an activity which lies at the heart of science. This significant omission has led to important shortcomings in the education provided *about* science. As it has given a false impression of science as an unproblematic collation of facts about the world, this creating controversies between scientists, whether historical or contemporary, about puzzling events (Geddis, 1991; Driver, Leach, Millar & Scott, 1996). Such disregard for the practice of argument has failed to empower students with the ability to examine critically the claims generated by the plethora of socio-scientific issues that confront them in their lives (Norris & Phillips, 1997). This theoretical paper examines the role and function of rhetoric within science and science education and explores how pupils may be offered educative experiences that provide some understanding of its function in contemporary science.

Science in schools is commonly portrayed from a 'positivist perspective' as a subject in which there are clear 'right answers' and where data lead uncontroversially to agreed conclusions. Essentially, science is 'taught as a nearly unmitigated *rhetoric of conclusions*, in which the current and temporary constructions of scientific knowledge are conveyed as 'empirical, literal and irrevocable truths' (Schwab, 1962, 24). Consequently, the stock-in-trade of the science teacher is knowledge that is 'unequivocal, unquestioned and uncontested' (Claxton, 1991).

Current research into the activities of scientists points to a different picture of science. Here, the contribution of discursive practices such as rhetoric and argument to the construction of scientific knowledge is portrayed as important. Practices, such as assessing alternatives, weighing evidence, interpreting texts, evaluating the potential viability of scientific claims, are all seen as essential components in constructing scientific arguments (Latour & Woolgar, 1986; Taylor, 1996; Weimer, 1977). Arguments concerning the appropriateness of an experimental design or the interpretation of evidence in the light of alternative theories, are seen to be at the heart of science and central to the discourse of scientists (Druker, Chen & Kelly, 1996). Furthermore, the work of scientists also includes argument in the public domain through journals, conferences and the wider media.

This change in perspective has major implications for pedagogy, requiring the opportunity to study discursive activities, especially argument and rhetoric. Especially of portraying empirical work as the major foundation of scientific practice - the "scientific method - it argued that practical work is valued by science teachers for the role it plays in providing evidence for knowledge claims and resources for argumentation - that is as a rhetorical device, that seeks to persuade and convince their students of the validity of the science teacher's world view. From this viewpoint, observation and experiment are not the bedrock on which science is built. Rather they are the handmaidens to the rational activity of generating arguments in support of knowledge claims advanced by science.

However, if the claim, to know science, suggests that one knows, not only *what* a phenomenon is, but also: *how* it relates to other events, *why* it is important and *how* this particular view of the world came to be, then the study and use of rhetoric and argument has a function in developing an understanding of the concepts of science; providing epistemological insights; developing a sense of science as a social practice and an understanding of the role of controversy in contemporary science along with its rhetorical function in the management of uncertainty.

The paper concludes by critically examining some of the materials and procedures that have been suggested for developing a better understanding of discursive practice in science and its use (Giere, 1984; Solomon, Duveen & Scott, 1992). Such study demands that students are offered plural interpretations of phenomena (Monk & Osborne, 1997; Siegel, 1989) which are the *sine qua non* of undermining the common view that science is an authoritative and unquestioned body of knowledge.

However, the inherent tensions placed on the science teacher by the inclusion of epistemic goals, which seek to show the limits of science and the intrinsic uncertainty which surrounds the new products of scientific endeavour, sit uncomfortably with the rhetoric of unequivocal and unquestioned conclusions that are used to persuade pupils of the validity and value of the scientific world view. Thus it may be that such features can only be incorporated by radical reconstruction of science education rather than minor surgery.

The rhetorics of school, science textbooks

Martins' recent work on rhetorical analyses of school science textbooks builds on earlier research on explanation in science in the classroom, which included analyses of the role of visual representation in explaining, (Kress et al 1998; Martins 1998) drawing on ideas from visual semiotics (Kress & van Leeuwen 1996) so as to emphasise the multimodal nature of classroom communication (Kress, Ogborn, Martins, 1998). A series of classroom observation studies documented a number of strategies used by teachers to: (i) create interest and need for understanding, (ii) construct new entities in the discourse, (iii) re-contextualise specialist knowledge and (iv) put meaning into matter through demonstrations. The types of work, which are necessary to all the discursive interactions described above to occur, are realised through relations within, between and across languages. Rhetorical devices are, consequently, more than identifiable textual structures. Martins' accounts of rhetorical devices go beyond questions of accessibility of "the language of science" to non-specialist readers and strategies of persuasion to include a discussion of how different texts attempt an engagement with readers through cognitive, affective and communicative channels. In doing so she draws upon insights provided in recent work on the rhetorics of science (Gross 1990, Bazerman 1991, Myers 1990) and on aspects of the rhetorics of science teaching (Millar 1989, Solomon 1989, Sutton 1992) such as the relationship between theory and experimental evidence, the processes of creating new semiotic objects (especially through metaphor) and the role of argumentative practices in the scientific community.

Textbooks are a reliable source of information and ideas for practising teachers and contain accepted re-contextualisations of scientific knowledge. Work developed by Martins seeks to characterise their rhetorical devices pointing to textual features, such as linguistic expressions and images and analyses of intentions and proposed interactions. It then relates these to larger patterns of text organisation and to sets of both the author's and readers' intentions and expectations. Through an inspection of both Brazilian and British secondary school science textbooks it was also possible to detect many of the changes in both conception and presentation that textbooks have undergone in the last decades, such as: an increasing reliance on visual communication; emphasis on establishing links between science content and the contexts of everyday life; interdisciplinary organisation and explicit addressing of students' misconceptions. Martins' analyses try and relate these changes to broader changes in education and in society and search for examples, describing them in terms of patterns of textual organisation.

Examples of rhetorics found to be present in science textbooks include, for example: (i) "science is about the real world as we know it", which deploys strategies of providing context and relevance; (ii) "it's only logical after all", where conclusions are reached through logic and argument; (iii) "surrender to irrefutable evidence" in which authority rests upon experimental or phenomenological evidence.

These issues are considered in relation to the design and organisation of textbooks in ways which help realise a number of functions such as, for example: conveying images of the nature of science and of the scientific enterprise; constructing authority of scientific knowledge and discourse; altering subjectivity in relation to a domain; reconceiving the world in terms of new scientific entities.

One example of her analyses refers to the ways teachers and learners are portrayed in textbooks, showing students and teachers as subjects in the classroom, helping to create a new set of expectations and attitudes towards this domain of knowledge. For example, learners are increasingly portrayed in the textbooks, usually actively engaged in some kind of science related task. It is common to find images of students wearing aprons, with their hair tied back, manipulating equipment or setting up apparatus in the science classrooms. These pictures actually do more than echo the metaphor of the "pupil as scientist". They relate a set of behaviours which are expected from students in science classrooms (e.g. obeying necessary safety procedures) and characterise the nature of activities they are expected to perform (e.g. conduct experiments). This rhetorical function of the image is reinforced by accompanying text, that stresses aspects, involved in learning science as, for example, in the stylistic forms, found in a Brazilian textbook, but also typically found on British materials. Here, the use of the pronoun 'we' blurs a distinction between scientists and science learners and brings teachers and students closer. For instance, *"in the study of physics, we often need to ... measure and interpret data ... formulate hypotheses and make predictions"*. This subtle induction into scientific activity is usually found in introductory chapters, at the beginning of a textbook and is often accompanied by other instances where science is either shown to be or brought closer to the learner. Here image and language articulate and help convey, together with, for example, applications of science in everyday life, a view of science as a field where methods and results can be made accessible to novices.

This example of rhetorical analyses of science textbooks seeks to highlight elements of an intentional organisation of discourse, exploring relations with larger organisations such as curriculum recommendations. Analyses of this kind could help raise awareness of different possibilities, styles and strategies for explaining science and would enable teachers to think more critically about the materials they use. They are also important in discussing how textbooks help both shape and construct the audience they are intended for.

Mediational tools and discourse interactions in science classrooms

In this section we explore methodological concerns involved in the understanding of discursive interactions in the classroom. Starting from a concern with the need to develop a language for describing discursive, rhetorical and semiotic features of science classrooms and textbooks (Lemke, 1990, 1998; Sutton, 1992; Halliday & Martin, 1993; Scott, 1996; Ogborn et al. 1996; Mortimer, 1998), Mortimer's analyses of science classroom interactions emphasise the role of both verbal interaction between teacher and students and cultural tools in the process of meaning making.

According to Mortimer, there are three main kinds of mediational tools characterising classroom interactions. They are: genre, theme and approach. His analysis of genre in classroom discourse is based upon Bakhtin's claim that each utterance is individual but each sphere in which language is used develops its own relatively stable types of utterance: the speech genre (Bakhtin 1986). If we look at the classroom as a space where at least two different social languages – the scientific and the common sensical – interact to generate new meanings, it is necessary to understand the genres by which these meanings are conveyed in the flow of the discourse. The second tool – theme – can be defined as the core content of the explanation or description for the phenomenon that is under discussion. The theme of a classroom discourse is part of the agenda that the teacher has pre-planned for the lesson. The third tool – approach – is clearly related to theme and refers to the level at which teachers and students deal with an explanation or phenomenon: macroscopic or microscopic. As the tension between macroscopic phenomena and microscopic explanation is a key for understanding science, this tool is very important and can frame the classroom discourse in several ways. Normally, science and chemistry teachers go from one level to the other quite unconsciously and automatically although they are not always followed by the students in these shifts.

Mortimer also emphasises a characteristic he calls the *flow of discourse*. When engaged in dialogue with students, teachers normally create discursive patterns, the most common of which is described as I-R-F (see, for example, Edward & Mercer, 1987), where **I** corresponds to the initiation of the dialogue by the teacher, normally with a question; **R** is the student's response; and **F** is the feedback from the teacher. His analysis is expanded by focusing on turning points in the episodes where a student initiates a discourse move and can have much more control over the process of generating meanings. Apart from initiation, response or feedback, other elements to characterise the flow of the discourse are introduced. Some of the students' utterances cannot be characterised as responses but as assertions prompted by follow-ups from the teacher. The process of authorising a student voice as "the" voice of scientific explanation is also discussed.

These theoretical categories were applied to analyses of classroom interactions. Some excerpts, selected from transcribed extended episodes of science and chemistry classrooms (grade 8 to 11) video recorded in different schools of Belo Horizonte, Brazil, are discussed next. The teachers are involved in a project of continuing professional development. The episodes selected, were those in which meaning is being made and/or broken in the interaction between teacher and students or between students. Using the sociocultural approach, these episodes can be characterised as showing microgenetic development occurring in the intermental plane of the classroom (Wertsch, 1991). Mortimer's scheme of analysis uses a graphic representation to present the episodes in which the sequence of speech turns is represented and quoted using the categories discussed earlier. The episodes are also presented using a table with three columns, the first showing the verbal discourse, the second some gestures and other non-verbal elements and the third showing the quotation of each utterance according to the categories discussed above. The graphic representation allows one to see different phases in the episodes. In one of these episodes, in which there is a discussion between the teacher and students about the spontaneous dissolution of potassium permanganate in water, there is one first phase - the attempt phase - in which the students did not succeed in getting the right approach, genre or theme the teacher was expecting. One student (S5) tried to explain the phenomenon in macroscopic terms (turn 2) - an inappropriate approach, as the teacher was expecting a microscopic explanation. Another student (S4) offered an apparently inappropriate genre, attributing a human behaviour to the substances (turns 5 and 8). With S4 the teacher establishes an evaluative IRF sequence (turns 8 to 13), questioning apparently only the student's genre.

In turn 21, S5 introduced the theme the teacher was expecting - motion of particles - inaugurating the second phase of the episode - the meaning phase. It is worthy of note that S5 offered her contribution by using a question, initiating a sequence in which her assertions were followed by prompts from the teacher. The teacher also contributed by posing new questions (turns 30 and 35), which allowed for the extension of the meaning being produced by S5.

Together with a third phase (not discussed), this analysis shows how, ultimately, the teacher was able to establish a new IRF sequence with S5, *"eliciting the generalisation* (Turn 46...) *that the particles, their energy and the space are related not just in the phenomenon of the dissolution of permanganate but in general, as part of the particulate model of matter."*

This way of analysing classroom discourse helps to make visible the way in which language is used in classrooms as an important rhetorical device, that helps students to make meaning of scientific explanations, genres and approaches. This kind of analysis can help the professional development of teachers in a subject-matter pedagogical expertise (Schulman, 1986) that seems to be very important but goes unnoticed in most teacher practices: the expertise of discursive management.

Rhetorical construction of entities in science classrooms

This paper analyses how teachers attempt to make pupils reconceptualise the world through the construction of entities and attempts to show how the various communicational modes are used, how pupils contribute to the construction of entities and the rhetorical framings that a teacher uses to shape pupils' views about an entity in a particular way.

This paper draws on work from the ESRC funded project "Rhetorics of the science classrooms: a multimodal approach" directed by Gunther Kress and Jon Ogborn. The researchers of the project were Carey Jewitt and Charalampos Tsatsarelis. The data include a series of three to six lessons from four different schools with pupils 12-15 years old. Lessons were video-recorded using two cameras; one focused on the teacher and the second focused on the pupils. What follows is Tsatsarelis' discussion of the analyses of the lesson transcripts which show how the entities of an 'orbit', and other entities were constructed through the identification of a general pattern.

A framework with the rhetorical framings that were identified in all the lessons was developed and then applied to analyse the construction of orbits (the subject of a series of (recorded) lessons. The construction of the entity "orbit" in the classroom was done gradually, in a series of steps. These steps are realized in the classroom through the use of various rhetorical framings. Here is a discussion of those which were foregrounded:

- *You know about their existence and I will tell you what is relevant to our story.* The teacher introduces the actors (e.g. earth, planets, satellites, people) without any need to give evidence for their existence. She shows pupils which of the actors will be used in the stories and which of the attributes they know are relevant – portraying them as material things that have masses – making some of these actors look similar. She ignores other features that are irrelevant to the explanatory stories and which make these actors very different (e.g. their volume, humans on earth etc.).
- *I will be able to show you and you will understand what it is.* What the teacher attempts to avoid throughout these lessons is pupils saying (or thinking) "we do not believe you". Thus, she is forced all the time to be clear and understandable. The teacher uses the framing to present the existence of a relationship between two actors (e.g. earth and people). Particularly, the teacher presents it as a fact that the attraction between earth and objects (for instance people) is due to their masses.
- *See the world in a different way.* Pupils are asked to imagine some features of relationships between actors in a different way from their everyday life. For instance, she suggests that pupils should imagine that not only does the Earth attract people but that people attract the Earth as well, that is, that all masses attract each other. Also, pupils are expected to give a different meaning in the science classroom for some terms, such as the weight, that they use frequently in everyday life.
- *You know this is true, but I have the knowledge to explain it.* The pupils know that planets orbit the sun in circles but the teacher has the knowledge to explain why. The teacher attempts to unsettle the reality, and presents orbits as the results of 'actions at distance'. This kind of interaction is not common in pupils' everyday life where pushes and pulls happen mainly between objects in contact. She does this through the presentation of some examples of circular motion.

- *It's a logical implication of fact.* The teacher establishes explicitly a close relationship between the entity orbits and circular motion. It is presented as the case that planets follow the same principles as the objects in circular motion.
- *Authority.* The teacher expressed his authority in a consistent manner and in various modes. Most of the time in the two lessons involved a teacher's monologue. In the case of the images and gestures the teacher's authority was realised through their modality. For instance, she has the power, through modality, to represent the earth as an empty circle, that is, to highlight what is relevant to the story (e.g. a circle has a centre) to ignore the features of the actors that are irrelevant to her stories (e.g. mountains, sea), abstraction and generic paths.

But what was the contribution of various communicational modes to the construction of the meaning of this entity? Both images and gestures contributed to the construction of the meaning of orbits. In this synopsis we present briefly some cases to which the images and gestures do not convey the same meaning as language, that is they play an important role to the construction of the meaning and they are not just a translation of what the teacher says. What the teacher does not do verbally, although it appears in her gestures, is to differentiate between orbits. With her hands, she makes circles in a horizontal plane whenever she talks about planets orbiting the sun, while she makes circles in a vertical plane when she talks about satellites that move around the earth. It is possible that the gestures in different planes (horizontal versus vertical) are informed by the place that the teacher chooses to look from. It seems that she views planets orbits from a place outside of our solar system (since they are orbiting the sun and not the earth) while she views satellites orbits standing on the earth (she is inside the orbits now and the satellites move 'over her head'). In this case images convey meanings visually that are not conveyed verbally. The teacher uses gestures together with images. She uses gestures to show the direction of the arrow (force). In the case with the person moving away from the circular earth the teacher used the gestures to present an imagined path from the person to the surface of the earth.

The shaping of the entity of an orbit is realised through the various rhetorical framings the teacher used. The entity orbit is constructed in a multimodal environment. The various communicational modes contribute to this construction and, in some cases, in a contradictory way. It seems that what the teacher avoided saying using the channel of language, namely that orbits can be seen as 'railway paths', appeared to be the case in her use of images. In summary, the teacher used images to emphasise the path of orbits whilst in the verbal channel of communication the emphasis was on the reason for the path (that is, gravity and high speed). Gestures and action are thus used to mediate between the channels of images and language, combining what is observable with how it can be explained, differentiating actions and presenting an object as an actor. In this way in this series of lessons the teacher constructed the theoretical entity using different forms of argumentation.

The example shows how the project aims to provide teachers with an account of what they are doing (rather than of what they are just saying) and that may help them to think about their choices at conscious level to improve their practice.

Final remarks: understanding how meaning is made in classrooms

According to Jimenez-Aleixandre, the perspective which runs through these papers is an exciting one and points to the need to understand how explanations are built *inside* classrooms and to explore the processes through which meaning is made *in* science classrooms. Classroom is a keyword here, because these questions are concerned with situations, interactions, and exchanges taking place in actual classrooms rather than with studies in a laboratory setting.

To explore the processes through which meaning is made adds a new dimension to our understanding of science learning which has been improved in the past two decades by the study of its products. The study of processes includes modelling (besides other studies about models), means of persuasion (besides the success or the lack of it in actually persuading others) and the building of explanations (besides the actual form and content of the explanation).

Another issue raised by these papers is that of uncertainty. These studies show that it is appropriate to talk about decisions that must be taken by teachers and pupils such as choosing one or another explanation, deciding about evidence and that, contrary to an extended discourse running through textbooks and classrooms, science does have a dimension of uncertainty. For on many occasions, there is more than one possible interpretation of a phenomenon and that is why scientists argue and construct arguments relating data to theories. To develop a perspective of science which incorporates this problematic nature is one of the goals of science education – to address its epistemic nature.

Meaning, discourse and argument are different aspects of communication in classrooms. The implication is not that these are the only aspects worth studying in them, but that they can help to illuminate our understanding about teaching and learning science.

References

Bakhtin, M.M. (1986). *Speech genres and other late essays*. Austin: UT Press.

Bazerman, C. (1991). *Shaping written knowledge*. Madison: University of Wisconsin Press.

Claxton, G. (1991). *Educating the enquiring mind: the challenge for school science*: Harvester: Wheatsheaf.

Driver, R., Leach, J., Millar, R. & Scott, P. (1996). *Young people's images of science*. Buckingham: Open University Press.

Driver, R., Newton, P., & Osborne, J. (in press). Establishing the norms of scientific argumentation in classrooms. *Science Education*.

Druker, S.L., Chen, C. & Kelly, G.J. (1996). *Introducing content to the Toulmin Model of Argumentation via error analysis*. Paper presented at NARST meeting, Chicago, 1996.

Edwards, D. & Mercer, N. (1987). *Common Knowledge*. London: Routledge.

Fuller, S. (1997). *Science*. Buckingham: Open University Press.

Geddis, A. (1991). Improving the quality of classroom discourse on controversial issues. *Science Education* 75, 169-183

Giere, R. (1991). *Understanding scientific reasoning*. (3rd ed.). Fort Worth, TX: Holt, Rinehart and Winston.

Gross, A. (1990). *The rhetoric of science*. Cambridge, MA: University of Harvard.

Halliday, M. & Martin, J. (1993). *Writing science*. London: The Falmer Press.

Kress, G. & van Leeuwen, T. (1996). Reading images: the grammar of visual design. London: Routledge

Kress, G., Ogborn, J., Jewitt, C. & Tsatsarelis, C. (1998). *Rhetorics of the science classroom: a multimodal approach*. Mid project consultative meeting texts. Institute of Education,

University of London.

Kress, G., Ogborn, J., Martins, I., & McGillicuddy, K. (1997). Multimodal rhetorics of the science classroom. A research proposal to the ESRC. Institute of Education. University of London.

Kress, G., Ogborn, J., & Martins, I. (1998). A satellite view of language: some lessons from the science classroom. *Language Awareness*, Special Issue: Metacommunication in Instructional Settings, Volume 7:2 & 3, 67-89.

Latour, B., & Woolgar, S. (1986). *Laboratory life: The construction of scientific facts*. (2cnd ed.). Princetown, NJ: Princetown University Press.

Lemke, J. (1990) *Talking science*. Norwood: Ablex Publishing Corporation.

Martin, J. R., & Veel, R. (1998). *Reading Science*. London: Routledge.

Martins, I. (1998, December). *Visual imagery in science texts*. Paper presented at the EARLI SIG Group The Psychology of Scientific Text Comprehension, Cuenca, Spain.

Millar, R. (1989). Bending the evidence: the relationship between theory and experiment in school science. In R. Millar, *Doing science: images of science in science education* (pp. 587-596). London: The Falmer Press.

Millar, R. (1989). *Doing science: images of science in science education*. London: The Falmer Press.

Monk, M. & Osborne, J. (1997). Placing the history and philosophy of science on the curriculum: a model for the development of pedagogy. *Science Education* 81(4), 405-424.

Mortimer, E. (1998). Multivoicedness and univocality in the classroom discourse. *International Journal of Science Education* 20(1), 67-82.

Myers, G. (1990). *Writing biology: texts and the social construction of knowledge*. Madison: University of Wisconsin Press.

Norris, S. & Phillips, L.M. (1997). Intellectual independence for nonscientists and other content-transcendent goals of science education. *Science Education* 81(2), 239-258.

Ogborn, J., Kress, G., Martins, I., & McGillicuddy, K. (1996). *Explaining science in the classroom*. Buckingham: Open University Press

Pera, M. & Shea William, R. (1991). *Persuading science, the art of scientific rhetoric*. Canton, USA: Science History Publications.

Roth, W. M. & Lucas, K. B. (1997). From "Truth" to "Invented Reality": a discourse analysis of high school physics students' talk about scientific knowledge. *Journal of Research in Science Teaching* 34 (2), 145-179.

Schulman, L. S. (1986). Those who understand: a conception of teacher knowledge. *American Educator* 10(1): 9-15, 43-44.

Schwab, J. J. (1962). *The teaching of science as enquiry*. Cambridge, MA: Harvard University Press.

Scott, P. (1996). Social interactions and personal meaning making in secondary science classrooms. In G. Welford, J. Osborne & P. Scott (Eds.), *Research in Science Education in Europe* (pp. 325-336) London: The Falmer Press.

Siegel, H. (1989). The rationality of science. Critical thinking and science education. *Synthese* 80(1), 9-42. In R. Millar, *Doing science: images of science in science education* (pp. 126-136). London: The Falmer Press.

Solomon, J. (1989). The social construction of school science. In R. Millar (Ed.), *Doing science: Images of science in science education* (pp. 160-179). London: The Falmer Press.

Solomon, J., Duveen, J., & Scott, L. (1992). *Exploring the nature of science: Key Stage 4*. Hatfield: Association for Science Education.

Sutton, C. R. (1992). *Words, science and learning*. London: Open Univ. Press.

Taylor, C. (1996). *Defining science: A rhetoric of demarcation*. Madison, Wisconsin: The University of Wisconsin Press.

Weimer, W. (1977). Science as a rhetorical transaction: Towards a nonjustificationist conception of rhetoric. *Philosophy and Rhetoric* 10, 1-29.

Wertsch, J.V. (1991) *Voices of the mind*. Cambridge: Havard University Press.

Development of Complexity through Dealing with Physical Qualities: One Type of Conceptual Change?

Stefan von Aufschnaiter
University of Bremen, Institute of Physics Education, Germany

Abstract

An important element of our theoretical framework for investigating individual learning processes, (von Aufschnaiter & Welzel, 1999) is the distinction between cognitive processes ("every student continuously creates cognitions according to his own actions and perceptions") and cognitive structures ("tools producing these cognitions") (ibid.). We then must distinguish between situated activity (construction of meaning) and changes of situated activity (changes in the way of meaning is constructed in similar situations). We pointed out that we do not interpret learning as the acquisition of "new" knowledge but as the development of "new" tools for the construction of this knowledge. Furthermore, we demonstrated that each of a student's cognitions can be assigned to one of ten levels of complexity and that one kind of learning can be understood as the increase in the average level of complexity of the cognitive processes.

In the first part of this paper, our approach is compared to similar approaches from other groups. It will be stressed that these approaches also distinguish between cognitive processes and cognitive structures and that learning is interpreted as the development of cognitive structures. In the second part, results of a laboratory study will be presented, that show how development of complexity occurs as one kind of conceptual change.

Distinctions with respect to cognitive processes and structures

Piaget's constructivist starting point is that knowledge does not exist outside the cognitive system, but that it is being constructed by individuals through their interactions with the environment (e.g. Piaget & Garcia, 1991). In this respect, Piaget distinguishes cognitive processes as interpretations of the world within the framework of existing cognitive structures (assimilation) from those processes in which parts of the cognitive structures change (accommodation). Piaget's term for these parts of the cognitive structure is schema.

Adey calls the latter kind of processes, in which new schema evolve, meta constructions as opposed to the construction of knowledge (Adey, 1999).

Asking "what changes in conceptual change", *diSessa and Sherin* have developed a theory about one type of concept by introducing coordination classes as systematically connected ways of getting information from the world. "The difficult job of a coordination class is to penetrate the diversity and richness of varied situations to accomplish a reliable 'readout' of a particular class of information." (diSessa & Sherin, 1998). A coordination class (a special type of concept) is the ability to see a particular class of information in the world and learning may be understood as developing (new) coordination classes.

In their phenomenographic approach, *Marton and Booth* stress that cognitive processes, as a way of experiencing the world, are constituted as an internal relationship between the real world "out there" and the subjective world "in here".

H. Behrendt et al. (eds.), Research in Science Education – Past, Present, and Future, 199–204.

They focus on variation in ways of experiencing phenomena and show that "the qualitatively different ways of understanding the content of a learning task are logically related to each other, because they represent a more or less partial grasp of the same complex of constituent parts." (Marton & Booth, 1997).

The differences and similarities between the approaches described above and our approach are summarised in the following table:

	Processes	Structures	Structural Changes
Piaget	assimilation	schema	accommodation
Adey	construction of knowledge	schema	metaconstruction of new schema
diSessa & Sherin	reading out information	coordination class (p-prim)	development of "new" coordination classes
Marton & Booth	experiencing the world	ways of understanding	development of new ways of understanding
von Aufschnaiter et al.	construction of meaning	cognitive tool	development of cognitive tools

In accordance with Marton and Booth, we hold that construction of meaning always constitutes a relationship between phenomena of the world and ways of understanding, developed so far. As a consequence, "internal" and "external world" cannot be seen as separate in ongoing cognitive processes.

Similar to Adey, we think that cognitive structures develop in special "meta"-processes. Development of meaning takes place in intervals ranging from seconds to several minutes, whereas development of tools takes considerably longer (from hours to months). In our model, different co-ordination classes (diSessa & Sherin, 1998) as well as different ways of understanding specific phenomena (Marton & Booth, 1997) can be assigned to different levels of complexity. We agree with diSessa and Sherin on the point that a change between such levels of complexity, while working on a task, can be interpreted as a kind of conceptual change. Furthermore, we stress, in accordance with Marton and Booth, that development of meaning and learning have to be modelled as bottom-up processes: "... learning proceeds from a vague undifferentiated whole to a differentiated and integrated structure of ordered parts." (Marton & Booth, 1997).

In order to match empirical data with theoretical accounts, we should bear in mind that all empirical data we can use, refer to changing meanings, reconstructed by an observer from learners' actions and utterances. Therefore, in our programme, the first field of interesting questions that need to be addressed has so far been: How does meaning develop with respect to content, complexity and time? We think that this development takes place in a 3-dimensional space (fig. 1).

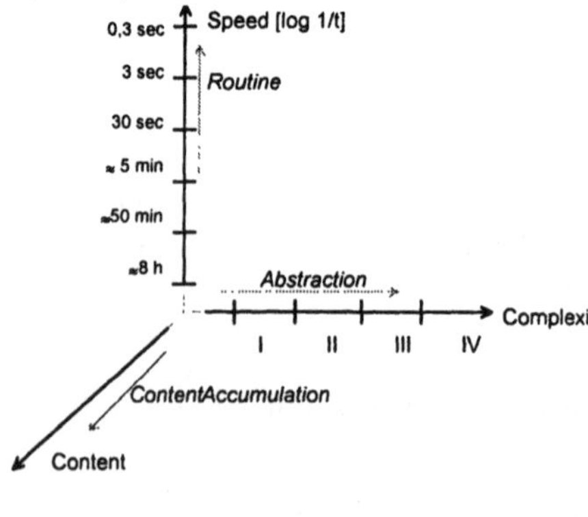

Figure 1

As *content,* one may imagine n-dimensions for n-different content areas out of physics, chemistry or biology. In this article, the content area, electrostatics will, be discussed as one dimension. In the study described below, we gave the students four context specific groups of tasks and instructions:

- chargeable objects and the attraction and repulsion of charged objects [19 tasks (T) / 22 instructions (I)]
- using a neon bulb as evidence of different types of charge (13 T/ 5 I)
- using the electroscope to measure amounts of charge (12 T/ 4 I)
- charging and discharging through induction and grounding (7 T/ 3 I)

Our main interest in the last ten years has been the *complexity* of cognitive process, and we have introduced a model of 10 levels for describing complexity development (cf. von Aufschnaiter & Welzel, 1999).

Here, I will combine two or three of them to one complexity area.

I Constructing meaning with respect to concrete objects and their interdependence (objects, aspects, and operations)

II Constructing meaning with respect to one invariant property of several objects and situations as well as interdependence between two or more invariant properties (properties and events)

III Constructing meaning with respect to single variable properties and co-variation between two variables (programmes and principles)

IV Constructing meaning with respect to co-variations of more than two variables (connections, networks and systems)

One can imagine complexity-development as the capability of thinking more and more abstract. One type of conceptual change can then be interpreted as changing cognition from a lower to a higher area of complexity.

In figure 1, a *logarithmic axis for 1/t* is used, because one main goal of the cognitive system is "to do and think things as quickly as possible". We could discuss, here, a time scale extending over more than the tenth power of 10 (from milliseconds to centuries), but for the development of meanings when solving tasks, only 3 are of interest.

- Up to 3 seconds for one conscious meaning or "Image-of-now" (Pöppel, 1994; Damasio, 1994)
- Up to 30 seconds for the development of several interconnected meanings with regard to finding out one way for a solution to a task
- Up to 5 minutes to find the best development for solving this task

On these time scales, we can study conceptual change as developing meanings situatively from lower to higher complexity, if conceptual change is necessary for solving the task and possible for the student.

A study of "conceptual change" in the content area electrostatics

In the content area electrostatics, students first have to build a concept 'charge' as an invariant property, which can be attributed to many different objects, and which can be either + or -. However, to understand Coulomb's law, they have to change their concepts of charge as being just a quality (invariant property) to a concept of charge as being variable (the quality of charge on a special object can increase or decrease over time). Such a variable property (charge) can vary with changes of other properties of the same or other objects (voltage of a power supply).

In the study presented now, we primarily investigated how students came from using the word 'charge' as a name for something "mysterious" to the concept 'charge' as an invariant property and how they learned to connect this property with the invariant property force between two charged objects. But we were also interested in studying the conceptual change of 'charge' from being invariant to being variable.

Our study was done in a special room at our university with three students at the same time working with given material on tasks without any interaction with a teacher. The design is very briefly described in the following table.

• 51 tasks and 34 instructions (written on cards) and material for experiments on electrostatics	
• mean time for working on cards and experimental material: 180 min	
• 9 groups of 3 students each, only student-student interactions	
• complete video documentation (c. 1600 min video taped)	
• c. 1200 min transcribed, c. 300 min analysed in detail	
• additional survey data about experience and interest	
We did	longitudinal studies of single students
	vertical studies of tasks / instructions
	coupling of video and survey data

Results Concerning Content: We ascribe meanings as being new concerning the content if single actions are sequenced differently, if mimic and gesture of actions change clearly and if words are combined differently in verbal and written expressions. Using these distinctions, we found that 27 students, as a whole, produced a large number of different content-specific meanings (ca. 150 to 250) for a single task. Each individual student constructed a large number of different content specific meanings (from 500 to 1000) for all tasks combined. Only 16 words or word combinations were used (maybe as concepts) more than ten times by individual students. We can show development of meanings, concerning their content, only with these words or concepts.

Results Concerning Complexity: Especially when the students started to work with "new" elements of the content (like a neon bulb or an electroscope) or at the beginning of a new session, meanings predominantly referred to concrete operations with objects. Often terms like 'force' or 'charge' for naming invariant properties were used only after the students carried out a number of operations. Each property was developed repeatedly by carrying out operations. The linking of this property to others seemed to be somewhat accidental at first. Only if a property proved to be viable in a variety of cases, did a high degree of certainty and stability in its use develop. The same was true for specific combinations of two properties. However, after a lapse of days or weeks, this certainty in using a property had to be redeveloped, in many cases, by carrying out operations with objects. Our students especially learned to develop important properties of the objects they used (i.e. neon bulb and electroscope) with respect to charge. However, none of the students managed to construct charge as a variable property and to use this property for explaining varying deflections of the needle of the electroscope. With the help of two tasks and one instruction, this is shown below.

The following tasks were given in the last of the three sessions. Until then, the students had had many opportunities to construct the properties 'charge' and 'force' when working on other tasks.

Task I:	"Rub an overhead transparency and hold it above the electroscope. Remove the transparency. What do you observe?"
19 stud.	carried out the experiment and discussed observations.
8 stud.	explained with respect to charge or force.
Task II:	"Why is the deflection and regression of the needle observable?"
9 stud.	did not work on the task before reading the following instruction.
7 stud.	repeated the experiment.
10 stud.	explained with respect to charge or force.
1 stud.	explained with respect to the relation between charge and force.
Instruction:	about the function of the electroscope (charge distribution, polarity of charge, repulsion of needle and mount).
11 stud.	dropped out after reading the instruction.
7 stud.	repeated the experiment.
6 stud.	explained with respect to charge or force.
3 stud.	explained with respect to the relation between charge and force.
Task III:	"Why is the deflection and regression of the needle observable?"
5 stud.	explained with respect to the relation between charge and force.

During task I, 19 students were participating actively in carrying out the experiment and talking about their observations (complexity area I, operations). Eight students additionally explained the result by referring to charge or force, that is, with respect to charge or force they reached area II (properties).

During task II, seven students repeated the experiment without attempting to explain it. 10 students referred to either force or charge in their explanations, that is, they reached area II (properties). One student linked charge and force, thus reaching area II (Events). Nine students started to work on task II only after working on the instruction.

When working on this instruction, eleven students dropped out after reading it (I, Aspects) or they did not participate in the ensuing discourse. Seven students repeated the experiment they had carried out earlier (I, operations) whereas six students commented on the instruction by referring to charge or force (II, properties). Only three students linked charge with force when explaining the mode of operation of the electroscope (II, events). When task II was carried out or repeated after the instruction, 5 students managed to find an explanation which linked charge with force.

Our interpretation of these dynamics is that most students experienced the instruction as being too complicated and, consequently, did not use it although it incorporates a link between charge and force.

Results Concerning Time Scales: Students were only seldom able to solve our tasks or instructions spontaneously (within 3 sec). This did not even happen if successive tasks were very similar. When students started working on tasks or instructions (which they did for nearly 100% of tasks but only for about 50% of instructions), they often needed several developments of meanings (each up to 30 sec) with slightly different content and/or different levels of complexity to solve the task successfully. Unless students found a solution, that was successful from their point of view, they stopped working on tasks/instructions after ca. 5 min (or even less). In such situations, dissatisfaction was very often expressed explicitly.

Summary

The presented investigation refers only to the content electrostatics. However, we have found similar results in this content area within other laboratory and field studies (e.g. von Aufschnaiter & Welzel, 1999):

- Fast access to information requires high familiarity with this information in different contexts.
- Students normally work on narrow content elements within less than 30 sec.
- While working on a task, students shift from content element to content element in an unsystematic fashion within less than 5 min.
- Development of meanings starts (nearly) always with manipulation of or discussions about concrete objects.
- Many similar operations concerning the same content area are the prerequisite for the construction of invariant properties.
- Connection of properties is only possible if every property can be constructed easily and quickly (within less than 3 seconds).
- In all cases there are two limiting time scales: 30 sec for the development of one way to solve the task and 5 minutes to work on a task.
- The step from one complexity area to the next can be interpreted as a kind of conceptual change.

We propose the analysis of learning processes and learning environments critically under the perspective of the presented theoretical framework and the empirical results.

References

Adey, P. (1999). Revisiting cognitive conflict, construction, and metaconstruction and discovering metaconstructivism. In M. Komorek et al., Eds., *Research in Science Education: Past, Present, and Future.* Proceedings of the Second International Conference of the European Science Education Research Association (E.S.E.R.A.), Vol. 1 (pp. 58-60). Kiel, Germany: IPN.

Damasio, A. R. (1994). *Descartes' error, emotion, reason and the human brain*. New York: Avon Books.

diSessa, A. A. & Sherin, B. L. (1998). What changes in conceptual change? *International Journal of Science Education* 20(10), 1155-1192.

Marton, F. & Booth, S. (1997). *Learning and Awareness.* Mahwah, New Jersey: Lawrence Erlbaum.

Piaget, J. & Garcia, R. (1991). *Toward a Logic of Meanings*. Hillsdale, New Jersey: Lawrence Erlbaum Associates.

Pöppel, E. (1994). Temporal mechanisms in perception. *International Review of Neurobiology* 37, 185-202.

von Aufschnaiter, S. & Welzel, M. (1999). Individual learning processes - a research program with focus on the complexity of situated cognition. In M. Bandiera et al., Eds., *Research in Science Education in Europe* (pp. 209-215. Dordrecht, The Netherlands: Kluwer.

On the Micro-Structure of Analogical Reasoning: The Case of Understanding Chaotic Systems

Jens Wilbers and Reinders Duit
Institute for Science Education at the University of Kiel, Germany

Abstract

Analogies are generally seen as powerful tools in facilitating conceptual change, i.e., to guide students from their pre-instructional conceptions to science concepts. However, many studies have shown that analogies may also deeply mislead students' learning processes. The study presented here is part of a project on investigating learning processes in the domain of limited predictability of chaotic systems (Duit, Komorek, & Wilbers, 1997). A pilot study on teaching this issue to grade 10 students clearly illustrated the potential power of the analogies used in our studies (Duit & Komorek, 1997). A closer examination however also revealed some pitfalls of analogy use (Duit, Roth, Komorek, & Wilbers, 1999). A micro-analysis clearly showed that students had severe difficulties in making use of various conceptual analogies to a magnetic pendulum. It appears that the processes of decoding and mapping are closely linked. It became apparent that students make their own sense of analogies provided and that, in some cases, the understanding of the random behaviour of the magnetic pendulum was even misguided by them. This study draws on these findings. The focus is on investigating the micro-structure of analogical reasoning in open inquiry settings. In brief, the results show that theories of analogical reasoning as developed, for instance, by Gentner (1989) have to be revised. They do not allow the description of analogical reasoning as observed in our study and hence may not be viewed as applicable frameworks for the instructional designs that make use of analogies.

A theory of analogical reasoning

--------------------context--------------------
analogical relation
base domain <========> target domain
--------------------context--------------------

Fig. 1: Analogy: Relation between base and target domain - embedded in a specific context

Analogies are commonly seen within a framework illustrated in figure 1. There is a base sharing certain features with the target of the analogy, i.e., there are similarities between base and target with respect to surface features and deep structural features. The analogical relation denotes these similarities. Gentner (1989) differentiates five sub-processes of analogical reasoning: *accessing, mapping, evaluation, storing, and generalization*. Access to an analogy is primarily facilitated by surface features. Mapping is conceptualized as *structure mapping* - the focus is on comparing similarities of propositional structures and on the carry over of certain sub-structures of the base to the target. Whereas in Gentner's structure mapping approach the focus is predominantly on carry over of propositional structures, the *pragmatic* approach of Holyoak (1985) also takes regard of contextual factors. However, both approaches share the view that mental representations of

H. Behrendt et al. (eds.), Research in Science Education – Past, Present, and Future, 205–210.

propositional structures of base and target are the starting point of analogical reasoning. Furthermore, mapping is exclusively conceptualized as a transfer from base to target. In other words, the symmetrical nature of the analogical relation (as given by similarities of structures of base and target) is not explicitly employed.

In our view, the context in which the analogical relation (figure 1) is embedded is essential. To understand learning by analogy, two perspectives need to be considered: that of the analogy provider (e.g., teacher or textbook author) and of the learner. When analogies are provided, the analogical relations between base and target are clearly defined within the particular context hold by the analogy provider. These analogies may be called post-festum-analogies. Usually, learners do not share this context. From their perspective they are merely confronted with objects. All they know is that these are in some way similar to each other and that they are expected to find out in which way this may be the case. Among the many potential relations between base and target they have to detect those that allow the construction of understanding. In other words, students use analogies in an heuristic manner. A theory of analogical reasoning has to differentiate post-festum and heuristic analogies. Such a differentiation is missing in present theories of this kind (Gentner & Medina, 1998). Gentner (1989), for instance, views analogical reasoning as a comparison of similarities between base and target. It appears that, in her approach, the post-festum-perspective dominates. The creative aspect of the heuristic perspective appears to be given only little attention. To use a base in order to facilitate understanding of a target requires the construction of an analogy and not simply the discovery of objective correspondence between certain features of base and target.

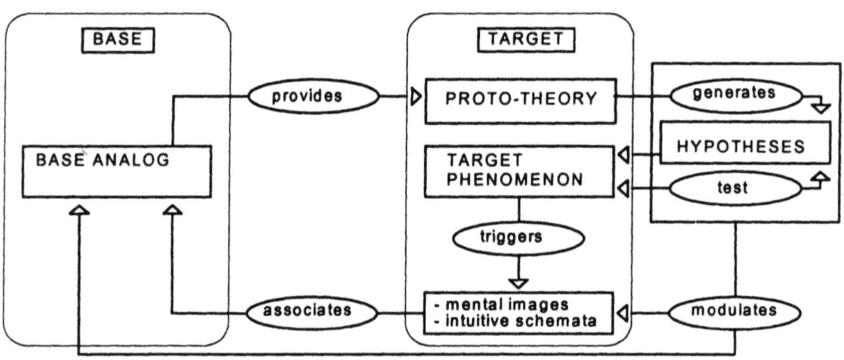

Fig. 2: A model of analogical reasoning

Both Gentners structure mapping and Holyoaks pragmatic approach consider propositionally based knowledge as a starting point for analogy use. As opposed to this, our revised model of analogical reasoning (figure 2) claims that intuitive schemata and mental models, spontaneously generated by the students when first confronted with the target phenomenon, are essential in analogy use. They lead to a preliminary associative link between target and base. The subsequent process of analogy construction is guided by these spontaneously generated associations. Or to put it the other way round: The analogy is a means of constructing (propositionally based) hypotheses on the basis of (image like) mental models and intuitive schemata triggered by the target phenomenon. This process of analogy construction, which serves an heuristical exploration of the target, draws on a better known base analog that provides some "proto-theory" for the yet unexplored target. This implies the epistemologically well known fact that analogies are more of a tool to bring about hypotheses instead of proving them. The proof of hypotheses is a matter of empirical testing and beyond analogy use (Bunge, 1973).

A study on analogical reasoning in the domain of limited predictability of chaotic systems

A variant of the *teaching experiment* proposed by Steffe and D'Ambrosio (1996) was employed. Teaching experiments draw on Piagetian critical interviews, where the interview situation is deliberately turned into a teaching situation from time to time. Instead of interviews with single students, a small group setting with four students each was employed. 12 groups of 4 students (German Grammar school; average age 16) were "interviewed" in two sessions of about 120 minutes each. All sessions were video-taped and transcribed.

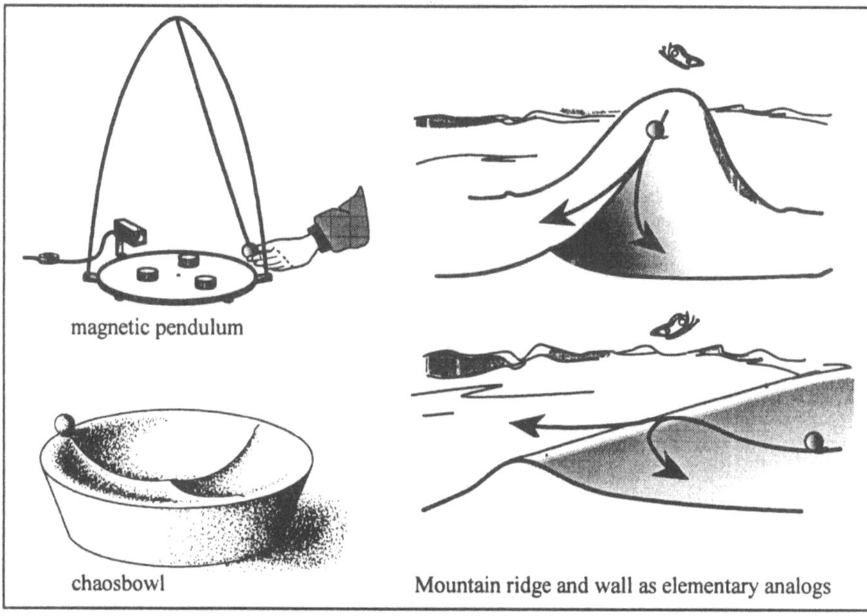

Fig. 3: The chaotic pendulum and analogies to explain its behaviour

The focus of the sessions was on investigating a magnetic pendulum as shown in figure 3. Releasing the pendulum bob again and again from the same starting position leads to diverging trajectories and one cannot possibly predict the magnet over which the bob will come to rest. In brief, this chaotic behavior is caused by the zones of unstable equilibrium (the Y-like figure between the magnets where the magnetic forces to the right and left balance out) that are passed several times by the pendulum ball. All chaotic systems possess such zones. The analogical relations investigated in this study, represent the unstable equilibrium in various artefacts like a chaos bowl or a dice. Students are given much time in quasi-open-inquiry sessions to construct their understanding of the magnetic pendulum. At certain times, the two elementary analogies shown in figure 3 (ridge and wall analogies) are presented as drawings on paper. Also, a computer simulation program is investigated that demonstrates the behaviour of the pendulum in an ideal world without any disturbances and with the possibility of always starting from exactly the same position.

The corpus of data comprises videotranscriptions and all artefacts produced by the students during class. There were also homework assignments between the two sessions. The data analysis is qualitative in nature. Standard methods of qualitative content analysis were used (Guba & Lincoln, 1989). A particular focus was on constructing students' mental models of the zones of unstable equilibrium that the chaotic pendulum passes many times on the one hand and students' ways of talking about chaotic motion (their dictions) on the other.

Results

The study confirms that analogies *can* be powerful tools in guiding students towards an understanding of the principle of limited predictability. The theoretical and methodological setting of the present study facilitates the construction of a fine-grained picture of analogy use as a micro-level description of the role of analogies in learning about chaos theory. In accordance with our theoretical considerations we focused our data analysis on the role of intuitive schemata and mental images in analogy construction and the heuristic use of analogies. Indeed the intuitive schemata that students hold of the notion of unstable equilibrium and the mental images they construct with respect to chaotic motion seem to play a crucial role in the course of analogy use. They both directed the ways the students in our study made use of presented conceptual analogies (e.g. Figure 3).

Space limitations, however, prevent the presentation of references from our data which substantiate the claims made in the subsequent section. Further details and abundantly discussed examples from our data are presented in Wilbers (1999).

Intuitive schemata of unstable equilibrium

In the course of the data analysis, we came to differentiate the following types of students' intuitive schemata with respect to unstable equilibrium. Every view includes a particular conception of explaining limited predictability.

- *Zones of decision.* When the ball comes into a sensitive zone it has to decide where to go so to speak. The decisions are random and hence may not be predicted.

- *Neutral zones.* There are no forces acting on the ball: Hence small disturbances, predominantly those occurring when the ball is in a neutral zone, determine its future path. As it is impossible to foretell the disturbances, the future path is unpredictable.

- *Dividing lines.* Many students are of the conviction that the lines of unstable equilibrium (forming the Y-like lines in the case of the magnetic pendulum) are basically borderlines for the fields of the single magnets, where the one field ends and the other begins. It is interesting that chaotic behavior does not need any explanation for the students holding this view. It is simply accepted as a fact and serves to explain the random sequence of target magnets. Accordingly the analogies provided are not used in the intended way.

- *Zones of topple over.* In such a zone the direction of an object's further motion is random. Chance is seen as a generating "force" for future behavior.

- *Sensitive zones.* This intended view includes the fact that small changes in the starting conditions and small disturbances result in small changes in the path of the pendulum bob over the zones of unstable equilibrium. If one compares two subsequent paths they deviate more and more when the sensitive zones are passed. At one stage they totally deviate. Only a small number of students in our study were able to proceed that far in their understanding.

Dictions of the systems' dynamic

Next, we analysed students' ways of talking about a system's dynamic. We identified the following patterns of diction:

- *Static.* Explanations include the starting point and the target magnet but not any discussion of the trajectory of the moving pendulum bob.

- *Animistic.* Especially in the beginning a remarkable number of students use animistic dictions. However, they do not appear to hamper understanding, but merely serve as a first heuristic.

- *Dynamic - local.* Among the dictions that include arguments concerning the motion of the pendulum bob there are several students focusing on the behavior at certain points, e.g. in zones of unstable equilibrium. Local arguments are either animistic or include force arguments (which are often not in complete accordance with the physics conception of the interplay of forces).

- *Dynamic - global.* This is the intended diction. Two or more trajectories are compared. Main emphasis is given to the significance of local changes for the development of the trajectories.

Processes of analogical reasoning

Particular intuitive schemata and dictions deeply influenced the processes of analogical reasoning observed. With regard to our attempts towards a revised theory of analogical reasoning, we would like to emphasize the following findings:

- Gentner's theory maintains that access to analogies provided as learning aids is facilitated predominantly by associations to surface features. There are a number of cases in our studies showing that access is also possible via deep structure similarities between base and target.

- Gentner also emphasizes the key role of the propositional structures of base and target in the mapping process. There is much evidence in our data that mapping often does not occur, even if students are familiar with the propositional structure of the base, i.e. understand it in the appropriate manner. The essential key to engaging in a mapping process, in the sense of Gentner, appears to be students' mental models and dictions of base and target. If students exhibit different frames of mental models regarding the unstable equilibrium and employ different dictions describing a system's motion for base and target a carry over does not take place.

- The mapping process from the perspective of Gentner's and Holyoak's theory of analogical reasoning is exclusively a process from base to target. In the present study, as well as and previous studies, students usually make use of the symmetrical nature of the analogy relation. In other words, they often switch the roles of base and target, i.e. view the base from the perspective of the target and vice versa.

- Approaches to analogy use in instruction usually hold that intimate familiarity with the base is essential (Duit, 1991). In accordance with findings by Corkill and Fager (1992) there are several cases in our data where students who are familiar with the base do not necessarily engage in the mapping process. As mentioned, this is the case if students view base and target within different mental models and dictions and, therefore, do not see the potential explanatory power of the base with regard to the target. This may also happen if base and target are seen from the perspective of the same (or at least similar) mental models and dictions. In this case, students may be pleased with the similarities between base and target that they constructed and, therefore, see no need for any further search for similarities.

Discussion

The model of analogical reasoning (figure 2) developed within the present study is based on data from 12 teaching experiments with 48 students in the domain of understanding chaotic systems. It is also supported by the findings of previous learning process studies in the same domain. However, it is preliminary in nature. Further studies are necessary to investigate whether the model is viable in general. If these studies support the model, significant changes to instructional strategies of analogy use like Glynn's (1989) "Teaching-with-Analogies Model" or the FAR-Guide by Treagust, Harrison and Venville (1996) are necessary.

References

Corkill, A.J. & Fager, J.J. (1992, April). *Individual differences in transfer via analogy*. Paper presented at the meeting of the American Educational Research Association, San Francisco, CA.

Bunge, M. (1973). *Method, model, and matter*. Dordrecht/Boston: Reidel Publishing Co.

Duit, R. (1991). On the role of analogies and metaphors in learning science. *Science Education* 75, 6, 649-672.

Duit, R. & Komorek, M. (1997). Understanding the basic ideas of chaos-theory in a study of limited predictability. *International Journal of Science Education* 19, 247-264.

Duit, R., Komorek, M., & Wilbers, J. (1997). Studies on educational reconstruction of chaos theory. *Research in Science Education* 27, 339-357.

Duit, R., Roth, W.M., Komorek, M., & Wilbers, J. (1999). Conceptual change cum discourse analysis to understand cognition in a unit on chaotic systems: towards an integrative perspective on learning in science. *International Journal of Science Education* 20, 1059-1073.

Gentner, D. (1989). The mechanisms of analogical learning. In S. Vosniadou & A. Ortony, Eds., *Similarity and analogical reasoning* (pp. 199-241). New York: Cambridge University Press.

Gentner, D. & Medina, J. (1998). Similarity and the development of rules. *Cognition* 65, 263-297.

Glynn, S. (1989). The Teaching-with-Analogies (TWA) model: Explaining concepts on expository text. In K.D. Muth, Ed., Children's comprehension of text: Research into practice (pp. 185-204). Newark, DE: International Reading Association.

Guba, E. & Lincoln, Y. (1989). *Fourth generation evaluation*. Beverly Hills, CA: Sage.

Holyoak, K.J. (1985). The pragmatics of analogical transfer. In G.H. Bower (Ed.), *The psychology of learning and motivation, Vol. 19* (pp.59-87). Orlando: Academic Press.

Steffe, L.P. & D'Ambrosio, B. (1996). Using teaching experiments to understand students' mathematics. In D. Treagust, R. Duit, & B. Fraser (Eds.), *Improving teaching and learning in science and mathematics* (pp.65-76). New York: Teacher College Press.

Wilbers, J. (1999). *Theorie und Praxis analogiebasierter Lernprozesse im Bereich des deterministischen Chaos* [Theory and practice of analogy-based learning processes in the domain of deterministic chaos]. PhD Thesis. Kiel: Institute for Science Education.

Treagust, D., Harrison, A., & Venville, G. (1996). Using an analogical teaching approach to engender conceptual change. *International Journal of Science Education 18*, 213-229.

Role-playing, Conceptual Change, and the Learning Process: A Case Study of 7th Grade Pupils

Pirjo-Liisa Lehtelä
Faculty of Education, University of Joensuu, Finland

Abstract

This study explores the teaching of the structure of matter in the science classroom in a way that aims at supporting seventh-graders' learning processes through the use of modelling and role-playing seldom used in science education. Through role-playing activities, pupils may, for example construct an understanding of how particles move and interact in different states of matter. The qualitative data used in the study were collected from classroom observations and the pupils' reflective writing during the ten-lesson period. In this article pupils' metacognitive conditions of learning processes are studied in a context where conceptual change is promoted.

Introduction

This study is a part of a larger project (Sormunen, Saari, & Lehtelä, 1999) seeking to develop alternative methods for the teaching of chemistry and physics to meet the particular needs of seventh-graders (aged 13 years). While pupils' understanding of the nature of matter has been focus of several earlier researchers (e.g. Andersson, 1990; Saari, 1997; Stavy, 1990), this study aims to clarify the nature of the pupils' learning process and their awareness of their own understanding in a context where conceptual change is promoted. The specially designed instruction used to achieve these aims was based on modelling and role-playing activities. Instruction consisted of ten lessons and its particular emphasis was on the pupils' learning of the nature of matter and physical change.

Two major theoretical ideas, conceptual change and metacognition, play a significant role in current science education. Conceptual change and metacognition are seen to be closely connected to the success of the learning process (Gunstone & Northfield, 1992). Pupils' ability to find the main idea from the lessons, to monitor and evaluate their learning and to use knowledge and skills comprise from the metacognitive dimension of the learning process. These metacognitive conditions are the focal point of this article and the issue of conceptual change is only briefly discussed.

From a perspective emphasising the role of conceptual change, the reconstruction of a pupil's knowledge is seen as a personally created, constructed and experienced process (Driver, 1989; Tobin, 1993). Furthermore, according to constructivist views of learning, the learning of science is not a process, in which students passively absorb information but one in which they actively construct for themselves an understanding of the events under observation (e.g., Driver, 1989; Tobin, 1993). Since the learning of many scientific concepts and phenomena requires changes in thinking, special instruction is often needed to guide the learner towards this goal. As this is the case, the students' own abilities to reflect on and

H. Behrendt et al. (eds.), Research in Science Education – Past, Present, and Future, 211–216.

monitor their learning process appear to play a significant role. In addition, when abstract and difficult topics (e.g., changes of the states of the matter) that are not visible as phenomena are studied, experiments and observations are not always possible. Therefore, in this study the activity role-playing is utilised to make abstract phenomena visible to the pupils and to assist in their reconstruction of knowledge. Resnick and Wilensky (1998) argue that role-playing activities, although rarely used in science classrooms, play a significant role in helping students to learn complex topics. I not only agree but also contend, that in certain teaching situations, role-playing activities promote pupils' understanding and conceptual change.

The role of metacognitive issues in achieving conceptual change is central in much research (Gunstone & Mitchell, 1998; Gunstone & Northfield, 1992). The person's ability to control and monitor his/her own learning and understanding is based on metacognition (Flavell, 1976). In the science classroom, the development of metacognitive skills should be studied in this particular learning context since it may give valuable information (Gunstone & Northfield, 1994). On the basis of these perspectives, this classroom-based study is thought to provide us with refreshing and significant information on the issues concerning how pupils reconstruct knowledge and take responsibility for their own learning during their study of the structure of matter.

Study design

The planned instruction was carried out in one secondary school in the city of Joensuu in February 1999. The subjects consisted of 18 seventh-graders, 10 girls and 8 boys. The ten-lesson teaching unit involved teaching the structure of matter through the use of modelling and role-playing activities. The teaching setting consisted of five sessions of 90 minutes each. The class teacher was familiar with the teaching approach, as she had used modelling and role-playing in teaching the structure of matter.

The idea of the instruction unit is described in Figure 1. All lessons were based on a similar procedure in which pupils are able to participate in role-playing through which they may construct their knowledge of scientific phenomena. The research questions of this study are:

1. How do the pupils find the main idea of the lessons ?
2. How do pupils monitor and evaluate their own learning process ?

Data collection and analysis

Data were collected by classroom observations and by pupils' reflective writing. Each pupil kept a learning diary during the ten-lesson period. After each lesson they were asked to answer questions supporting the monitoring and evaluation of their learning process. The following questions were asked: (A) What was the main idea of the lesson?, (B) Was there anything you didn't understand?, (C) How did you succeed in the task you did in class?, (D) Which events help you to understand the issues you mentioned? The aim of reflective writing is to follow and understand the reconstruction of the pupils' knowledge. The classroom observations were collected by recording the teaching activities used during the lessons, time dividing between these activities and also the general course of events in the classroom.

The data were analysed qualitatively and categorised according to the principles of "grounded theory". To evaluate the metacognitive conditions of the pupil's learning processes, his/her answers to questions (A, B, C), as presented in the learning diary, were ranked and divided into three groups (ranging from level 1 to level 3). The ranking was based on the metacognitive level of the pupil's answers. A mean score of the levels of the sessions was calculated for each pupil. On the basis of these levels, it was possible to place each pupil in a learning category (I, II, III). The category describes a typical learner, with certain metacognitive skills, implying the nature of the pupil's learning process in that perspective (see Table 1).

The answers to the question "Which events help you to understand the issues you mentioned?" (D) were analysed separately. Table 1 shows what pupils in each group (categories I-III) found most helpful in the learning process.

	Theme of the sessions		***Activities***
I N S T R U C T I O N	MODEL AND STATES OF MATTER GAS SOLID LIQUID APPLICATIONS	THE PROCESS OF LEARNING	INDIVIDUAL TASKS • writing notes • learning diary • tests GROUP WORK • group discussions CLASSROOM DISCUSSION TEACHING THROUGH PERFORMANCE • modelling by computer • experiments • role-play

Fig. 1: Teaching activities used in the science classroom.

Findings

As some of the pupils' written answers were short, their analysis was problematic in some cases. The section below will discuss the pupils' answers to each question (A-D) separately. Examples of their answers will also be given.

Understanding the main idea of the lesson (A)

The three levels related to the pupil's understanding of the main idea of the lesson are the following:

(1) only the title of lessons mentioned:
- *The word model.* (Girl 21)[1] [[1]Code for the pupil]
- *Liquid.* (Boy 33)

(2) one issue learnt mentioned:

- *I learnt that matter has the continuance model and the particle model.* (Boy 33)
- *I learnt the liquid state and the motion of particles.* (Girl 35)

(3) several ideas from the lesson mentioned:

- *The structure of gas or what it is formed from. Structure particles and the motion of particles, collision and interaction.* (Boy 33)
- *That the particles of gas spring back from the wall and that's why you can smell perfume farther away straight in front of you.* (Girl 26)

A mean score of understanding the main idea of each lesson was calculated for each pupil. 11% of the responses were categorised as belonging to level 1, 66 % to level 2, and 22 % to level 3.

Monitoring own understanding (B)

The pupil's level of being able to monitor his/her own understanding is based on his/her answer to the question "What did you not understand during the lesson?" The levels identified were the following:

(1) responses classified as belonging to this level give the impression that the learner was not able to think about his/her own level of understanding during the lesson:

- *Nothing.* (Girl 38)

(2) clearly written answers claiming that everything was understandable:

- *I understood everything that was talked about during the lesson.* (Boy 29)
- *I think that there was nothing unclear, nothing in my view.* (Girl 35)

(3) responses introducing that something related to the studied topic remained unclear:

- *Gas particles' movement in different temperatures.* (Boy 33)
- *For me the structure of gas is still a little unclear. If somebody asked me questions about the structure of gas I would not be able to give an answer.* (Girl 35)

Again, the mean score of monitoring own learning was reported to each pupil; 61% of the pupils' responses belong to the first level, 22% to the second level and 17% to the third level.

Evaluation of own success in the lesson (C)

The pupils' evaluation of their success during the lessons formed three levels presented below:

(1) a general answer without any explanations:

- *Well.* (Girl 21)
- *Excellently.* (Girl 38)

(2) answers claiming that they had succeeded because the respondent was beginning to understand the issues dealt with in the classroom:

- *Quite nicely and everything was clear.* (Girl 34)
- *I think that I was pretty good, because I learnt the issues taught in the class.* (Girl 31)

(3) answers giving explanations why s/he succeeded during the unit's lessons:

- *I'm very pleased, I understand everything because everything was discussed well.* (Girl 33)
- *Well. I participated actively and listened carefully.* (Boy 32)

Again, the mean score of evaluation of own learning was reported to each pupil: 39% belonged to level 1, 11 % to level 2%, and 50 % to level 3.

Types of learners

On the basis of their ability to monitor and evaluate their learning pupils were categorized into three different groups (Table 1). This data makes it possible to form three different types of learners.

Categories	Types of learners	Codes formed from the levels: (A,B,C)	Percentage %	Understanding assisted most by
I	Active analytical learner	(3,3,1) (3,3,3) (3,2,3)	22	• teacher • models
II	Average learner	(2,1,3) (2,1,2) (1,2,3) (2,2,3)	44	• models • different activities
III	Low-skilled learner	(2,1,1) (1,1,1)	33	• role-plays • teacher

Table 1: The metacognitive conditions of pupils' learning processes

Active analytical learners were able to process their understanding during lessons and to evaluate their learning outcomes. These pupils were aware of their level of understanding. They also seem to have metacognitive skills with which to process conceptual change in an active manner. If some information had remained unclear to them, they were able to talk about it. They usually mentioned some unclear abstract issues discussed in class. This supports further the impression that good pupils also need such classroom activities, that make abstract phenomena visible to them. *Average learners* try to understand the keywords of each lesson. They do have some metacognitive skills, that they were able to use in their learning. They were also able to monitor and evaluate their learning. *Low-skilled learners* were not aware of their own learning processes; nor did they have good skills to monitor their understanding. While they did not have a very active attitude, their attitude to learn science in these lessons was positive.

Issues that assisted learning

Factors that the pupils considered as helpful to the development of their understanding were gathered from the learning diaries. The answers to the questions were calculated by the mean average and can be found in Table 1. Pupils in all learning categories (I-III) reported that role-playing activities or modelling helped the construction of their knowledge. Pupils were aware of the role of classroom activities in learning and how these impinge on their own learning. Nonetheless, the teacher had a central role in their learning processes.

Discussion

The results indicate that 22 % of the pupils are able to evaluate their learning outcomes and understand well the main idea from the lessons. They give analytical answers and seem to be able to use different types of learning strategies during the lessons. The ways in which pupils think, select, and, for example, evaluate information are "cognitive styles". Learning occurs most effectively when learning strategies/styles are designed to suit the learner's particular needs. In this study 33 % of pupils seem to have some difficulties in using or changing their learning strategies, although the teaching approach was designed to help the pupils towards more advanced learning. Modelling and role-plays helped pupils' understanding of issues dealt with and the approach seemed to work rather successfully in this learning context. While there were some negative reactions, these focused mainly on the pupils problems of writing down their thoughts.

If we want pupils to take responsibility for their own learning process, the metacognitive dimension should also be taken into consideration while designing teaching aiming for conceptual change. In addition, the fact that role-playing activities may be associated with wide-ranging learning gains should also be taken into account. However, we still need more information dealing with pupils' learning processes and reactions to instruction, since the approach presented may differ from traditional science teaching in some ways. The pupils' conceptual change process and its relationship with the metacognitive dimension in this particular learning situation will be discussed in further research reports.

References

Andersson, B. (1990). Pupils' conceptions of matter and its transformation (age 12-16). *Studies in Science Education* 18, 53-59.

Driver, R. (1989). Students' conceptions and learning of science. *International Journal of Science Education* 11, 481-490.

Flavell, J.H. (1976). Metacognitive aspects of problem solving. In: L. B. Resnick (Eds.), *The nature of intelligence*. Erlbaum, Hillsdale, New York.

Gunstone, R.F. & Northfiled, J. (1994). Metacognition and learning to teach. International *Journal of Science Education* 16(5), 523-537.

Gunstone, R.F. & Mitchell, I.J. (1998). Metacognition and conceptual change. In J. Mintzes, J., J.H. Wandersee, and J. D. Novak, (Eds.), *Teaching Science for Understanding. A Human Constructivist View* (pp. 133-165). New York: Academic Press

Resnick, M. & Wilensky, U. (1998). Diving into complexity: developing probabilistic decentralized thinking through role-playing activities. *The Journal of the Learning Sciences* 7(2), 153-172.

Saari, H. (1997). *Learning the structure of matter*. Unpublished licentiate thesis. Department of Physics, University of Joensuu. (In Finnish).

Sormunen, K., Saari, H., & Lehtelä, P.-L. (1999). *Learning the structure of matter: Models and analogies, learners' metacognitions and epistemologies*. Paper presented at the Finnish Symposium on Mathematics and Science Education Research, 25-26.9.1998. Åbo Akademi, Vaasa, Finland.

Stavy, R. (1990). Children's conception of stages in the state of matter: from liquid or solid to gas. *Journal of Research in Science Teaching* 27(3), 247-266.

Tobin, K., Eds. (1993). *The Practice of Constructivism in Science Education*. Washington, DC: American Association for the Advancement of Science Press.

Concept Mapping as a Tool for Research in Science Education

Helmut Fischler, Jochen Peuckert, Free University of Berlin, Germany
Helmut Dahncke, Helga Behrendt, University of Kiel, Germany
Priit Reiska, University of Educational Sciences, Tallin, Estonia
David B. Pushkin, Wilmington College of Delaware, USA
Milena Bandiera, Roma Tre University, Rome, Italy
Matilde Vicentini, La Sapienza University, Rome, Italy
Hans E. Fischer, Lorenz Hucke, Kristin Gerull, Univ. Dortmund, Germany
Jenny Frost (discussant), University of London, UK

Abstract

In recent years concept mapping has become a powerful tool that is frequently applied in different contexts in science education. There are still, however, questions and problems that arise when researchers in science education begin to design concept map formats and search for appropriate methods of analysis. In a symposium, five research groups from different countries shared experiences of using concept maps and hence provided some answers to the questions and some illumination of the problems.

Concept maps: Report on a symposium

Many research activities reveal a vast amount of details concerning general and special conditions for applying concept maps (Liu & Hinchey, 1996; Markham & Mintzes, 1994; Novak, 1990; Fischler & Peuckert, 2000; White & Gunstone, 1992; Ruiz-Primo & Shavelson, 1996; Novak, 1998; McClure, Sonak, & Suen, 1999; Rice, Ryan, & Samson, 1998). However, the literature still reports many unsolved problems concerning the application and evaluation of concept maps. Therefore, it is necessary to carry out further research studies and to present research results obtained in different contexts. In the symposium, five research groups presented their different approaches to a graphic representation of knowledge structures by means of concept maps. These contributions showed a great variety of goals connected with concept maps, of modes in applying them and of age groups who used the maps. This diversity reflects how flexibly concept maps can be used by adjusting their formats to given situations. In reporting the symposium, therefore, it was important to find a common structure that would allow description and identification of characteristic features of concept maps as used in the different research projects. The following questions were used to catch the different characteristics and contexts highlighted by the presenters:

- Which research questions guide the project; which problems does the project focus on?
- What are the purposes of the application of concept maps?
- Are there other research tools complementing the use of concept maps?
- Which types of concept maps are being applied?
- Which task formats are used?
- What are the methods of analysis?

H. Behrendt et al. (eds.), Research in Science Education – Past, Present, and Future, 217–224.

- Which features describe the power of concept maps, what are the limitations of applying concept maps?

Research projects, research questions and the purpose of the maps

Three of the five projects use concept maps exclusively as tools for analysing students' conceptions. Reiska et al. (1999), working in secondary schools, compare the effect of teaching about energy and energy supply through traditional physics lessons and in STS lessons (Science Technology Society). They use students' knowledge and ability to take action as indicators of effect. In this research, concept maps are used to provide information about breadth and structure of students' knowledge.

Fischer et al. (1999), in their study of first year physics undergraduates at University, use concept maps to detect the knowledge of students before and after the performance of a laboratory experiment. The researchers ask under what conditions students use physical and experimental knowledge for action regulation during traditional and computer based labwork and, furthermore, whether students acquire physical and experimental knowledge by performing a lab experiment. The concept action regulation (e.g. Hacker, 1986; Dörner, 1993) describes how human action always involves cognitive processes occuring on different levels of complexity (Fischer, 1993). For example, students' observations and actions during a physics laboratory experiment can be regulated by physical concepts (concept driven action regulation) or by describing events and moving objects (object oriented action regulation) without relating to physical concepts.

Peuckert and Fischler (1999), working in lower secondary schools, investigate how students' conceptions of particles change over a period of time, in which this topic is not part of the science education they are experiencing. Concept mapping *by students* is expected to elicit their conceptions about particle models. Concept maps constructed *by researchers* summarizing all data including students' maps are used as an assessment tool for intra-individual and inter-individual comparisons of conceptions as well as for the description and comparison of groups.

In two research projects concept maps serve as research tools and as means to support teaching-learning processes. Pushkin (1999a), working with introductory physics students at a College, asks if there is an alternative form of concept mapping that can teach us more about novice physics students' problem solving approaches (Pushkin, 1999b). Can we evaluate *maps of equations*, rather than terms? Might this technique enhance student problem solving? In Pushkin's project, students are encouraged to draw equation webs as a means to foster metacognitive self-awareness of how they approach problems and solve them, successfully or unsuccessfully.

Bandiera and Vicentini (1999) regularly incorporate concept mapping into pre-service and in-service secondary science teacher-training courses. They take concept maps into consideration as a product (significant representation of „here and now" knowledge) and as a process (an activity which helps recovery of concepts and understanding and induces their ordered organisation). They analyse and compare three task formats, drawing distinctions between the methodologies used and identifying the contexts most appropriate for each format. The project presented is aimed at formalising different types of concept maps as educational tools, to help achieve the aims of science education.

Concept maps in the context of other research tools

In all research groups concept maps are regarded as an essential instrument. Do researchers totally rely on the diagnostic power of this tool, or are concept maps complemented by other tools?

Reiska et al. use computer simulations, combined with the method of thinking aloud, in addition to the concept maps. Thus, they gather additional oral data that have to be interpreted according to criteria fitting the goals of the project. Fischer et al. complement the mapping process by an open interview, in which students are asked to explain the whole concept maps in their own words and in which the researcher asks for a more detailed explanation of the links that had not been explained clearly in writing. In addition to this, the researchers video tape students' performing of the experiment and analyse their action regulation during laboratory work by means of a category based video analysis, using the same category base for both methods. The results of the concept map analysis are linked to the results of the video analysis in order to study the links between the students' knowledge acquisition and action regulation during laboratory work.

Peuckert and Fischler ask students to draw a concept map during interviews conducted at the beginning and at the end of a period of nearly one year. This mapping aims at addressing the cognitive structures, in an alternative way, in order to get mutual validation with other results from interview transcriptions.

These descriptions of the contexts of methods demonstrate that concept maps are a part of a set of instruments, applied in order to achieve an acceptable degree of mutual validation. Pushkin (1999a) and Bandiera and Vicentini (1999), in their projects, focus almost entirely on concept maps because these maps are the objects of their investigation.

Concept mapping tasks

In most cases, concept mapping, as a task for learners, refers to the diagnostic power of concept maps. Nevertheless, even for instructional purposes, not only complete maps can be presented to learners (Bandiera & Vicentini) but also active drawing can help to meet certain objectives of the learning process, e.g. concept clarification (Bandiera & Vicentini) or fostering of metacognitive self-awareness (Pushkin). What also distinguishes these two projects from the other three is that the maps are being used for formative assessment - i.e. the outcomes of the maps inform subsequent learning and teaching.

Focussing on diagnostic mapping, two types of mapping tasks can be distinguished: Firstly, *students'* mapping as an externalisation of knowledge is a means that is used by all research groups. Secondly, some projects include *researchers'* mapping as a reconstruction of knowledge of individuals or groups (Reiska et al., Fischer et al., Peuckert & Fischler). To some extent, the format of this type of map reflects properties of students' mapping tasks.

Formats of students' mapping

In considering formats of maps, it seemed sensible to separate the maps researchers asked students to make, which are dealt with in this section, and the maps the researchers made for themselves, which are dealt with in the next section. The illustration in Fig. 1 shows a part of a student's map, in which a student linked given words at fixed positions by short phrases on a sheet of paper (Peuckert & Fischler). This is only one of many possible approaches. The specific function of the mapping procedure and, especially the intended way of map analysis, leads to a great variety of what is expected to be done by the students.

When designing the task with regard to the purpose of mapping, some important aspects that have to be considered are:

- given concept words (and if so, which words? and how many?) versus free choice of concept words
- fixed positions of concepts versus choice of arrangement of concepts by the mapper
- given relations (and if so, which relations? and how many?) versus free choice of description of associations by the mapper
- a hierarchical structure versus an associationist structure

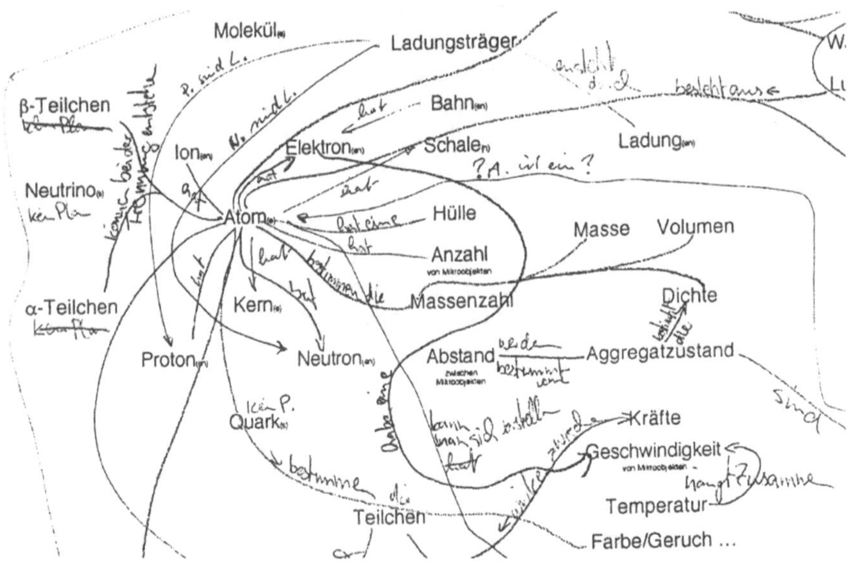

Fig. 1: Concept map of a student - an example from a German study on particle models

The only common feature in the presented studies in the symposium was the basic idea of mapping: i.e. some concepts connected to a specific domain being linked explicitly in a way that makes sense to the student. Students are expected to externalise their individual view on a topic. One important aspect of mapping was emphasised by Bandiera and Vicentini in their explicit instruction to the teachers: „It is important that you make explicit what is in your mind without searching for the correct map". In most cases the „concepts" are simply represented by words, but can also be mathematical terms and equations (Pushkin).

Some researchers provide a set of words to reflect normative conditions (Fischer et al.: laboratory guide, Reiska et al., Peuckert & Fischler: objectives of secondary science teaching) or expert rating (Reiska et al., Peuckert & Fischler). For some, an additional reason for using given concepts is to simplify the comparison of maps with each other or with an external reference. In the case of providing the words, the relevance and suitability of these is tested in pre-studies. The number of words differs from 20 (Fischer et al.), 39 (Peuckert & Fischler) to 51 (Reiska et al.). Where a large number of words is used (especially at fixed positions: Peuckert & Fischler), the problem of reliability arises because the graphical barriers on the sheet inhibit all connections being made and there is an element of chance as to whether or not a concept is considered. Reiska et al., therefore, began with a phase of familiarizing the students with concept maps. Before concept mapping was initiated, students were explicitly introduced to the procedure by means of an example concerning animals. For the procedure of concept mapping itself, the researchers reduced the number of concepts and relations - which was originally quite substantial - to 51 concepts and 10 relations. The acquisition and subsequent evaluation of data indicated such a relatively large number could be coped with. The students did not have any problems with the relations and concepts presented and completed their maps in a relatively short time. No student required more than 45 minutes of task time.

As a result of variations encountered with concept map formats, in two studies the students' maps are not analyzed separately but complemented by other data (Fischer et al., Peuckert & Fischler). Some researchers allow students to work freely without any restrictions or further guidelines (Bandiera & Vicentini, Pushkin) or at least allow students to add concepts freely to the list given (Fischer et al., Peuckert & Fischler). The same methodological difference can be found for the linking task: relations between the concepts can be given (Reiska et al.: 10 relations) or left to students' ideas; the latter implies a paper and pencil approach.

Since both given and freely chosen links enable students to express order and hierarchy (if necessary at all), the result of students' mapping is an associationist „network concept map" (Ruiz-Primo & Shavelson, 1996). An exception is hierarchical instructional maps for concept clarification used in biology teachers' training (Bandiera & Vicentini), where the promotion of a structured, logical and rational thought is a relevant aim. This science domain is probably more closely linked to a Novakian map format that is not used in the studies concerning physics education.

Formats of researchers' mapping

Since researchers' maps are based on students' maps, the format of these maps is not free, but depends on the way in which students' maps are integrated. In two studies (Fischer et al., Peuckert & Fischler), students' maps are complemented by oral data from interviews or other sources, that have to be transformed into a propositional format. This reconstruction of maps needs a categorisation of links, if the students were free to describe their associations between the concepts. This is, of course, not necessary, if relations are provided in the mapping task and if oral data are not included (Reiska et al.).

By means of these more or less reconstructed individual maps, generalising or averaging maps (reference maps: Fischer et al.; modal maps: Reiska et al., Peuckert & Fischler; field graphs: Reiska et al.) can be developed that allow further interpretation of data. All of these reconstructed maps are arranged systematically. They include the same number of words as the students maps in those studies where students were not allowed to add their own words (Reiska et al., Peuckert & Fischler). Reconstruction can, however, lead to much bigger maps of e.g. 195

propositions (Fischer et al.), if additional concepts of all students are taken into account. As an example, the following figure shows a small extract of a researcher's map that is intended to describe the knowledge of a student with respect to a reference map. The reference map takes into account concepts and propositions of all students and, therefore, permits comparison of student maps to each other. In Figure 2, the different shapes describe different types of concepts (physical, methodological, objects, etc.). Empty shapes and propositions without description indicate that certain individual students did not mention them in their concept map, while other students did. Therefore, such concepts and propositions will appear in the reference map.

Since compilation and reconstruction procedures are time consuming for larger samples, data processing of researchers' maps is usually done by means of special software based on data matrices, which are isomorphic to the graphs of nodes and links.

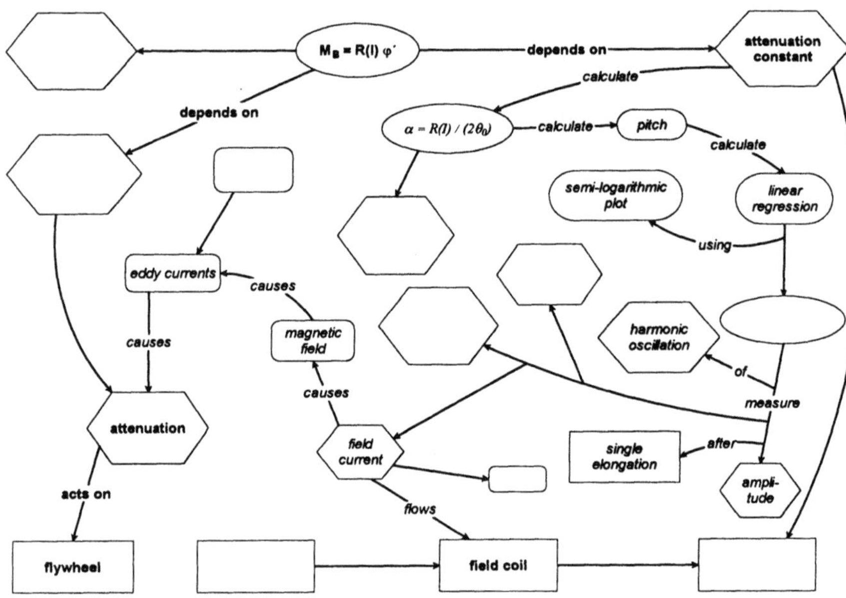

Fig. 2: Reconstructed student map based on a reference map (Hucke & Fischer, 2000).

Analysis of concept maps

In three of the projects, evaluation is carried out by analyzing the students' maps directly. This analysis has features of a learning situation in those cases, where mapping meets instructional purposes. Concept maps, then, are analyzed jointly between teacher or researcher and student (Bandiera & Vicentini, Pushkin). This analysis is mainly qualitative and aims at the elicitation of deficiencies or the development of self-awareness. In a pure diagnostic context (Reiska et al.), students' maps are analyzed quantitatively by means of about 50 numerical variables that are to describe the maps - and, therefore, students' knowledge - in terms of structure (e.g. point centrality, degree of embedding), size (e.g. number of concepts), correctness (number of correct concepts and relations) and content (e.g. number of concepts from physics terminology, (for details see Reiska, 1999). This analysis is carried out by researchers with the aid of suitable software (see Dahncke et al., in this volume). Likewise, generalizing maps are analyzed, that are drawn from the individual maps of the students.

In the two other studies (Fischer et al., Peuckert & Fischler), students maps are not analyzed, apart from being standardised and compiled with other information from interviews. Evaluation is then based on the maps that are reconstructed by the researchers. These maps are analyzed qualitatively and quantitatively by researchers. Quantitative analysis focusses on getting evidence for development or stability of knowledge and on estimating the impact of external objectives. Qualitative analysis aims at identifying content related features of the maps. Furthermore, results from the mapping analysis are correlated with results obtained by other research methods.

Summary: Power and limitations of concept maps

In general, all researchers who have contributed to this symposium agree that concept maps are a powerful and flexible instrument for analysing differences and changes in the knowledge of learners in science. One of the main characteristics of concept maps is that they are easily adaptable to different research topics. They unfold their most convincing advantages when linked to other methods used in the same project. This makes possible to obtain results that may go beyond the results which can be obtained using the methods separately.

In processes of teaching and learning, concept maps help to clarify concepts, to define conceptual schemes, to single out misconceptions, to overcome cognitive obstacles, to plan and to test educational activities and to promote self-assessment. This latter purpose (self assessment) was most evident in those projects that had a strong teaching purpose and necessarily involved the learner in making some assessment of their own concept maps. This was one of the issues raised in the questions at the end of the symposium.

The research groups using concept maps as a research tool do not rely on this method alone and this indicates one of the limitations of the maps. The strength of concept maps becomes more evident once they are linked to other methods. Another limitation, identified by one of the research groups, is that because a concept map is a representation of knowledge and seems to share correspondences with knowledge organisation structures, it can be considered exhaustive and correct, as an objective and definitive construct of implied concepts. In order to prevent researchers and teachers (and learners) from developing such a view, it is necessary to support the use of a diversity of methods for understanding learners' conceptual networks.

Both the presentations and the questions focussed on how concept maps were being used, and this focus has been followed in this summary. The findings from the individual projects have consequently not been reported.

References

Acton, W.H., Johnson, P.J. & Goldsmith, T.E. (1994). Structural knowledge assessment: Comparison of referent structures. *Journal of Educational Psychology* 86, 303-311.

Bandiera, M. & Vicentini, M. (1999). Drawing teaching-learning tools from concept mapping. *Proceedings of the Second International Conference of the European Science Education Research Association, Kiel. Vol. 1* (pp. 266-268). Kiel, Germany: IPN.

Dahncke, H. (1997). Science learning and competence to act – a report on a study on computer simulations and STS-teaching. In R. Panwar, Ed., *Science and Technology Education – Responsible Citizenship and Economic Development*. Vol 1 (pp. 24-31). Edmonton: University of Alberta, Faculty of Education.

Dörner, D. (1993). Wissen, Emotionen und Handlungsregulation oder die Vernunft der Gefühle. *Zeitschrift für Psychologie* 2, 167-202.

Fischer, H.E. (1994). Physiklernen: Eine Herausforderung für Unterrichtsforschung. In D. Nachtigall, Ed., *Didaktik und Naturwissenschaften, Band 3*. Frankfurt: Lang.

Fischer, H., Hucke, L. & Gerull, K. (1999). The interpretation of concept maps using a reference. *Proceedings of the Second International Conference of the European Science Education Research Association, Kiel. Vol. 1* (pp. 269-271). Kiel, Germany: IPN.

Fischler, H. & Peuckert, J., Eds. (2000). *Concept Mapping in fachdidaktischen Forschungsprojekten der Physik und Chemie*. Berlin, Germany: Logos.

Hacker, W. (1986). *Allgemeine Arbeits- und Ingenieurpsychologie*. Bern: Huber.

Hucke, L. (1999). *Handlungsregulation und Wissenserwerb in traditionellen und computergestützten Experimenten des physikalischen Praktikums*. PhD thesis at the University of Dortmund.

Hucke, L. & Fischer, H.E. (2000). Wissenserwerb und Handlungsregulation im physikalischen Praktikum. In H. Fischler & J. Peuckert, Eds., *Concept Mapping in fachdidaktischen Forschungsprojekten der Physik und Chemie (pp. 57-89)*. Berlin, Germany: Logos.

Liu, X. & Hinchey, M. (1996). The internal consistency of a concept mapping scoring scheme and its effect on prediction validity. *International Journal of Science Education* 18, 921-937.

Markham, K.M. & Mintzes, J.J. (1994). The concept map as a research and evaluation tool: further evidence of validity. *Journal of Research in Science Teaching* 31, 91-101.

McClure, J.R., Sonak, B. & Suen, H.K. (1999). Concept map assessment of classroom learning: reliability, validity, and logistical practicality. *Journal of Research in Science Teaching* 36, 475-492.

Novak, J.D. (1990). Concept mapping: a useful tool for science education. *Journal of Research in Science Teaching* 27, 937-949.

Peuckert, J. & Fischler, H. (1999). Concept maps as a tool for investigating and analyzing the development of students' conceptions. *Proceedings of the Second International Conference of the European Science Education Research Association, Kiel. Vol. 1* (pp. 263-265). Kiel, Germany: IPN.

Pushkin, D.B. (1999a). Concept mapping and students, physics equations and problem solving. *Proceedings of the Second International Conference of the European Science Education Research Association, Kiel. Vol. 1* (pp. 260-262). Kiel, Germany: IPN.

Pushkin, D.B. (1999b). Post-formal thinking and science education: how and why do we understand concepts and solve problems? In J.L. Kincheloe & S.R. Steinberg, Eds., *The Post-Formal Reader* (pp. 449-467). New York: Garland.

Reiska, P. (1999). *Physiklernen und Handeln von Schülern in Estland und in Deutschland*. Frankfurt a.M.: Lang.

Reiska, P., Dahncke, H. & Behrendt, H. (1999). Concept maps in a research project on "Learning Physics and Taking Action". *Proceedings of the Second International Conference of the European Science Education Research Association, Kiel. Vol. 1* (pp. 257-259). Kiel, Germany: IPN.

Rice, D.C., Ryan, J.M. & Samson, S.M. (1998). Using concept maps to assess student learning in the science classroom: must different methods compete? *Journal of Research in Science Teaching* 35, 1103-1127.

Ruiz-Primo, M.A. & Shavelson, R.J. (1996). Problems and issues in use of concept maps in science assessment. *Journal of Research in Science Teaching* 33, 569-600.

White, R.T. & Gunstone, R. (1992). *Probing understanding*. New York: Falmer.

The Need for and the Role of Metacognition in Teaching and Learning the Particle Model

Peter Buck
University of Education, Heidelberg, Germany
Philip Johnson
University of Durham, UK
Helmut Fischler, Jochen Peuckert, Silke Seifert,
Free University of Berlin, Germany

Abstract

The research focus of this symposium on teaching and learning the particle model (discussant: Robin Millar) addressed the following questions: How can misconceptions on atoms be avoided and how can we gain insight into and describe the learning processes involved? Three major influences were identified: metaconceptional awareness, the order in which concepts are learnt and the metaphoric content of language.

Starting points and targets of the symposium

There are two main situations in which particle models are usually introduced in chemistry and physics education: the microscopic description (explanation) of the solid, liquid, and gaseous state and the description (explanation) of chemical change. Chemistry instruction frequently uses multiple models ranging from simple balls through to the mathematical quantum model. Often, such multiple models are used in unsystematic ways leaving students believing, at different times, that different models are 'right' rather than contextually appropriate.

A significant area of research - past, present and with a continuing need for the future - centres on the various "strategies" used to teach particle concepts and students' resulting mental models of particle phenomena.

Of course, in this synopsis, only a few facets can be illuminated and findings can only be reported in a very foreshortened manner. An important outcome for the contributors to the symposium, however, was to realize that they were working in mutual basic agreement.

State of research - conceptual shortcomings and teaching approaches

The conception of matter as a collection of particles is one of the most powerful ideas developed by science. Indeed, it is axiomatic to modern science. It is not 'by accident' that it has been part of the school curriculum, worldwide, for many years. However, research into students' understanding has shown that they have many difficulties with respect to this model of matter, despite a great deal of teaching effort in our schools (cf. Pfundt 1981, Novick & Nussbaum 1981, Brook et al. 1984, Andersson 1990, Duit 1992, Lee et al. 1993, Johnson 1998a):

H. Behrendt et al. (eds.), Research in Science Education – Past, Present, and Future, 225–234.

- The central misconception that influences many students' ideas can be described as a transfer of macroscopic properties to particles; e.g. particles have a specific temperature and expand when heated up.
- Only a few students apply the particle model by themselves to describe macroscopic phenomena.
- There is often a mixture of continuous and discontinuous conceptions.
- The empty space between atoms and molecules is not accepted.
- Particles stop moving at some time.

Thus one of the major problems lies in the difficulty to achieve an acceptable understanding of the particle world, i.e. to acknowledge that most of the properties that we are so familiar with from the macroscopic world, are *lacking* in the micro world. In other words, that particles are "alien", "andersweltlich".
Different teaching approaches have been developed which explicitly try to overcome these shortcomings:

- The teaching scheme used by the Children's Learning in Science Project (CLIS 1987, Scott 1992) begins with students' identifications of the various properties of different kinds of matter. Students are encouraged to make their own ideas explicit and to develop a pattern of properties in the behaviour of solids, liquids and gases. At some stage, students are introduced to the basic elements of theory making, in order to enable them to develop their own theories for the nature of solids, liquids and gases. In the case that they do not meet the scientists' view, the teacher will help the students to move towards conceptions held by scientists. An important means for achieving such a shift is a cognitive conflict strategy, in which students are exposed to phenomena they cannot explain using their existing theories.
- In an approach which he calls "Jumping to the atoms", Buck (1987, 1990) relies on the concept of systems and components, which he considers at length, before he uses it to discuss textbook representations of particle models. He concurs with Millar (1990) that particles "can only be taught by ostentation" and emphasizes that any visual, touchable notion of particles like balls and sticks should be avoided as much as possible. He stresses metacognitive reasoning as being a major concern in his approach of "epistemic teaching" (cf. Buck 1997)
- Vollebregt (1998, 1999) questions *both* the teaching phase of CLIS, in which students are expected to find theories about the nature of matter that are in line with the scientific view *and* Buck's "aporetic situation" which makes "jumping to the atom" neccesary. Therefore, her "problem posing approach" offers an initial particle model to the students and helps them to come to a better understanding of the behaviour of gases, liquids, and solids by linking macroscopic variables (for instance air pressure) to model variables (for instance force per area exerted by particles by means of their collisions).

Common to all approaches is an awareness of the need for explicit metacognition in teaching the particle model.

Gaining insight into learning and teaching processes

In a longitudinal research project *"Pathways to concepts of atoms"* Fischler, Lichtfeld & Peuckert observed students from a Berlin Gymnasium from grade 7 through to grade 11. The development of students' cognitive structures and their understanding of the topic *'microworld'* were described. Using different research methods (e.g., analysis of outcomes for all pupils out of four parallel courses, observing single students in detail, video-recording lessons and interviews of lessons) the following questions were addressed: What influences do physics and chemistry lessons have on science thinking and learning? What is the influence of students' alternative frameworks within the cognitive field of particles and atoms?

The design for this long term study follows two main lines (for more details see Lichtfeldt 1996): a sample of 24 students from four different courses were observed in detail and every half year all the students (100) from all four different courses completed the same questionnaire and were video-recorded during their lessons. Verbal data gathered from interviews and videotaped lessons deliver information about students' performance, in given situations that address particle ideas. Concept maps and responses to questionnaires provide a less context dependent view. The combination of these methods allow identification of individual as well as group developments.

Single student – results and examples

The following drawings and concepts were made by a female student in the questionnaires. Before lessons in physics and chemistry had begun, at the beginning of the long-term-research, she was not able to draw a picture of an atom. She said she could not imagine *what* an atom looks like. But her concept map shows ideas of an atom (Fig. 1): the nucleus is surrounded by a "shell". This is called an atom, and the atom is a "tiny little body".

Fig. 1: A student's "drawing" and her concept map before lessons in chemistry and physics (grade 7)

To illustrate Buck's "epistemic teaching" approach (Buck, 1997), he would rejoice in encountering such a student in his class. Indeed, scientists also "have no idea what an atom looks like", as she puts it (e.g. Heisenberg, 1970). Buck would encourage the student to keep her general notion and go on to discuss with the class, what people intend to convey when they draw pictures or build tangible model objects even though "nobody knows, what an atom looks like". In most cases, however, this attitude is (as yet) not to the taste of the average science teacher, who prefers to build up a pictorial image, i.e., a misleading concept which has to be *deconstructed* later in order to overcome the "central misconception" mentioned above. All the same, the concept map shown in fig. 1 (produced after more insisting questions) shows a concept of atoms in terms of "tiny little bodies".

The influence of physics and chemistry education is shown by the concept map and drawing constructed in the middle of grade 9 (Fig. 2). The electron was introduced in grade 8. For her, it behaves like a classical particle moving on a trajectory. The atom is still a tiny little body and has the shape of a ball. The videotaped lessons in grade 8 and 9 clearly show that the strengthening of already existing conceptions mainly occurred in chemistry education where electron clouds are offered to the students but not integrated into the knowledge structure represented by the propositions in the concept map.

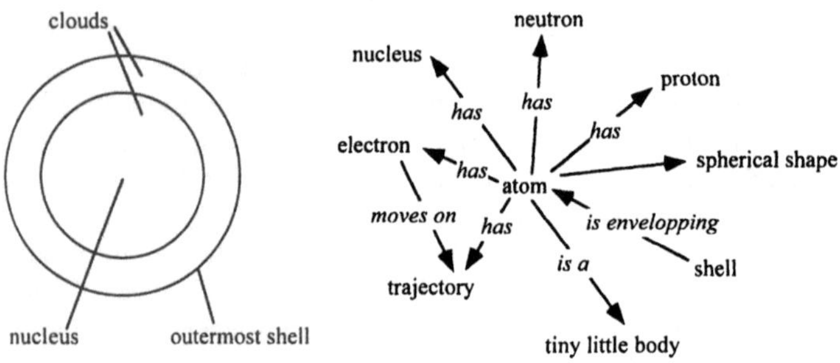

Fig. 2: The same student's drawing and her concept map in grade 9

All students – results and examples

In the questionnaire, all students had to comment upon a set of statements that reflect an understanding of particles and an atom's structure. One example is given (Fig. 3) to demonstrate the development of students' concepts of 'atom' in relation to microscopic and macroscopic properties (which was central to the analysis of the single student's concept maps and drawings).

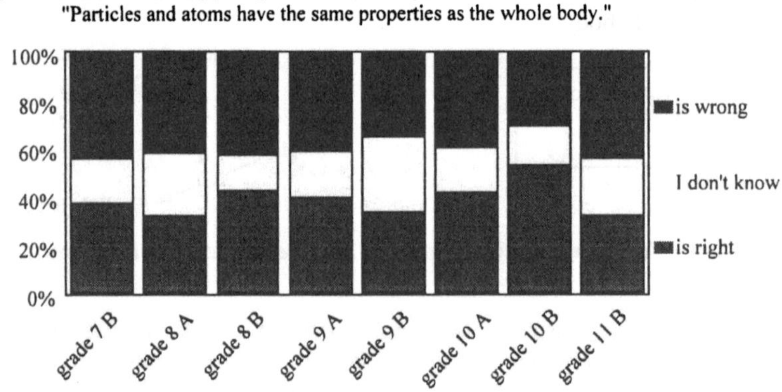

Fig. 3: Students' comments on a statement about properties of atomic particles

The diagram (Fig. 3) does not clearly show a definite development. But it seems that from the middle of grade 9 until the end of grade 10 instruction reinforces a macroscopic view. Models, at this stage of teaching, were introduced without discussing their characteristic features. It is interesting to note that in grade 11 the number of correct answers increases although particle models are not a topic of the syllabus in both subjects physics and chemistry.

On stability and structure of students' conceptions

Peuckert & Fischler's research project "Stability of students' conceptions concerning particle models" focuses on how students' particle conceptions change over a period of time when particle topics are *not* the primary focus of their current science education experiences. Among others, this question aims to describe the stability of students' knowledge: i.e., the efficiency of instruction in terms of types of learning behaviour. To satisfy this last point, the study focused on two types of learner: students *with a metaconceptual awareness* of at least some concepts within the field of particle models and "context-bound" students, who adapt scientific conceptions (e.g., by rote learning) but do not consciously relate them to real world phenomena. These two groups were selected by analyzing video transcriptions of former science lessons and by using suitable questionnaires.

Findings suggest that metaconceptual awareness (found in 15% of all students) is accompanied by a comprehensive, precise, and coherent knowledge, especially in the stability of the majority of conceptions and by some stability of microscopic thinking. However, this metacognitive orientation does not protect students from misunderstandings. In addition, it turns out that differences between the groups of students are smaller than were expected. This can be understood if we assume that metaconceptual awareness is related to very restricted parts of understanding but is not a comprehensive characteristic of the learner as anticipated at the beginning of the study.

For instance, some students were found to be explicitly aware of consequences that would result from thinking in terms of the continuum theory but had no idea about how to interrelate macroscopic and microscopic properties. Others, who showed epistemological sensitivity, had problems in applying particle concepts to everyday phenomena. Nevertheless, data analysis reveals interdependences between the nature of conceptions.

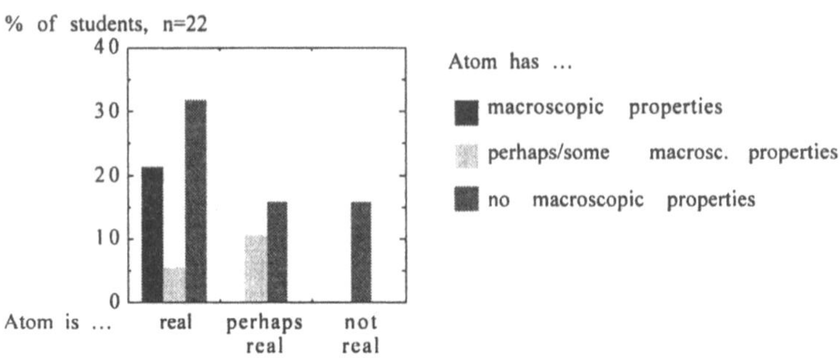

Fig. 4: The relationship between ontological positions and attitudes towards the properties of an atom

Both groups held only a limited number of alternative conceptions, but students without metaconceptual awareness often argued less systematically and used simultaneously held additional conceptions that were elicited by the situational context. Fig 4 is an example of the relationship between the ontological positions concerning atoms and the attitude towards the properties of an atom.

As to macroscopic thinking and (microscopic) understanding of models: applying continuum theory or assigning macroscopic properties to microscopic particles is *not* a matter of being a member of one of the groups. It seems to be the result of the way students were selected, which was based on the assumption that a sum of scattered indications of metaconceptual awareness should allow a comprehensive judgement on the student. Findings suggest that, above all, ontological and epistemological views are important parameters for macroscopic thinking.

In general, data suggest that science education has a greater and more stable impact on students with metaconceptual awareness. Furthermore, the knowledge of these students is more stable even if it is not the result of science education (cf. Peuckert 1999).

A multidimensional approach: level system of representations

Peuckert's research (1999) shows that metaconceptual awareness is a key qualification for overcoming misconceptions about the microworld. The Berlin working group, therefore, developed a "level system of multiple representations" both as a multidimensional curricular approach to the understanding of particle models and as a means for analysing teaching and learning processes. The curricular intention in using the "level system" is a flexible modelling of the microworld (in which metaconception means an active dealing with the two worlds: real world and model world), and developing an awareness of the existence of these two worlds. Thus, macroscopic and microscopic phenomena should be distinguished systematically. The "level system" as a descriptive tool allows for such metaconceptual distinction.

The system contains two areas: the world of reality and the world of models. Within these areas, different sections can be distinguished, that in the following representation are called levels. The arrangement of the levels, however, does not express any hierarchy. The superordinate level of metaconceptual reasoning reflects explicit argumentations about individual levels or combinations of them. The system, in essence, is a hypothetical construct derived from teaching experience. The system, as shown in fig. 5, can be used in different ways. When planning lessons or constructing a curriculum, the "level system" can serve a guidance function. E.g. every level should be touched during the treatment of a phenomenon in a series of lessons. When teaching according to the "level system", it can be used as an explicit or implicit point of reference. When analysing lessons, the "level system" is a tool for characterising the teaching and learning processes.

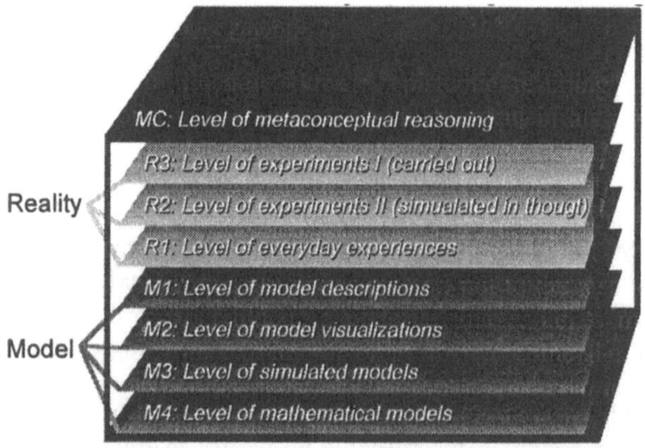

Fig. 5: Level system

Seifert developed a teaching sequence addressing all of the levels step by step. The teacher initialised the problem by showing a poster distinguishing between 'reality' and 'models':

Similarly, if applied to Buck's teaching approach (e.g. as described by him in 1986), a level diagram would reveal that his approach is more or less a "top-down" approach, beginning with metaconceptual reasoning and omitting the levels R2, M3 and M4. Research on the benefits of such a level system both for curriculum design and description is still in progress.

Major factors in understanding the particle model adequately

As mentioned the *ability of metaconceptual reasoning* is one major factor in understanding the particle model adequately (for more detail see Peuckert, 1999). Another factor concerns the *order in which concepts are learnt.* Johnson's findings in a three year longitudinal study, which explored the development of children's understanding (from ages 11 to 14) of a number of ideas relating to the basic concepts of chemistry, indicated that children's understanding of chemical change is dependent on their understanding of particle ideas. Only those pupils who understood the properties of state adequately as *collective* properties of particles also developed an acceptance and understanding of chemical change. Given that research shows a widespread failure of pupils to grasp particle ideas, the extremely poor understanding of chemical change (also well established, cf Andersson 1990, Garnett, Garnett & Hackling 1995, Johnson 1999) is only to be expected.

Johnson placed the interviewed pupils in three broad 'chemical change' categories:

Category A: The pupil shows very little grasp of the idea of chemical change as a phenomenon and, therefore, by default, has no understanding of how it is that it can happen. This is characterised by only considering copper oxide in terms of its perceived ingredients and by considerable uncertainty in the application of particles ideas, if at all, to the composition of water.

Category B: The pupil appreciates what a chemical change is 'supposed to be', but still shows signs of difficulty in coming to terms with the idea and/or has yet to fully grasp a basic understanding of how it can happen. This is characterised by some doubt surrounding the conception of copper oxide as a substance in its own right and/or a non molecular particle view of the composition of water.

Category C: The pupil seems to have fully accepted the idea of chemical change as a change of substance and has a sound, basic understanding of how it is that it can happen. This is characterised by an understanding of the reaction between copper and oxygen as a combination of the two substances, and a firm conception of copper oxide as a substance in its own right. The pupil is also able to relate a molecular view of the composition of water to its singularity as a substance.

For each of these chemical change categories, 11 testees' responses were then correlated with the categories he had formed for the pupils' 'basic' particle ideas.

Model X: Continuous substance.

Particle ideas have no meaning. Nothing that resembles having particles of any description is drawn.

Model A: Particles *in* the continuous substance.

Particles are drawn, but the substance is said to be between the particles. The particles are additional to the substance. There can be varying degrees of 'profile' for the particles (weak to strong) and of association with the substance (none to close).

Model B: Particles *are* the substance, but with *macroscopic* character.

Particles are drawn and are said to be the substance. There is nothing between the particles. Individual particles are seen as being of the same quality as the macroscopic sample - literally small bits of it.

Model C: Particles *are* the substance, properties of state are *collective*.

Particles are drawn and are said to be the substance. The properties of a state are seen as collective properties of the particles.

Fig. 6 shows this correlation:

	Chemical change category		
	A N = 11	B N = 11	C N = 11
Particle Model A	5		
Particle Model B	5	9	
Particle Model C	1	2	11

Fig. 6: Basic particle model against chemical change category at interview 5.

The coincidence of pupils in both category Cs, of itself, is unremarkable - these are pupils who made good progress in all aspects of the study. Equally, it is not surprising to find pupils in A:A. The interest lies in the absence of any pupil with a particle model B in chemical change C. These pupils did not use ideas of bonding between atoms to explain chemical change and were still uncomfortable with the phenomenon. This raises the question of whether a particle model B *inhibits an understanding* of chemical change. There was evidence to suggest this might be the case. Particle model B is linked with the idea of three different *types* of substance - 'solids', 'liquids' and 'gases'. A passage from an interview concerning the reaction of copper with oxygen is given below:

I ..... what do you think we mean by reacts?
P that .. um .. like makes it change state or makes it change .. change in some way
I yes .. makes it change in some way .. so what do you think becomes of the oxygen in this reaction
P the .... where it's got hot .. I think it's turned into a .. um solid or .. liquid or something like that
I what the oxygen has
P yes .. don't think it can be a gas mixed in with the solid
I you can't have a gas mixed in with a solid
P no I don't think so .. I think it's turned into a solid or liquid
I the oxygen does
P yes
I and what becomes of the copper
P um the copper's still there just been broken down

Here the pupil talks of the oxygen turning into a solid or liquid so that it can mix with the copper: there is no sense of an interaction to form a new substance. An interpretation of his response is as follows. He has experiences of solids and liquids mixing together, but he cannot see how 'a gas' could mix with 'a solid'. Therefore, he speculates that the oxygen might turn into a solid or liquid, although there is no reason why this might be (especially on heating). This idea of mixing, at least, was some advance on his response at an earlier interview, where the oxygen simply 'went off' after it had changed the copper. He had started to think of oxygen as a substance and 'knew' that the copper and oxygen were together in the copper oxide, but he could not see how this could be. Pupil 7 was a very interesting pupil, because he seemed to embody many of the difficulties experienced by other pupils and he also reflected at great length. He was interviewed again (in a "teaching" interview) and after first establishing that he still had a particle model B and still the problem of understanding how a 'gas' could combine with a 'solid', the interview moved into its intervention phase. Here the pupil was encouraged to think in terms of invariant particles for a change of state (boiling) and different inherent attractions between particles accounting for different substances in different states. It ended thus:

P I was getting like (yes) the particles mixed up with the actual substances itself (yes exactly) .. the state .

I that's right .. now .. can we see - now does that help us to think how . um .. what do you think this copper oxide is like

P ..... cop .....

I how can it .. so we don't have the problem to think of it [oxygen] being a gas because in a sense all we just think of - these are .. these are particles (ah ha) they're just a bit further apart .. it doesn't

P the heating up of the ox (yes) the oxyen and the copper reacting (yes) has just made the par .. the par - the attraction between the copper particles and the oxygen particles strong (yes) which has formed a solid - solid powder (right . yes) the oxygen and the copper (yes) have like stuck together and attracted to each other and that's how the oxygen's in the copper oxide

I exactly .. does that make sense

P yes .. that helps a lot

There is still the issue of copper oxide as a substance in its own right to be resolved, but it seems quite clear that the idea of different types of substance, and hence particles, was hindering progress with the idea of chemical change.

Understanding, in such a discussion, is influenced by the "linguistic environment", i.e., the metaphoric (connotations) by load of the words and phrases used in it. The metaphoric content of a word might be different in different languages. Buck, in his contribution to the symposium, compared (among others) the English, German and Nguni word for 'oxygen' and analysed its (impeding or supporting) effects on understanding the basic concepts of chemistry: In English it is "only" a question of grasping a difference between 'the substance oxygen' and 'the element oxygen' (which, as Johnson has shown, is difficult enough). In German, however, due to the very deplorable translation of Lavoisier's 'principe oxygènique' into 'Sauerstoff' ('acid substance') students ever since have great difficulties to distinguish between 'der Stoff Sauerstoff' ('the substance sour-substance') and the quite conflicting 'das Element Sauerstoff' ('the element sour-substance'). In Nguni, due to Dlodlo's translation which keeps close to Lavoisier's idea of naming (Dlodlo, 1999), 'iMpiliso' ('which sustains life') is closer to an adjective and can be; therefore, combined with the noun 'substance' equally well as with the noun 'element'. Thus, alleviating the very difficult learning process of understanding chemical change.

References

Andersson, B. (1990). Pupil's conceptions of matter and its transformations (age 12-16). *Studies in Science Education* 18, 53-85.

Brook, A., Briggs, H. & Driver, R. (1984). *Aspects of secondary students' Understanding of the particulate nature of matter.* Leeds: Centre for Studies in Science and Mathematics Education. The University of Leeds.

Buck, P. (1987). Der Sprung zu den Atomen. *physica didactica* 14, 41-45.

Buck, P. (1990). Jumping to the atoms. In P.L. Lijnse, P. Licht, W. de Vos & A.J. Waarlo, *Relating macroscopic phenomena to microscopic particles: a central problem in secondary science education* (*pp*. 212-219). Utrecht, The Netherlands: CDß-Press.

Buck, P. (1997). About pitfalls in the road to scientific literacy. In W. Gräber & C. Bolte, Eds., *Scientific literacy* (*pp*. 217-246). Kiel, Germany: IPN.

CLIS (1987). *Approaches to teaching the particulate theory of matter*. Leeds: Centre for Studies in Science and Mathematics Education. The University of Leeds.

Dlodlo, T.S. (1999). Science Nomenclature in Africa: Physics in Nguni. *Journal of Research in Science Teaching* 36, 321-331.

Duit, R. (1992). Teilchen- und Atomvorstellungen. In H. Fischler, Ed., *Quantenphysik in der Schule* (*pp*. 201-214). Kiel, Germany: IPN.

Fischler, H., Lichtfeldt, M. & Peuckert, J. (1998). Wege zum Atombegriff I - III. In H. Behrendt, Ed., *Zur Didaktik der Physik und Chemie* (*pp*. 349-357). Alsbach, Germany: Leuchtturm.

Garnett, P., Garnett, P. & Hackling, M. (1995). Students' alternative conceptions in chemistry: A review of research and implications for teaching and learning. *Studies in Science Education* 25, 69-95.

Heisenberg, W. (1970). *Das Naturbild der heutigen Physik*. Hamburg, Germany: Rowohlt.

Johnson, P.M. (1998a). Progression in children's understanding of a 'basic' particle theory: A longitudinal study. *International Journal of Science Education* 20, 393-412.

Johnson, P.M. (1998b). Children's understanding of changes of state involving the gas state. Part 1: Boiling water and the particle theory. *Int. J. Sci. Ed.* 20, 567-583., part 2: Evaporation and condensation below boiling point. *Int. J. Sci. Ed.* 20, 695-709.

Johnson, P.M. (1999). Children's understanding of substances, Part 1: Recognising chemical change. Paper submitted to the *International Journal of Science Education*.

Lee, O., Eichinger, D., Anderson, C., Berkheimer, G. & Blakeslee, T. (1993). Changing Middle School students' conceptions of matter and molecules. *Journal of Research in Science Teaching* 30, 249-270.

Lichtfeldt, M. (1996). Development of pupils' ideas of the particulate nature of matter - Long-term research project. In G. Welford, J. Osborne & P. Scott, Eds., *Research in Science Education in Europe. Current issues and themes* (*pp*. 212-228). London, Washington, D.C.: Falmer Press.

Millar, R. (1990). Making Sense: What Use are Particle Ideas to Children? In P.L. Lijnse, P. Licht, W. de Vos & A.J. Waarlo, *Relating macroscopic phenomena to microscopic particles: a central problem in secondary science education* (*pp*. 283-293). Utrecht, The Netherlands: CDß-Press.

Novick, S. & Nussbaum, J. (1981). Pupils' Understanding of the Particulate Nature of Matter. *Science Education* 65, 187.

Peuckert, J. (1999). Stability and nature of students' ideas concerning particle models. In M. Méheut & G. Rebmann, Eds., *Theory, methodology and results of research in science education (pp.* 150-156). Paris: European Science Education Research Association.

Pfundt, H. (1981). The atom: Ultimate piece in a division process or the first building bloc? *chimica didactica* 7, 75-94 (in German).

Scott, P.H. (1992). Pathways in learning science: A case study of the development of one students' ideas relating to the structure of matter. In R. Duit, F. Goldberg & H. Niedderer, Eds., *Research in physics learning: Theoretical issues and empirical studies* (*pp*. 203-224). Kiel, Germany: IPN.

Vollebregt M.J. (1998). *A problem posing approach to teaching an initial particle model*. Utrecht, The Netherlands: CDß-Press.

Vollebregt, M., Klaassen, K., Genseberger, R. & Lijnse, P.L. (1999). Inzichtelijk een deeltjesmodel leren. *Tijdschrift voor didaktiek der ß-wetenschappen* 16, 12-26.

Evolving Mental Models of Electric Circuits

Melvin S. Steinberg
Physics Department, Smith College, Northampton, USA
John J. Clement
University of Massachusetts, Amherst, USA

Abstract

This is a case study of a 16 year old student, alias "Susan", discussing her thinking with a tutor during hands-on investigations of electric circuits. Susan had not studied any physics. To facilitate complex learning, her experiments were designed to foster model building in a series of small steps. Each step revised only one of Susan's alternative conceptions, so that most of her model was always available to support reasoning during an episode of conceptual change

Strategy of instruction

In this short report we outline our model of Susan's learning process.[1] Figure 1 shows the role of three conceptual change facilitators in each step of the process.

D = Discrepant Event: challenges a single conception in the existing model

A =Analog Conception: explains the event when transferred into the model

O = Observational Link: constrains revision and supports the revised model

The large white circles in Figure 1 labelled M1, M2, M3, M4, M5 represent stages of Susan's evolving explanatory model. The small grey circles in M1 represent four initial alternative conceptions that were revised during the course of her tutoring interviews. A colour change from grey to black indicates revision to a more expert-like conception. The arrows indicate contributions to model revision, that we have inferred from video taped tutoring interviews. One consideration in designing Susan's experiments was to introduce discrepant bulb lighting events, that would conflict with her alternative conceptions. The first two discrepant events were provided by inserting a capacitor in battery-and-bulb circuits, in order to reveal the presence of a *non*-battery origin of charge and a *non*-battery current driving agent. Conflicts with Susan's alternative conceptions are represented in Figure 1 by jagged lines.

The experiments were also designed to introduce observational links that would constrain the direction of model revision by supporting the plausibility of a particular analogy. These links were provided by deflections of a magnetic compass needle that was placed under each of the circuit wires. Susan interpreted the deflections as indicating the directions of movement and rates of flow for "charge" moving through the wires of an operating circuit.

---

[1] A more detailed analysis, which includes Susan's development of imageable models, is given in Clement and Steinberg (to appear).

H. Behrendt et al. (eds.), Research in Science Education – Past, Present, and Future, 235–240.

Diagram of model building

The video tapes of Susan's investigations and discussions with the tutor show her generating the four steps of model construction shown in Figure 1. The descriptions of these steps are expanded below, using quotes from Susan's transcript.

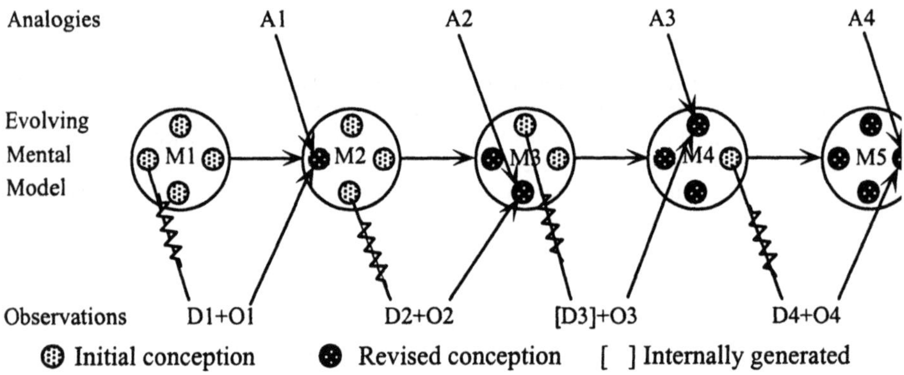

Figure 1

M1: Charge moving through the circuit wires originates only in the battery. –> **M2**: The moving charge comes also from metal in the capacitor plates.
M2: The battery is the only causal agent that can make this charge move. –> **M3**: Pressure in compressed charge can also make the charge move.
M3: NORMAL pressure in a capacitor plate cannot push any charge out. –> **M4**: NORMAL pushes charge into LOW pressure in a battery terminal.
M4: Flow rates into and out of a wire will be the same in all circumstances. –> **M5**: Inflow and outflow may differ, which will alter pressure in the wire.

STEP #1 – Capacitor charging experiment (task set by tutor)

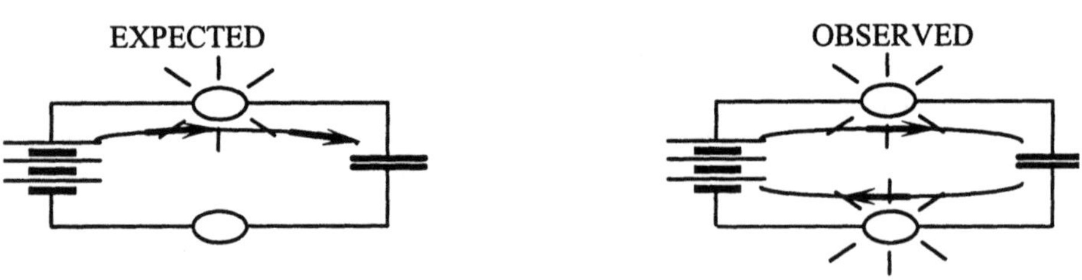

Figure 2

D1 Bottom bulb lights without a conducting path to it from the top of the battery.
O1 Flow direction implies origin in the capacitor "where it can't be coming from".

STEP #1 – Concept Transfer (analogy discovered by student)

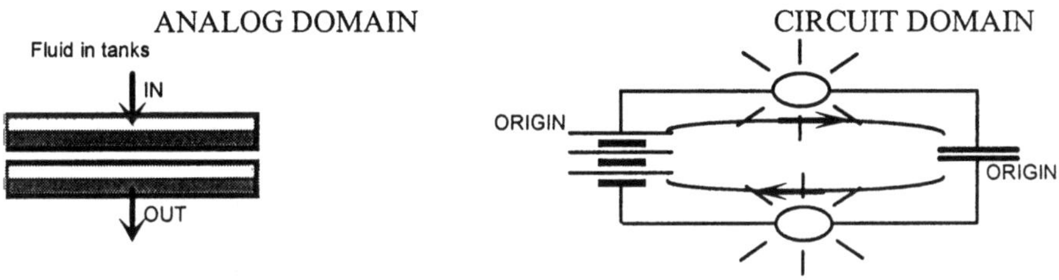

Figure 3

A1 Outflow from the lower tank can occur because *the tank contains a fluid.*
M2 Outflow from the lower plate can occur because *the plate contains charge.*

STEP #2 – Capacitor discharging experiment (task set by tutor)

Figure 4

D2 Motion occurs without a battery to cause it: "there's no place to get it to go".
O2 Movement is "in the other direction", in comparison to capacitor charging.

STEP #2 – Concept Transfer (analogy introduced by tutor)

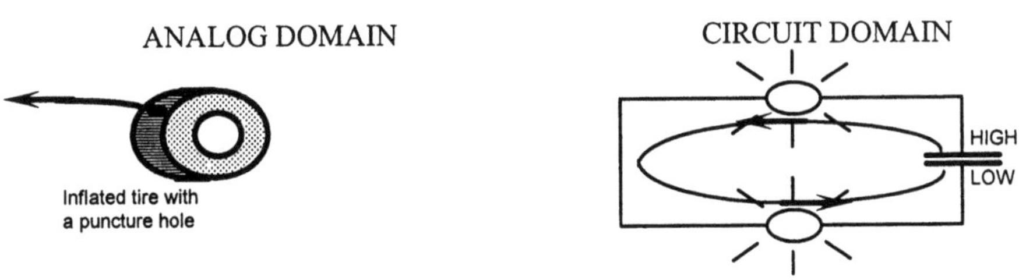

Figure 5

A2 *"Pressure" in an inflated tire* will push air out through a puncture hole.
M3 HIGH *"pressure" in compressed charge* pushes charge toward LOW.

STEP #3 – Capacitor charging revisited (no intervention by tutor)

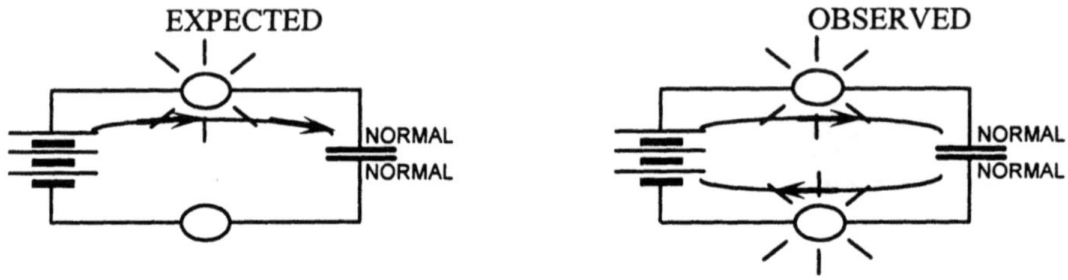

Figure 6

[D3] Lower bulb lights without HIGH pressure in lower capacitor plate causing it.
O3 Flow direction implies agent makes flow "leave here and go to the battery".

Susan's surprise in step #3 was generated internally. We think of it as precipitated by "mentally observing" a circuit representation that now includes pressure as well as visual elements and by failing to attribute causal agency to NORMAL pressure.

STEP #3 – Concept Transfer (analogy introduced by tutor)

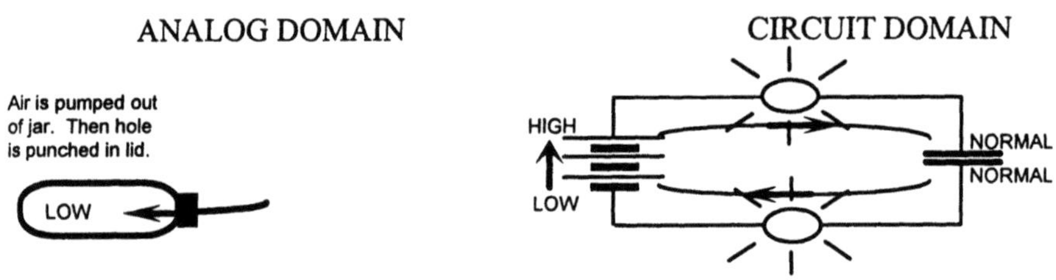

Figure 7

A3 *LOW pressure in the jar* will allow NORMAL outside pressure to push air in.
M4 *LOW in bottom end of battery* lets NORMAL in bottom plate push charge out. Battery moves charge upward, causing LOW in bottom and *HIGH in top end*.

This conceptual change episode began with Susan sensing a lack of coherence in her causal model. It ended with her adopting a revised conception of the battery as a device that maintains a pressure difference in its terminals – leading to a unified model of current propulsion in circuits based on the "electric pressure" concept.

STEP #4 – Bulb paradox experiment (task set by tutor)

Non-identical bulbs are connected in series as in Figure 8. The tutor introduced resistance by telling Susan that bulb A is "difficult" for charge to get through while bulb B is "easy" to get through. The tutor also provided a color code for pressure values, in which yellow is quantitatively midway between red and blue.

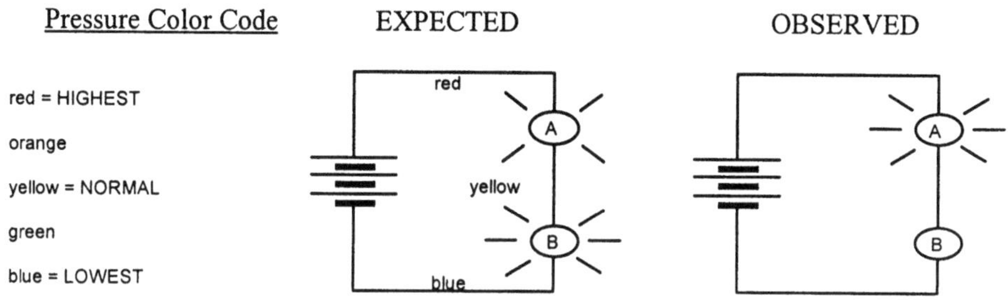

Figure 8

D4 Bulb B being out conflicts with same pressure difference across both bulbs.
O4 Same flow through A and B suggests middle wire pressure is below yellow. Susan had to overcome her intuition that inflow *always* equals outflow for a wire.

STEP #4 – Concept Transfer (analogy discovered by student)

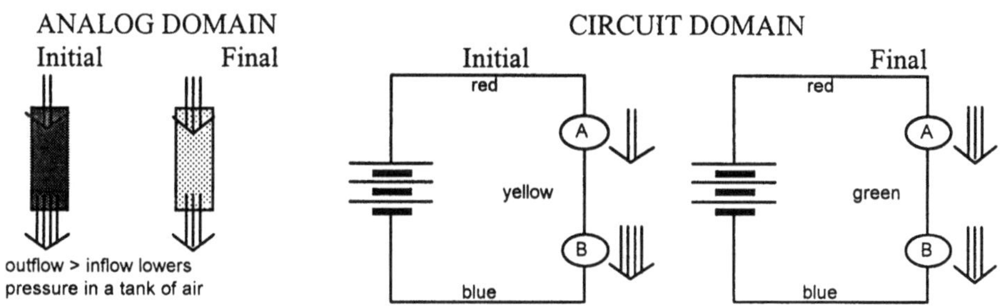

Figure 9

Asked what equal pressure differences across A and B would do, Susan replied:

"Less is coming in [to the middle wire] than is going out".

The tutor drew a narrow arrow by bulb A and a wide arrow by bulb B (see Figure 9).

"Would that really change the color of that wire then?"

The tutor invited Susan to say more about this.

"Whatever is coming through here [bulb A] would turn into orange. But there is more of it leaving [through bulb B]. So it overcompensates and gets rid of what would make it orange, but also takes even more away, which would turn it green. So I think that would make it green"

The tutor asked what will that do to movement in the circuit.

"You're going to have a larger push through here [A] and a smaller push through there [B]. Your arrows are going to change [to same width]."

A4 A wire is like a tank that can have *unequal rates of inflow and outflow*. Compression or depletion will raise or lower the pressure in a wire.
M5 Pressure change modifies inflow and outflow *until they become equal.*

Susan has now begun using electric potential – in the concrete prototype form of "electric pressure" – as *a causal-agent property of wires*, the value of which is *altered by a transient process* whenever a circuit parameter changes. She went on to use this conception to solve difficult transfer problems (Clement & Steinberg, to appear).

Discussion

Susan never lost confidence in her ability to modify her ideas when faced with anomalies. This suggests to us that she experienced "optimal dissonance" — i.e. the anomalies were discrepant enough to motivate conceptual change, but not so discrepant as to seem unexplainable and discouraging. We believe dissonance was made optimal by three qualities of Susan's instruction:

(1) Each discrepant event was designed to falsify only one feature of her model, maximizing chances to criticize and modify the model, at least partially, on her own.

(2) The observational link in each experiment was designed to constrain the search for a corresponding feature of a new model that can explain the discrepant event.

(3) The learning environment valued incremental model modification, rather than asking her to suddenly revise her ideas all the way to the final target model.

Research on discrepant events in the seventies did not provide sufficient strategies for revising models after their limitations are exposed. Later work on revision has focused on transfer of a critical causal relationship from an analog domain. But a single analogy can be insufficiently complex. White (1989), Steinberg et al (1995) and Niedderer and Goldberg (1996) have attempted to remedy this by introducing multiple analogies for electric circuits. The present paper provides an example of how model evolution, in small steps, can exploit the power of discrepant events to motivate complex learning, while avoiding the confrontational burnout or reduction of motivation that Stavy (1991) and Smith, diSessa, and Roschelle (1993) have expressed concern about in discussing the use of dissonance in instruction.

References

Clement, J. & M. Steinberg (to appear). Step-wise evolution of models of electric circuits: A "learning-aloud" case study. *The Journal of the Learning Sciences*.

Niedderer, H. & F. Goldberg (1996, April). Learning processes in electric circuits. Paper presented at the annual meeting of NARST. St. Louis, MO.

Smith, J., DiSessa, A. & Roschelle, J. (1993). Misconceptions reconceived: A constructivist analysis of knowledge in transition. *The Journal of the Learning Sciences* 3, 115-163.

Stavy, R. (1991). Using analogy to overcome misconceptions about conservation of matter. *Journal of Research in Science Teaching* 28, 305-313.

Steinberg, M. et al (1995). *Electricity Visualized – The Castle Project*. Roseville, CA: PASCO Scientific.

White, B.Y. (1989). The role of intermediate abstractions in understanding science and mathematics. In *Proceedings of the Eleventh Annual conference of the Cognitive Science Society*. Ann Arbor, MI.

Two Models for a Physical Situation: the Case of Optics. Students' Difficulties, Teachers' Viewpoints and Guidelines for a «Didactic Structure»

Philippe Colin
Laboratoire de Didactique des Sciences physiques, Université Denis Diderot Paris 7, France

Abstract

This paper focuses on difficulties linked to physical situations implying two models: geometrical optics and wave optics. The starting point is an investigation of students' difficulties, at university level, in this respect. A content analysis is presented, that underlines two main features required in order to deal with such situations: «backward selection» of paths of light and awareness of the status of the drawings. Teachers' viewpoints, collected through ten interviews, concerning students' difficulties as well as various possible presentations of the contents, make it possible to infer some likely obstacles and possible facilitating factors in the communication of an innovative teaching process in line with our content analysis. The paper ends in a discussion of such a «bridging» approach for the design of research-based teaching programmes.

Subject / Problem

Our research question concerns the difficulties that are involved in the analysis, of situations at university level that imply both geometrical and wave optics. Connecting these models of optics seems to be very difficult for students. All the more so because many teachers and textbooks do not explicitly deal with this point.

Our study starts with a content analysis of physics. This constitutes the frame of subsequent investigations bearing respectively on students' difficulties (Colin & Viennot, 1999), textbooks (Colin, 1997) and teachers' viewpoints (Colin, 1999).

We aim at providing some guidelines for the design of an innovative teaching process that could be realistically negotiated with normal teachers. The new proposals must not only achieve internal coherence (see for instance Gilbert 1998) but maximize accessibility to undergraduate students and acceptability by teachers. Some studies have already underlined the decisive part played by teachers, as potential transformers of research-based innovations (Hirn, 1998; Hirn & Viennot, 1999; Chauvet et al., 1999). We think the time is ripe to take into account teachers' main trends and views at the outset of the design of a sequence and not only after (possibly «disappointing») implementation. As learners, teachers are not empty pots to be «trained». Their own views are worth documenting thoroughly, if the suggestions made by researchers are to be adopted appropriately. Keeping this perspective in mind, in the communication of our didactic intentions to teachers, we looked for potential obstacles as well as for facilitating factors.

Very few published paper are available, specifically concerning the topic of physics. Several studies have shown (Galili, 1996) that many difficulties may occur when using the light ray model in teaching. Some authors (Dunin-Borkowski, 1993)

H. Behrendt et al. (eds.), Research in Science Education – Past, Present, and Future, 241–246.

stress the lack of connections between geometrical and wave optics without further development. Very few proposals try to build coherent bridges between these two fields of optics. L. Maurines dealt with students' «spontaneous» conceptions about diffraction (Maurines, 1999a). By contrast, we chose to work on classic situations that arise during traditional teaching, i.e. a priori far from being envisaged in a spontaneous register. In a more recent work, Maurines (1999b) seems to denounce some students' responses that we find acceptable according to our content analysis. For instance, she considers as incorrect students' comments which state that, when a non diffracting diaphragm is placed in front of a lens and illuminated by a plane wave, this diaphragm has no optical image. Such disagreements bring out the complexity of this field, and the lack of consensus on the way in which simple models, such as geometrical optics and Huyghens-Fresnel diffraction, can be integrated in a coherent frame.

Content analysis: the decisive part of « backward selection »

We took the notion of «backward selection» (Viennot, 1996; Colin, 1999; Colin & Viennot, 1999) as a pivot for our content analysis. This makes it possible to ascribe an appropriate status to what is often called, with no further specification, «a ray». We consider that a given path of light has no intrinsic status. It is the grouping of such paths that allows a given status to be ascribed to one of them.

For instance, if a diffracting object is illuminated by a monochromatic plane wave with a lens and a screen placed behind it, the screen can be located in the conjugate plane of the object plane (situation 1) or in the back focal plane of the lens (situation 2).

If the screen is placed at the conjugate position (in the absence of other diffracting devices), we get the image of the object on the screen, in a point to point replication. The grouped paths of light used in this analysis are those starting at a given point on the object and arriving at its image. They correspond to a unique spherical wave. Then, a particular path can be seen as a route for energy. More precisely, a flux of density of power is co-linear to this path. Situation 1 can be dealt with in the frame of geometrical optics.

This is not the case in situation 2. We get the diffraction pattern of the object on the screen: a point on the screen selects the paths of light of the same angular direction, starting from different points on the object. Then, the superposition of different spherical waves makes it impossible to see each of the concerned paths as a route for energy, because this is not compatible with the existence of dark fringes. This kind of selection is far from being a point to point replication. Note, however, that geometrical optics is still used to analyse what happens to a given spherical wave when crossing the lens.

It is noteworthy that a given line, if grouped with those starting from the same point on the object (situation 1), does not have the same status as if it were grouped with those starting with a given direction from different points on the object (situation 2).

Thus, it is the position of the screen that determines the relevant grouping of paths of light and accordingly their status. Numerous situations, common in higher education contexts, raise this problem.

Design / Procedure

To investigate students' difficulties, we used three questionnaires. Two of them were parts of an exam (third year of university) on Fourier optics and the third one was about the understanding of a prototypical experiment, i.e. the Young's holes.

The investigation of teachers' viewpoints is based on 10 semi-directive interviews, that consist of the following: open conversation about light rays in geometrical optics, confrontation with excerpts from textbooks about the use of «rays» in wave optics, comments on selected students' answers to the questionnaires and on certain elements of new proposals. To facilitate communication, we chose to have the teachers themselves act as judges. The interviews turned out to be between two and three hours long.

The data processing is based on the same lines that we used in our content analysis of physics: status of the paths of light and backward selection.

Data analysis and findings

Students' difficulties

The students' difficulties seem to be strongly connected with misreading of the drawings and with a lack of awareness about their status (Colin & Viennot, 1999). The students' reading often seems to be just from left to right, it is the history of a given object (light ray, or plane wave) before and after a diffracting screen. Backward selection cannot be present with this kind of reading and, consequently, the complexity of the diffracted wave seems to be completely ignored. In any situation, parallel straight lines are often thought to represent light rays of geometrical optics, which themselves constitute a plane wave. No polysemic reading of the drawings is envisaged, i.e. that a line might represent different things according to their grouping is not accepted.

One instance of such difficulties concerns a plane wave diffracted by Young's holes, followed by a converging lens and a screen placed in its back focal plane. Although a large number of students succeeded in tracing the paths of light and calculating the resulting amplitude at a point on the screen, many of them made surprising comments. Claiming, for example, that, when passing a diffracting hole, an incident ray is «deviated», as if the same entity was following its individual fate. A story-like analysis is used, from left to right, the opposite of a view in terms of backward selection of particular paths, linked with the choice of a particular point on the screen.

This example also shows that vigilance is needed concerning some textbook images, where the same arrow is put on a line before and after such a diffracting hole, thus reinforcing a common trend (Colin, 1997).

Teachers' trends: obstacles and facilitating factors

We used this example of an answer and others, together with our own suggestions, in order to provoke teachers' reactions. We focused on interrelated questions of the status of paths of light, and of backward selection to analyse the transcripts.

The former aspect seemed somewhat irrelevant to the teachers consulted. When asked whether a line on a diagram is a route of energy, the teachers considered the question to be very problematic, especially when diffraction is taken into account.

For instance, after a diffracting slit full agreement could not be reached on the status of a given line and it was still more difficult to envisage an explicitly possible multiplicity of status for such an entity, according to the grouping that is used.

To analyse teachers' ideas on the latter aspect - backward selection - we defined three levels:

- Level 1 involves of considering the arrival point of the light and selecting some of the emitted «rays» accordingly.
- Level 2 involves of accepting that, for the same device (see above situations 1 and 2), the kind of selection (point to point correspondence or not) changes with the location of the arrival point.
- Level 3 involves of explicitly putting the kind of selection in relation to a given model of optics - geometrical optics or wave - and consequently with the status of the selected paths.

The analysis of the interviews showed that:

- Backward selection is used diversely: level 1 concerns 7 out of 10 teachers; level 2 concerns two teachers; none of them reach level 3.
- Backward selection is not a common way of analysing situations in optics.
- Backward selection, even when explicitly mentioned by a given teacher for some situations, may be forgotten by the same person in other cases.

Two main lines of resistance were found, that are very similar to what was observed in students:

- story-like reading and analysis of the situation: for instance, after a diffracting slit, the lines drawn in a given direction were seemingly considered as the only protagonists of the story.
- one-to-one linkage between a set-up and a model: a lens appears as an imaging device with all the corresponding properties; what is seen on a screen may convey only one meaning - it is an image - .

As in the case of students, these obstacles were often linked with a superficial reading of the diagrams: parallel lines were said to represent a plane wave, converging lines were seen as a spherical wave, the point of convergence of such lines was, in any case, considered as a point source.

To overcome these obstacles, we decided to consider some points that are easily accepted by teachers as «facilitating factors» in the communication of our suggestions.

For instance, in the case of three Young's holes, calculation of the difference of phases corresponding to the overlapping waves convinced the teachers that parallel lines drawn after the holes do not constitute plane waves. In this case, teachers accepted that a path of diffracted light no longer represents travelling energy because of the existence of dark fringes in the interference pattern. However, when a continuous distribution of diffracting elements was envisaged, as in a rectangular slit, the same kind of awareness was not frequently observed.

Another point easily accepted by teachers involves the link between the kind of grouping of paths of light (point to point correspondence or not) and the location of the screen.

On the basis of this analysis, several versions can be submitted for our proposals, ranging from one which is likely to be easily accepted by teachers to one which would need a very sizeable previous effort of persuasion.

A hierarchy of acceptability for elements of an innovative lesson

Teachers agreement with innovative proposals is dependent on the level of need demanded concerning the use of backward selection. We reserved some aspects, that seemed difficult for teachers to accept, for more demanding formats of our proposals. For instance, the multiplicity of possible status for a given line, according to the grouping of paths used, was not often explicitly accepted. Consequently, our first (easier) version does not offer an explicit analysis of this point. In this version, backward selection is simply connected with the kind of grouping (wave or not) based on calculation of optical paths. In the second version, the status of the paths of light and the kind of grouping are linked. As explained above, the Young's experiment (two or three discrete sources) might be a facilitating factor for introducing this version of our proposals. After this first step, the case of continuous distribution as slit, grating or any kind of slide, can be introduced as an extension of the preceding situation.

For sake of brevity, we limit this account to the two first versions. More elaborated versions are proposed in (Colin, 1999).

Outcomes and general interest

We started from the viewpoint that, when implementing innovations, teachers are not passive transmitters. This is why, when designing a new research-based teaching process, it is reasonable to document teachers' views on the teaching of the corresponding domain. This approach makes it possible to a research-based teaching strategy in a «bridging» spirit, that uses possible «facilitating factors» positively as well as avoiding the most probable stumbling blocks. The first attempt described, above, leaves no doubt concerning the difficulties to be expected in the particular case dealt with and some likely facilitating factors were also found.

The outcome of such an investigation is not really the proposal of a given - best possible - sequence. We would rather say that this research provides a content analysis and some didactic elements that might be used by various people to construct, implement and evaluate their own sequence, as well as to try and convince their colleagues, a major problem especially at university level.

Our unusual focus on teachers might give an impression of pendular motion. Working with teachers, did we forget the students? In fact the whole project started from students' difficulties. But the question might be posed: are we loosing sight of the students when trying to design some innovation acceptable to the teachers? It might be the case, although we have not found proof of it, that things that sound very strange to teachers may be much easier for students and that the scale of acceptability may be different for both populations.

Such questions, obviously not addressed in this work, deserve attention and further research.

References

Chauvet, F., Colin P. & Viennot L. (1999). Images in optics and corresponding learners' difficulties : awareness and decision-making in teachers. In M. Komorek, H. Behrendt, H. Dahncke, R. Duit, W. Gräber & A. Kross (Eds.), *Proceedings of the Second International Conference of the European Science Education Research Association, Vol 2* (pp. 626-629). Kiel, Germany: IPN.

Colin, P. (1997). *Passage de l'optique géométrique à l'optique ondulatoire : l'idée de sélection par l'aval de l'information lumineuse.* Unpublished report «Mémoire de tutorat», Université Denis Diderot Paris 7 (on request from LDSP).

Colin, P. (1999). *Deux modèles dans une situation de physique : le cas de l'optique. Difficultés des étudiants, points de vue des enseignants et propositions pour structurer des séquences d'enseignement.* Unpublished thesis, Université Denis Diderot (Paris 7).

Colin, P. & Viennot, L. (1999). Les difficultés d'étudiants post-bac pour une conceptualisation cohérente de la diffraction et de l'image optique (to be published in *Didaskalia*).

Dunin-Borkowski, J. (1993). Crooked paths of straight rays. In L.C. Pereira, J.A. Ferreira & H.A Lopes (Eds.). *Proceedings of the International Conference on Physics Education GIREP'93* (pp. 333-337). Braga (Portugal): Minho University.

Galili, I. (1996). Student's conceptual change in geometrical optics. *International Journal of Science Education* 18, 7, 847-868.

Gilbert, J.K., Boulter, C. & Rutherford, M. (1998). Models in explanations, part 2 : Whose voice ? Whose ears ? *International Journal of Science Education* 20, 2, 187-203.

Hirn, C. (1998). *Transformations d'intentions didactiques par les enseignants: le cas de l'optique élémentaire en classe de Quatrième.* Unpublished thesis, Université Denis Diderot (Paris 7).

Hirn, C. & Viennot, L. (1999). Transformation of didactic intentions by teachers: the case of geometrical optics in grade 8 in France. In M. Komorek, H. Behrendt, H. Dahncke, R. Duit, W. Gräber, & A. Kross (Eds.), *Proceedings of the Second International Conference of the European Science Education Research Association, Vol. 2* (pp. 447-450) Kiel, Germany: IPN.

Maurines, L. (1999a). Students and the wave-geometrical model of the propagation of light in a three dimentional medium. In M. Banderia et al (Eds), *Selected papers from the First International Conference of the European Science Education Research Association (ESERA), Rome* (pp. 103-112). Dordrecht: Kluwer Academic Publishers,.

Maurines, L. (1999b). Spontaneous reasoning on light diffraction and coherent illumination optical imaging, In M. Komorek, H. Behrendt, H. Dahncke, R. Duit, W. Gräber, & A. Kross (Eds.), *Proceedings of the Second International Conference of the European Science Education Research Association, Vol. 1* (pp. 92-94) Kiel, Germany: IPN.

Taylor, C. (1986). L'optique retrouvée. In *New trends in Physics Teaching, Vol IV* (pp. 162-177). Paris: UNESCO.

Viennot, L. (1996). *Le rayon lumineux en optique géométrique et en optique ondulatoire.* Internal report, LDSP, Université Denis Diderot Paris 7.

The Influence of a Historically Oriented Course on the Content Knowledge of Students in Optics

Igal Galili and Amnon Hazan
Science Teaching Department, The Hebrew University of Jerusalem, Israel

Abstract

We developed an experimental course in geometrical optics that heavily incorporated historical ideas regarding the understanding of light and vision. Its design and content were guided by findings from a study, that explored the knowledge of students after a regular optics course. In a year-long experiment, we assessed the effectiveness of the new course's instruction and materials, using a facets-scheme structure of the students' knowledge. Clear advantages of knowledge after the experimental instruction provide evidence in favour of the adopted rationale.

Theoretical background

Structure of knowledge in learners

Researchers are still attempting to monitor and understand the multidimensional process of learning. Its extreme complexity caused scholars to speculate on the internal structure and constitutional elements of human cognition (diSessa, 1993). One empirical approach to infer how people learn is to study students' explanations of related physical phenomena, followed by the elicitation of the structure of such knowledge. This approach parallels inferences of the constructivist perspective on education (von Glaserfeld, 1992; Staver, 1998), which anticipates and explains differences between students' understanding of a subject and scientific knowledge about it provided in formal instruction.

Learning is knowledge construction, the establishment of a cognitive web of conceptions. During this process, the old structure is not merely expanded but essentially reconsidered, in light of both new and previously existing knowledge. One may ask whether this complex process can be described in terms of a reasonably simple design. diSessa (1993) suggested that simple, cognitive regulative forms, labelled phenomenological primitives ("p-prims"), are spontaneously created by an individual and form personal knowledge. Minstrell (1992) described students' knowledge in terms of *facets of knowledge,* that represent patterns of reasoning or strategy used by students in addressing *particular* situations, thus representing operational ideas and beliefs of children in making sense of reality. *Schemes of knowledge* were suggested as more abstract units of organization (e.g., Galili & Kaplan, 1996). Each scheme corresponds to a cluster of related facets, all sharing the same idea or explanatory mechanism, modified to match specific traits of a particular situation. Knowledge of schemes is valuable, as it can guide instruction, that efficiently addresses the fundamentals of alternative knowledge, rather than treating each particular erroneous view constructed by students, which we know can be numerous and multifaceted, as can be observed browsing the collection prepared by Pfundt and Duit (1994).

H. Behrendt et al. (eds.), Research in Science Education – Past, Present, and Future, 247–252.

Using HPS learning materials

Many science educators consider the history of science to possess a great potential for a multifaceted improvement of the learning process and its results (Brush, 1989), especially within the constructivist perspective (Matthews, 1994). Besides the importance of cultural and social perspectives, the history of science uncovers the scientific process, instead of focusing solely on final products, as is common in regular instruction. Thus, exposure to the "kitchen" of science can aid students, facilitating the meaningful adoption of currently practised scientific knowledge that they have to master. For teachers, knowledge of the historical background of the material they present in class, can elucidate its complexity. For example, the particular difficulties people had in establishing that knowledge, including unavoidable controversies, instead of the image of an exact truth. Such knowledge culturally enriches and enhances the appreciation of individual difficulties learners have. In particular, the exposure of historical "turning points" in science history can bring collisions of ideas to a class, subjective as well as objective, individual and social, that can grab the minds of a much wider variety of students and in a wider perspective, than rote problem solving. We believe that this approach, despite being less focused on traditional content, will appear as a more powerful means to facilitate students' construction of a deeper and genuine conceptual understanding of the pure scientific content (*content* knowledge) embedded in an introductory physics course.

Optics, traditionally taught in high school, was chosen as a suitable subject matter to introduce History and Philosophy of Science (HPS) content. In fact, scientific knowledge in optics is highly anti-intuitive. The particular reasons for this (Galili & Hazan, 2000), can explain the impressive abundance of students' alternative conceptions with regard to optical phenomena. These very same factors of complexity, which hinder students' learning, explain the extremely rich chronicle of optical conceptions that replaced each other during 2500 years of the documented history of science. In addition, the elicited schemes of knowledge, that students hold after regular instruction in optics, mentioned in our earlier study, guided us in choosing adequate HPS materials and also served as assessment references in our teaching experiment. This article presents the results of the follow up assessment of the new kind of instruction, focusing on the impact on students' content knowledge in optics.

Design and procedure of the experiment

Teaching resource

A specially prepared textbook served as a learning resource for the instructors and students in our experiment. Though substantially different from a regular textbook in several aspects, the text covered the whole content of a standard curriculum for high school geometrical optics. The most obvious innovation was a parallel exposure to the historical growth in the understanding of vision and the nature and behaviour of light itself. These two, interwoven trends established a permanent focus on the relationship within the triad: object, light and eye - the actors in the vision scenario. The line of instruction followed the major trend of historical progress, interweaving the inquiry of light with that of vision, appear to be equally relevant and important for the novice learner of optics.

The previously discussed (*ibid.*) historical similarity of collective (scientific) and individual (students') understanding of optical phenomena has an important implication within the constructivist perspective. Besides providing insights into the history of science, it establishes a bank of potentially useful instructional materials. The schemes of students' alternative knowledge elicited with respect to vision, nature of light, optical imaging and shadow (*ibid.*), guided us in such a search through history. Thus, we located (e.g. Ronchi, 1970; Lindberg, 1976) and incorporated in our course ancient Greek and medieval Arabic theories of vision, ideas regarding the nature of light, its expansion in space, visual and light rays, shadows and their use, reflection and refraction of light, mirror images and their nature, and speed of light. These contents, besides creating a fascinating story, can be viewed in correspondence with the "alternative" knowledge students commonly develop while learning optics (Table 1). This parallelism made the contents relevant, and thus more interesting to the students, positively influencing their learning.

Historical opus (practised in the past in science)	Scheme of knowledge (practised by students)
Pythagorean theory of vision	"Active" vision
Euclidean visual and light rays	Rays of sight, Rays of Light [Rays Reification]
Atomists' theory of "Eidola"	Image Holistic Scheme
Biblical-Medieval Dichotomy of light: Lumen-Lux	Static Light located in Light Sources, Halos, Bright Sky, Illuminated Surfaces [Light Reification as static entity]
Al-Hazen theory of vision by means of light rays	Image Projection Model

Table 1: Parallelism of views suggesting pedagogical relevance

Study chart

1. Our study comprised four stages: An introductory study, which investigated high-school students (N_C=102) knowledge immediately after taking a standard course in geometrical optics. A facets-schemes, hierarchical structure of knowledge was identified (Galili & Hazan, 2000).
2. A new course was constructed implementing the relevant HPS contents that could address the schemes of alternative knowledge.
3. The innovative course was applied in four randomly chosen 10th grade classes (N_e=141). Three regular instruction classes of an equivalent composition served as a control group (N_C=93). The study spanned one academic year, four hours a week.
4. The knowledge of the experimental and control groups was assessed and analyzed in the same way as in stage 1.

Data processing

Students accounts for different situations involving light, vision, mirror and lens images were elicited. The data were analyzed, in several steps, to identify facets and schemes of knowledge. In the first categorization, similar accounts for each particular situation were identified. These patterns of explanatory responses constitute facets of knowledge. A particular facet may result from responses to separate questions containing various contexts. After the facets were elicited, more general constructs (schemes) were identified. Each of the facets, united in a cluster, shared the same explanatory idea. A scheme usually presents a theoretical model, which can be recognized in each of the facets associated with that cluster and represents the students' understanding of reality in a more abstract form. Quantitative assessment included the coefficient FA_c for the control group and FA_e for the experimental group, that provided frequencies of each identified facet. Similarly, the popularity of schemes of alternative knowledge and of scientific conceptions, was characterized by the coefficient SA (scheme abundance) and SCA (scientific conception abundance). They were determined by taking into account the FA's contributions to each of the facets associated with the particular scheme or conception. Finally, coefficients of Scheme/Scientific Conception Abundance Difference (SAD/SCAD) reflect the difference of quality of knowledge in terms of frequency of schemes/conceptions.

Findings

Knowledge of vision

The most pronounced, the Spontaneous Vision Scheme, implies that vision is performed naturally (spontaneously and subconsciously) by the mere presence of eyes. No mechanism or agent mediating between eyes and the observed object is included. Instead, there might be a recognition of a necessity to turn the face towards the observed object, "to aim the eyes" at it, "to focus" on it. The mere location of the object in the "field of vision" is considered to be a sufficient condition for the object to be seen. Students also don't show any physical connection between the eye and the object or image in their sketches. Light is perceived as a bright object, either stationary and filling the space, or travelling in it and observed "from the side". At best, light is conceived as a necessary background, a factor helping and improving vision. The rate of facets of this Scheme in the experimental group, was considerably lower (SAD= -32%, Table 2). The scientifically correct conception of vision was also observed in a variety of facets. Yet, students of the experimental group exhibited, with a much higher frequency, an understanding of vision which included "light reflection from bodies, its rectilinear expansion and refraction within the observer's eye", and "beams of light which travel from the object and cause formation of an image on the retina".

Knowledge of the nature of light

Regarding the nature of light, we previously discussed the strong scheme of light reification, "Light is Corporeal". Within this scheme, students comprehend light as an external object, a passive subject of observation, not related to the observer. Facets in this scheme present light as "composed of light rays," a literally understood expression common in textbooks and instruction. Apparently, the "Light is Corporeal" knowledge of light is often accompanied by the understanding of vision within the Spontaneous Vision Scheme. Thus, both schemes share the facet "Light moving through space, or being stationary and filling the space, can be seen from the side". It simultaneously testifies to the understanding within both schemes of knowledge, reiterating the interdependence between knowledge of vision and nature of light. Circular definitions - light as that which causes vision, and vision as that caused by light, often in practice, may be congenial to the proximity of these schemes.

All facets of this scheme substantially decreased in the experimental group. The strongest facet, the conviction that light is a composition of material rays, was 55% lower. At the same time, this group demonstrated closer approach to the scientific conception of the nature of light. Not only naive ideas about rays received the desired refutation, but also the adoption of the idea of stationary light greatly decreased. Notions like "Light expands in the environment with a decreasing intensity until it strikes opaque objects" and "Light is the energy propagating in space in the form of beams/vibrations/waves" appeared. In all, the difference in frequency of views considered as close to the scientific one was very clear (CAD=+76%, Table 2).

Knowledge of optical imaging

Imagery presents a central subject in all optics school curricula. Its contexts include reflection, refraction, shadows and illumination. Among the strongest schemes of alternative knowledge about optical image is the Image Holistic Scheme. This knowledge is especially strong in the pre-instructed students (Rice & Feher, 1987; Bendall et al., 1993). Whereas, after instruction, the Image Projection Scheme was found very popular, seemingly replacing the Image Holistic Scheme (Galili et al., 1993; Galili & Hazan, 2000). Our data confirmed the wide use of both schemes, but especially the Image Projection Scheme, in post-instructed students.

The Image Holistic Scheme conceives the image as an entity, which replicates an object and, as a whole, can move, stay or revolve in the environment. This view commonly lacks details of how the image was formed and moved in space. The image might be observed when it reaches a screen or observer. The frequency of this scheme (seven facets found in the control group) was 17% lower in the experimental group (Table 2), where four of the facets disappeared entirely. Within the Image Projection Scheme, the formation of an image is understood as a result of mapping object-image performed by means of a single light ray per object point. Commonly, that ray travels in the "relevant" direction and "carries" the image point. The abundance of this scheme considerably decreased (-27%) in the experimental group, and some of the facets associated with the scheme totally disappeared. As to the scientifically correct conception of image formation, it appeared in a number of facets, each presenting a certain fragment of the desired knowledge. Regarding the differences between the groups, the ambiguous term "light ray" used by the control group regarding the nature of light, was "replaced" in the experimental group by the use of scientifically preferable terms "light", "light beam" and "light flux". The fundamental advantage of the scientific conception of the real image vs. naive ideas about it, is the understanding of an image as a collection of light spots, each obtained due to a converging light flux. This understanding appeared only in the experimental group.

Subject	Conceptions	Sa_c or SCA_c	Sa_e or SCA_e	SAD or SCAD	SAD_r or $SCAD_r$
Light and Vision	Spontaneous Vision Scheme	41	9	-32	-0.78
	Facets of the scientific conception	22	29	+7	+0.32
Nature of Light	Reified Light Scheme	47	7	-40	-0.85
	Facets of the scientific conception	0	76	+76	pure gain
Optical Im-agery	Image Holistic Scheme	36	19	-17	-0.47
	Image Projection Scheme	39	12	-27	-0.69
	Facets of the scientific conception				
	overall regarding optical images	9	33	+22	+2.4
	- by a converging lens:	0	44	+44	pure gain
	- by a plane mirror:	19	23	+4	+0.21

SAD_r - Relative Difference of Abundance of holding a Scheme of Knowledge (SAD_r = SAD/SA_c); $SCAD_r$ - Relative Difference of Abundance of holding a Scientific Conception ($SCAD_r$ = $SCAD/SCA_c$)

Table 2: Comparison of knowledge between the sample groups.

Regarding the plane mirror, most control group students apparently divide the process of image observation into two separate processes: image formation and subsequently and independently, image observation. This approach is incompatible with the scientific idea of virtual image. In contrast, many students of the experimental group demonstrated a more mature knowledge by defining the mirror image as an optical illusion, and elaborated on its formation on the retina (Table 2).

Conclusion

Our study provides evidence that instruction based on historical materials can be applied in a regular tenth grade class. It apparently has a positive and multidimensional effect on students' learning of physics. The abundance of schemes of alternative knowledge decreased following such instruction, and new knowledge, conforming to the scientific one appeared with much greater frequency. This shows that instruction of the suggested kind may be appealing to regular students, although we registered a certain feeling of unfulfilled expectations amongst those students who prefer a traditional style of learning emphasizing problem solving. The success of the experimental teaching can be interpreted as confirming the rationale implied in its design regarding the learning materials. The theoretical inferences may be that; firstly, representation of students' content knowledge (in optics) is possible in terms of a facets-scheme hierarchical structure, as it was previously suggested regarding other domains of knowledge (weight and gravitation, Galili & Kaplan, 1996, seasons and illumination, Galili & Lavrik, 1998); secondly, the support for the assumption that science educators should, in their activities, address the schemes of students' knowledge. This is preferable to challenging the numerous situated pieces of alternative knowledge; thirdly, repudiation of the undesirable schemes of knowledge is possible, deliberately chosen relevant historical materials are capable of prompting the construction of knowledge, that is closer to the scientific one.

References

Brush, S. J. (1989). History of science and science education. *Interchange* 20(2), 60-70.

Bendall, S., Goldberg, F. & Galili, I. (1993), Prospective elementary teachers' prior knowledge about light. *Journal of Research in Science Teaching* 30(9), 1169-1187.

diSessa A.(1993).Toward an epistemology of physics. *Cognition and Instruction* 10, 105-225

Galili, I., Goldberg, F. & Bendall, S. (1993). Effects of Prior Knowledge and Instruction on Understanding Image Formation. *Journal of Research in Science Teaching* 30(3), 271-303

Galili, I. & Hazan, A. (2000). Learners' knowledge in optics: interpretation, structure and analysis. *International Journal of Science Education* 22(1), 57-88.

Galili, I. & Kaplan, D. (1996). Students' operation with the concept of weight. *Science Education* 80 (4), 457-487.

Galili, I. & Lavrik, V. (1998). Flux concept in learning about light. A critique of the present situation. *Science Education* 82(5), 591-614.

Glaserfeld, E. von. (1992). A constructivist view of learning and teaching. In R. Duit, F. Goldberg & H. Niedderer (Eds.), *Research in physics learning: Theoretical issues and empirical studies (*pp. 29-40). Kiel: IPN.

Lindberg, D. C. (1976). *Theories of vision from Al-Kindi to Kepler*. Chicago: The University of Chicago Press.

Matthews, M. (1994). *Science teaching: The role of history and philosophy of science*. Routledge, New-York.

Minstrell, J. (1992). Facets of students' knowledge and relevant instruction. In R. Duit, F. Goldberg & H. Niedderer (Eds.), *Research in physics learning: Theoretical issues and empirical studies (*pp. 110-128). Kiel: IPN.

Pfundt, H. & Duit, R. (Eds.) (1994). *Bibliography: Students' alternative frameworks and science education*. IPN, Kiel.

Rice, K., & Feher, E. (1987). Pinholes and images: children's conceptions of light and vision. I. *Science Education* 71, 629-639.

Ronchi, V. (1970). *The Nature of Light*. Harvard University Press, Cambridge.

Staver, J. R.. (1998). Constructivism: Sound theory of explicating the practice of science and science teaching. *Journal of Research in Science Teaching* 35, 501-520.

Using Everyday and Scientific Conceptions for Developing Guidelines of Teaching Microbiology

Catja Hilge
University of Oldenburg, Germany

Abstract

The task of the study presented here is to investigate students' and scientists' conceptions to obtain a solid basis for teaching of microbiology subjects at school. Accordingly the study is conducted within the model of Educational Reconstruction, including three components of research: Investigation of Students' Conceptions, Scientific Clarification and Construction of Instruction. The mutual comparison of the students' and the scientists'conceptions shows different kinds of correlations. Besides particularities and limitations, similarities and congruities between students'and scientists' conceptions are identified. On the basis of the correlations, guidelines for teaching microbiology as well as some basic principles for teaching biology are formulated.

Subject

The motivation to develop a project dealing with microbes as part of instruction in biology was caused by the fact, that microbiology and microbial processes concern everyday life as well as modern scientific research. However, microbiology plays only a subordinate role in German curricula so far, although there is no doubt about the fundamental role that the microorganisms play within the economy of nature. There is a marked demand to meet students' conceptions in an adequate and fruitful way to make possible a fundamental understanding of contents of microbiology. Therefore, one of the most important tasks of the project was to find out which conceptions of microorganisms students have and how they understand processes, in which microorganisms are involved (cf. Bayrhuber & Stolte, 1997).

Design

The model of Educational Reconstruction was used to develop guidelines for a better understanding of microbiology as it brings together empirical investigations of students' conceptions and the analysis of the scientific content structure (Gropengießer, 1997; Kattmann et al., 1998). Students' conceptions as well as scientific conceptions were both used in order to develop principles and guidelines for a better understanding of the subject. In accordance with this research model the project consists of three interrelated components: the Scientific Clarification, the Empirical Investigation of Students' conceptions and the Construction of Instruction.

H. Behrendt et al. (eds.), Research in Science Education – Past, Present, and Future, 253–258.

Procedure

a) Empirical investigation of students' conceptions

Empirical investigation of student conceptions aims to discover student conceptual frameworks (notions and principles) regarding their ideas on different aspects of the microbes and issues in which they are involved. The following questions were, therefore, relevant:

- What idea do the students have of micro-organisms?
- How do they explain processes caused by micro-organisms?
- How do they connect processes and organisms?
- How do they argue different aspects with each other?
- Which role do they assign to the micro-organisms in the economy of nature?

Ten students of grades 11 to 13 (16- to 18-year-old) from German Grammar Schools (Gymnasium) were interviewed individually. The interviews were carried out in an open and problem centred format. A guide for the interviewer contained several main questions, that determined the rough course of the interview. The interviews were audio taped. A transcript was made and revised in a five-step-process, following the qualitative content analysis method. The students' notions and principles, as elements of the students' conceptions, were identified named and put into a structured sequence.

b) Scientific clarification

Scientific Clarification deals with an analysis of the scientific content with a view to detecting the basic qualitative ideas of the scientists and their relationships. The leading questions were the same as in the empirical investigation of students' conceptions.

The following publications have been relevant: Monographs by Christian Gottfried Ehrenberg (1838), Justus Liebig (1848), Louis Pasteur (1861), and Ferdinand Cohn (1866, 1872, 1876). "Mikrobiologie" by H.G. Schlegel (1994) and "Biology of microorganisms" by T.D. Brock served as leading up-to-date scientific textbooks.

Qualitative content analysis (Mayring, 1990) was adapted for the Scientific Clarification of monographs as sources of the history of science and of leading scientific textbooks. Scientific theories and concepts on microbiology and microbial processes, their genesis, function and meanings were of great importance to this aim besides, the use of terms and their meaning in learning processes. In this way the main theories in the history of science and related notions and principles have been identified and analyzed.

c) Construction of Instruction

After the Empirical Investigation of Students' Conceptions and the Scientific Clarification had been carried out, the conceptions from both sides were brought to the same level of complexity, namely notions and principles. This allowed comparisons between both sides. Thus, relations could be drawn between the results of the Empirical Investigation of Students' Conceptions and Scientific Clarification. Students' and scientific notions and principles were correlated to each other.

Consequently, the conditions that have to be arranged in order to support learning microbiology were discussed in this part of the investigation (Baalmann et al., 1999).

Within the third step, the main result of the project was the production of seventeen guidelines for teaching microbiology and biology in general.

Findings

Processes of decomposition

Different phenomena of decomposition were mentioned during the interviews. A considerable proportion of the notions and principles, concerning decomposition of organic matter, found in the interviews, was on an abiotic level. Even if micro-organisms were included in the students' conceptions, notions concerning abiotic processes still seem to be dominant. These results support the findings of Helldén (1999). Accordingly, the students explained the corresponding phenomena mainly by physical and chemical processes, e.g. by mechanical destruction or oxidation.

Also in the scientific works of Justus von Liebig and Gottfried Christian Ehrenberg abiotic patterns of explanation play a major role in processes of decomposition. Gottfried Christian Ehrenberg developed the conception of mechanical decomposition, while Justus von Liebig explained decay by processes of oxidation. In this way similarities were clearly observable between the students' and the historical scientific conceptions. With regard to the aim of the project, namely to improve instruction of biology, the following guideline was deduced from this observation:

To develop purely abiotic patterns of explanation to biotic ones within the interpretation of processes of decomposition.

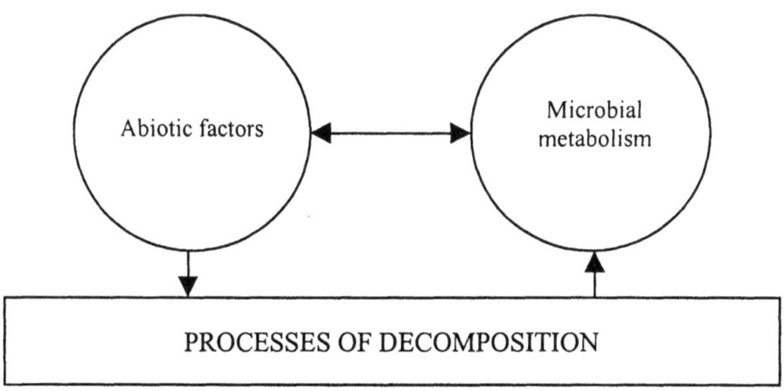

***Fig. 1:* The development of patterns of explanation from purely abiotic ones to those, that also include aspects of the microbial metabolism.**

In figure 1 those components are underlayed white, which were observed in the interviews with the students. Mainly abiotic factors were mentioned regarding the decomposition of organic matter. From that point of view, the aim of teaching microbiology is to give the opportunity to students to develop also conceptions for the biotic interpretation of those processes. This aim is marked grey in figure 1. The abiotic factors can be used in a fruitful way by connecting them with living organisms. In this way a new meaning is assigned to the students' conceptions, so that they do not have to be rejected as wrong or incorrect. Instead, a broadening and new assignment of meaning takes place.

Microorganisms as living beings

All students interviewed described micro-organisms as microscopic organisms, that are invisible to the naked eye. This conception led many of the students to the conclusion that there is a close relation between the size and complexity of the organisms. Because of the micro-organisms' small size, the students characterized them as less complex than macroscopic organisms.

This conception has a guiding function towards notions that propose that bacteria are invariably simple-structured organisms. Some students assume that they have only an insignificant metabolism and that certain structures are even missing which exist in macroscopic organisms.

The view of micro-organisms as simple-structured organisms also has fundamental consequences in the fields of microbial processes and the economy of nature. Especially the simplicity of metabolism is seen by many students as a reason for the only possible function of the micro-organisms to be decomposers. This is expressed by another relation between the structure and the function of the micro-organisms. In this way, most students describe the microbial metabolism as limited to the decomposition of organic matter. They cannot think of bacteria as simple organisms building up complex organic compounds. Bacteria are seen as organisms that are always and only tearing down structures, built up by macroscopic multicellular organisms before. This picture is very simple, of course, and does not at all reflect the role that micro-organisms actually play in the economy of nature.

In comparison, a similar relation between the size and the degree of complexity exists in the scientific world, too. In most scientific sources one will mainly find characteristics, in which structures of micro-organisms are described as simpler or smaller than those of bigger organisms, or even missing. However, a differentiation seems important, made in both modern textbooks by Brock and by Schlegel. They both appose morphological uniformity to physiological versatility.

This contrast characterizes the micro-organisms in a very special way and seems to be fundamental for an adequate understanding of the microbes as living beings and their role within the economy of nature. Accordingly, the following guideline arose from this observation:

To make clear fundamental forms of life of micro-organisms and their ecological functions.

As presented in figure 2, for the students, the ecological role of micro-organisms consists mainly of the decomposition of organic material. Therefore, the aim of the instruction of microbiology should be - and this is marked grey in figure 2 - to offer an extension of their conceptions in the direction of a comprehensive picture of the microbes. To this aim, the ability of the micro-organisms to build up biomass, to live in most varied habitats and also to form their habitats by their physiological activity has to be particularly pointed out. The search for contexts, in which the micro-organisms' ability of assimilation plays a major role, seems to be fundamental here.

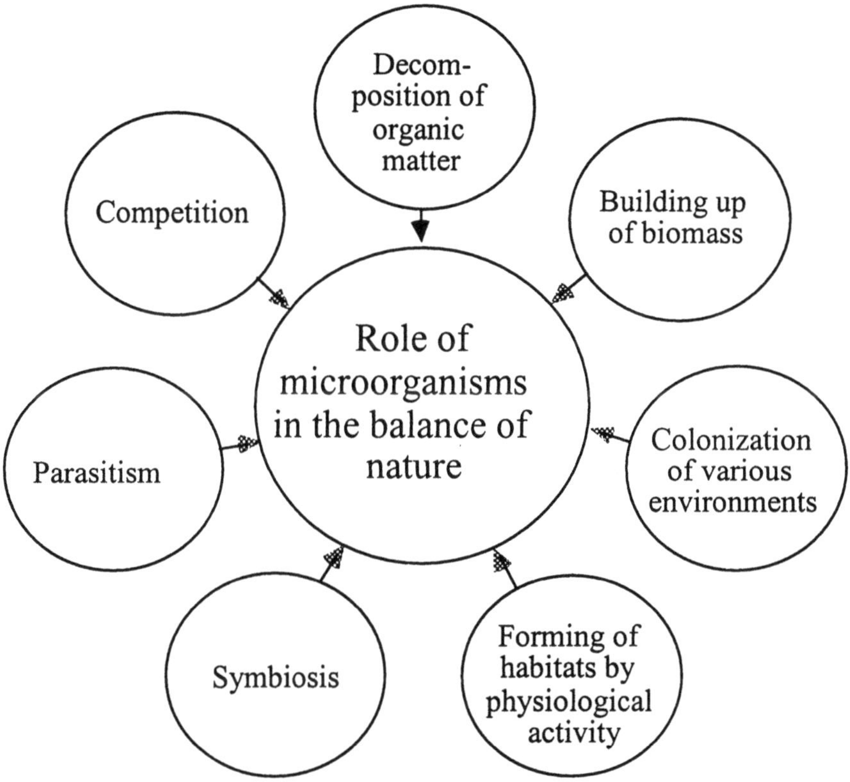

Fig. 2:* *The development of conceptions, which consider micro-organisms only as decomposers, to a more comprehensive view.

Furthermore, different relationships to other organisms, e.g. forms of symbioses, parasitism and competition are basic aspects of the picture of micro-organisms. The presentation of these relationships helps to create a more comprehensive picture of the micro-organisms as members of the economy of nature. In this way, the instruction of microbiology could contribute to the extension of the limited concept of micro-organisms as mere decomposers.

The guidelines, of which two have been presented here, form the basis for planning and performing teaching of microbiology. In this way, they introduce the next step of the investigation, the implementation of concrete units of instruction. Then the acceptance of the guidelines with respect to teaching microbiology will have to be proved. The final aim will be the access to practical work of teachers, according to the third step in the model of Educational Reconstruction, the Construction of Instruction.

The presented project was supported by the *Studienstiftung des deutschen Volkes*.

References

Baalmann, W., Frerichs, V. & Kattmann, U. (1999). How the gorillas became dark – research in students`conceptions leads to a rearrangement of teaching genetics and evolution. In O. de Jong, et. al. (Eds), *Bridging the gap between theory and practice: What research says to the science teacher* (pp. 171-190). Hongkong: ICASE.

Bayrhuber, H. & Stolte, S. (1997). Schülervorstellungen von Bakterien und Konsequenzen für den Unterricht [Students' conceptions of bacteria and consequences for teaching]. In H. Bayrhuber et al. (Eds.), *Biologieunterricht und Lebenswirklichkeit* (pp. 311-315). Kiel, Germany: IPN.

Brock, T. D. et al. (1994). *Biology of micro-organisms*. Seventh Edition. Prentice-Hall International.

Cohn, F. (1866). Über Bakterien, die kleinsten lebenden Wesen [About bacteria, the smallest living beings]. In R. von Virchow & F. Holtzendorff (Eds.), *Sammlung gemeinverständlicher wissenschaftlicher Vorträge*. Berlin, Germany.

Cohn, F. (1872). Untersuchungen über Bakterien [Investigations about bacteria]. *Beiträge zur Biologie der Pflanzen* 1 (2), 126-224, (3),141-203.

Ehrenberg, G. C. (1838). *Die Infusorientierchen als vollkommene Organismen – ein Blick in das tiefere organische Leben der Natur* [Infusoria as complete organisms - a view into the deeper organic life in nature]. Leipzig, Germany: Leopold Voss.

Gropengießer, H. (1997). *Didaktische Rekonstruktion des "Sehens"*. [Educational reconstruction of seing]. Oldenburg, Germany: Zentrum für pädagogische Berufspraxis,.

Helldén, G. (1999). A longitudinal study of pupils' understanding of conditions for life, growth and decomposition. In M. Bandiere et al. (Eds), *Research in science education in Europe* (pp. 23-29). Dordrecht, The Netherlands: Kluwer.

Kattmann, U., Duit, R. & Gropengießer, H. (1998). Educational Reconstruction - Bringing together issues of Scientific Clarification and Students' Conceptions. In H. Bayrhuber, & F. Brinkman (Eds.), *What, Why, How? Proceedings of the First European Conference on Didaktik of Biology* (ERIDOB) (pp. 252-262). Kiel, Germany: IPN.

Liebig, J. von (1840). *Die organische Chemie in ihrer Anwendung auf Agrikultur und Physiologie* [The organic chemistry and its application to agriculture and physiology]. Hildesheim, Germany: Gerstenberg.

Mayring, P. (1990). Qualitative Inhaltsanalyse [Qualitative content analysis]. Weinheim, Germany: Deutscher Studien Verlag.

Pasteur, L. (1871). *Die Alkohol-Gärung* [Alcoholic fermentation]. Augsburg, Germany: Lampart.

Pasteur, L. (1862). *Die in der Atmosphäre vorhandenen organisierten Körperchen, Prüfung der Lehre von der Urzeugung* [The organized corpuscles in the atmosphere - examination of the doctrine of sponaneous generation]. Leipzig, Germany: Akademische Verlagsgesellschaft.

Schlegel, H. G. (1992). *Allgemeine Mikrobiologie* [General microbiology]. Stuttgart, Germany: Thieme.

Teaching and Learning the Concept of the Model in Secondary Schools

Heikki Saari and Jouni Viiri
Department of Physics, University of Joensuu, Finland

Abstract

The main purpose of this project was to discover the conceptions that grade 7 pupils have of models. Another purpose was to construct a research-based teaching unit for teaching the concept of the model. The concept of the model was taught while learning the structure of matter in grade 7 physics (secondary school). Data was collected by means of questionnaires, tests and interviews both before and after the teaching unit. The pupils' conceptions of models were very limited before the teaching unit. The analysis of the interviews shows that there was progress in pupils' understanding of the concept of the model during implementation of the teaching unit.

Introduction

Models are an important part of physics. We can even say that in order to understand science, pupils have to know how scientific models are constructed and validated (Hestenes, 1992). Research has, however, shown that it is hard for pupils to understand the concept of the model (Grosslight et al., 1991).

The use of models and analogies in physics teaching has received considerable attention (see, e.g., Clement, 1998; Duit, 1991; Gilbert & Boulter, 1998). The particulate nature of matter has also been the subject of many studies (see, e.g., Andersson & Bach, 1995; Johnson, 1997). However, there are fewer studies of pupils' conception of models. To teach models and modelling, we would need to know what kind of conceptions pupils have of models. Moreover, we would also need to pay attention to how pupils' concepts of the model can be affected. In this study we present a description of ways in which pupils' conceptions of models were changed through a teaching unit where different kind of models were applied.

Aim of the study

Some studies already exist concerning learners' conceptions of models (Finegold & Smit, 1993; Gilbert, 1991; Grosslight & al., 1991). The objects of these studies have mainly been university students. However, the effect of teaching on their conceptions has not been studied.

The aim of this study was to investigate pupils' understanding of the concept of the model and the effect of teaching grade seven pupils this concept by means of models and modelling. We chose pupils who were about to begin learning physics at school because we considered that models and modelling should be taught right from the beginning of physics teaching. The specific aims of the study were to investigate the following questions:

1. What are the typical categories of pupils' concepts of the model?
2. Does teaching by means of models affect pupils' concepts of the model?

H. Behrendt et al. (eds.), Research in Science Education – Past, Present, and Future, 259–264.

Framework of the study

The research findings, described in this paper, form part of a larger research project, whose aim is to develop alternative methods for teaching and learning physics. In the following, we first describe the context of the study and then the structure of the teaching unit.

Models in teaching and learning physics

Gilbert and Boulter (1998) classify the models according to the following presentation. A model can be:

1. A *mental model,* that is a personal, private representation of a target.
2. An *expressed model,* that is a version of a mental model expressed by an individual through action, speech or writing.
3. A *consensus model,* that is an expressed model that has been subjected to testing by scientists and which has been socially agreed by some of them as having some merit.
4. A *teaching model*, that is a specially constructed model used by teachers to aid pupils' understanding of a consensus model.

Our aim in this research project was to teach pupils how to use models in learning about the structure of matter and to find out how this will affect their concept of the model. We used teaching models to portray the most important aspects of the topic to the pupils. A teacher already knows the structures and properties of the consensus model and, for this reason, he/she plays an essential role in designing the teaching model. The purpose of the teaching model is to influence a pupil's mental model of the subject under study. The teaching itself takes place in a context where the pupils can test their models in practice. We could assume that both the teaching of the structure of matter by modelling and the testing of those models might, in practice, affect a pupil's mental models (Grosslight, Unger & Jay, 1991; Saari, 1997).

The teaching unit

The teaching unit, which was based on an earlier study (Saari, 1997), was designed for 7th grade physics (13 year-old pupils). The duration of the teaching unit was eight hours. The pupils' pre-interviews were used in designing the unit, that started with a "black box" experiment, where the pupils had to make models of what was inside the box without opening it. After the experiment, and even after the conclusion of the whole teaching unit, the teacher did not tell the pupils what was inside the box. By means of this experiment we wanted to simulate Nature, since Nature does not tell the researcher her secrets. The pupils were subsequently required to classify things according to their state of matter.

The pupils were told that the structure of matter could be portrayed by means of different models. Two of them, *the model of continuous matter* and *the particle model of matter* were introduced. Both of these models are scientifically valid. It is not always necessary to take the particulate nature of matter into account.

In the next phase, the pupils modelled the gaseous, solid and liquid states of matter using role-play simulations and computer models. Each state of matter was studied through, approximately, the same phases (Fig. 1).

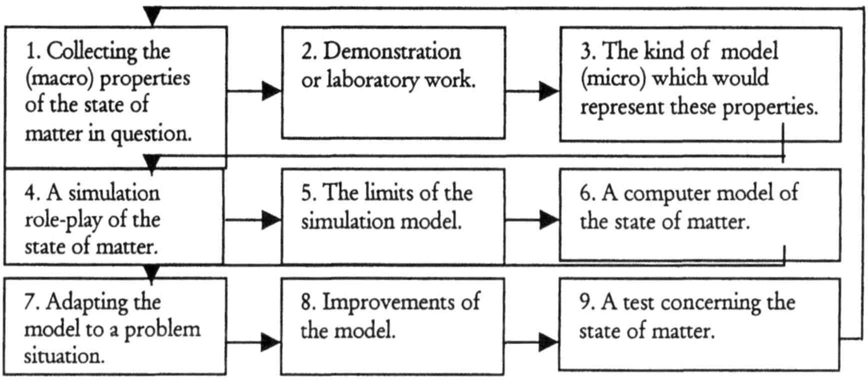

Fig. 1: The structure of the teaching unit

Method

Subjects

Fifteen pupils participated in the first phase. In the second phase, the teaching unit was presented at another school, with 18, 13 year-old pupils. The researcher was also the teacher of the first phase. The researcher holds a permanent teaching position at the school, but, at the time of the study, he was a researcher at the University of Joensuu. In the second phase, the teacher was the normal teacher of the class.

Data gathering

The data was collected by means of questionnaires, tests and interviews, both before and after presentation of the teaching unit. In the interviews, we used, in part, the questions of Grosslight, Unger and Jay (1991) and we also used a prepared list of questions, although every pupil was not asked exactly the same questions. In this paper we describe the results of the interviews.

Data analysis and findings

We wanted to find out about the ways in which the pupils think about the model concept. For this purpose, the pupils were interviewed both before and after the teaching unit, the interview lasting about half an hour. The interviews were recorded and afterwards transcribed. The transcribed data was, firstly, divided into four classes, based on the contents of the interviews. The classes were: *(a) The definition of the model, (b) the use and purpose of the model, (c) the preciseness of the model and (d) the reason for changing a model.*

The pupils' names were removed from the written answers during the process of categorisation. The pre- and post interviews were united into a single database, so that we would get all the categories that exist in the data. Next, each of the four classes was coded into qualitatively different subcategories. No stated categorisation was used, but it emerged from the data itself, with the result that we gained the following subcategories.

a Definition of the model	b Use and purpose of the model	c Preciseness of the model	d Reason for changing a model
a=1. A model is an object or function which is to be copied.	b=1. To help in doing and identification.	c=1. A concrete model. Complete preciseness.	d=1. The failings of the model or makers necessitate its change.
a=2. A model is an engineer model.	b=2. To help in learning and teaching.		
a=3. A model is an expressed model that expresses a target that cannot be seen.	b=3. To help in illustrating, understanding, forecasting and testing.	c=2. An abstract model. Preciseness to some degree.	d=2. New knowledge, target of research, target of use or comparing.
a=4. An advanced model concept. (This subcategory demands that a pupil may also adapt his/her knowledge in use).	b=4. To help all-inclusively. (This subcategory demands that a pupil may also adapt his/her knowledge in use).		

Table 1: Subcategories of the concept of the model

Each of the four categorisations forms a hierarchical structure. The greater the number of the subcategory, the more advanced an idea it represents. The four categorisations were then combined so that we could form the main categories.

The main categories were formed so that they differed from each other qualitatively. Each of them included certain combinations of the subcategories (a, b, c, d). We acquired three main categories:

- Main category A (a basic understanding of the concept of the model)
- Main category B (a moderate understanding of the concept of the model)
- Main category C (an advanced understanding of the concept of the model)

Each pupil can be described by a row (a, b, c, d) in which **a** describes one of the numbers of subcategory (**a**) etc.

To the main category (A) belong all pupils who do not provide answers to each of the four classes. Pupils who have the combination (a=1, b, c, d) or (a=2, b=1, c, d) also belong to main category (A). A pupil belonging to main category (A) thinks that:

A model is a thing, an act or a so-called engineer model which is to be copied. The preciseness of the model depends on who makes it but it has to be as accurate as possible. We can change the model if it contains mistakes or if its maker wishes to do so.

Within the main category (B) belong all those pupils with the combinations (a=2, b=2, c, d) where either c and/or d has the value of 1 and also the combination (a=3, b=3, c=2, d=2). A pupil belonging to main category (B) thinks that:

A model represents a target that is known or unknown. The main purpose of the model is to help in learning and teaching. The preciseness of the model depending on the nature of the model and changes in it depend on the researcher's willingness or research.

Within the main category (C) belong all those pupils with the combinations (a=4, b=4, c=2, d=2), (a=4, b=3, c=2, d=2) and (a=3, b=4, c=2, d=2). A pupil belonging to the main category (C) thinks that:

A model represents a target that is known or unknown. The purpose of the model is to give an idea of the target and to help in conceptualising it. A model also provides the vocabulary for representing the target. The preciseness of the model depends on its use and its changing is founded on research.

Pupils belonging to the main category (A) possess a very concrete concept of the model. The pupils in the main category (B) differ from those in the main category (C) in that they have a brief notion of one of the four classes. Furthermore, their ability to apply and widen their concept of the model is limited.

After forming the main categories, the pre- and post-interviews were reread and each pupil was given a combination (a, b, c, d), that described their understanding of the concept of the model. Then each of the pupils was allocated to one of the three main categories. Hence each pupil will belong to one main category before the start of the teaching unit and potentially to another after its conclusion. The results of our analysis are presented in Table 2.

Main category	**Number of pupils prior to presentation of the teaching unit (n)**	**Number of the pupils after presentation of the teaching unit (n)**
A	30 (1, 1, 1, 1) n=14; (1, 1, 1, 2) n=4	1 (1, 1, 1, 1)
B	3 (2, 3, 1, 2), (2, 3, 1, 1), (3, 3, 2, 2)	15 (3, 3, 2, 2) n=3; (3, 2, 2, 1) n=2
C		16 (4, 4, 2, 2) n=8; (3, 4, 2, 2) n=3; (4, 3, 2, 2) n=5

Table 2: Pupils belonging to each main category before and after presentation of the teaching unit. The most common combinations are shown.

Analysis of the pre-interview shows that the pupils had very limited perceptions of the models (Table 2). Thirty pupils out of 33 belong in the main category (A) and almost half of them (14) have the most concrete combination (1, 1, 1, 1). They think that a model is an object which has to be copied perfectly. Only three pupils had the idea that a model represents something and that the model does not have to be an object.

In the course of the teaching unit, the pupils developed and used the models that they were given (the model of continuous matter and the particle model of matter). They used role-plays in modelling and also studied computer models. After completion of the teaching unit the pupils were interviewed again. Analysis of the post-interviews shows that improvement has taken place concerning their conceptions of models (one pupil moved to another school before the post-interview). At this point, only one of the pupils now belongs to the main category (A) and as many as 16 belong to the main category (C) (Table 2). This means that the vast majority of the pupils have gained the idea that a model represents a target that is either known or unknown and we can claim that teaching with models and modelling has affected our pupils' conceptions of models. In sum, the differences between the post- and pre-interviews are statistically very meaningful (Jonckheere-Terpstra test stat. 6.885, Exact Sig. (1-tailed) $p = 0.000$ in three decimals).

Discussion

The aim of this study was to discover the typical categories of pupils' conceptions of models and whether teaching with models affects pupils' concept of the model. This was investigated by means of interviews both before and after the presentation of the teaching unit. Pupils' ideas of models could be classified into three qualitatively different categories. Those belonging to the first category have a very concrete idea of models, those in the second category have a more abstract idea of models which could represent an unknown target, while those in the third category have an even more abstract and general idea of models.

Before the presentation of the teaching unit most of the pupils belonged to the first category, during the presentation pupils used different models to study the properties of matter. At the conclusion of the presentation of the unit, roughly half of the pupils had moved into the second category and the other half into the third. We may, therefore, conclude that the teaching unit noticeably changed their ideas. We can also conclude that the categories can be used in describing pupils' concepts of the model and also in describing their actual learning.

In this case study, pupils' understanding of the modelling idea of science improved when modelling was taken as a real goal of the teaching unit. However, we need further research. We should, for instance, investigate more thoroughly how teaching units containing modelling ideas affect pupils' ideas both by studying individual pupils' ideas and using control groups.

References

Andersson, B. & Bach, F. (1995). *Att utveckla naturvetenskaplig undervisning. Examplet gaser och deras egenskaper.* Rapport NA-SPEKTRUM, 14. Göteborg: Göteborgs universitet, Institutionen för ämnesdidaktik.

Clement, J. (1998). Expert novice similarities and instruction using analogies. *International Journal of Science Education* 20 (10), 1271-1286.

Duit, R. (1991). On the role of analogies and metaphors in learning science. *Science Education* 75 (6), 649 - 672.

Finegold, M. & Smit, J.J.A. (1993). Learning in science as affected by perceptions of the nature and functions of models. In *The Proceedings of the Third International Seminar on Misconceptions and Educational Strategies in Science and Mathematics.* Ithaca, NY: Misconceptions Trust.

Gilbert, J.K. & Boulter, C.J. (1998). Learning science through models and modelling. In B.J. Fraser & K.G. Tobin (Eds.) *International Handbook of Science Education* (pp. 53-66). Dordrecht, The Netherlands: Kluwer Academic Publishers.

Gilbert, S.W. (1991). Model building and a definition of science. *Journal of Research in Science Teaching* 28(1), 73-79.

Grosslight, L., Unger, C. & Jay, E. (1991). Understanding models and their use in science: conceptions of middle and high school students and experts. *Journal of Research in Science Teaching* 28 (9), 799-822.

Hestenes, D. (1992). Modelling games in the Newtonian world. *American Journal of Physics* 60 (8), 732-748.

Johnson, P. (1997). Progression in children's understanding of a 'basic' particle theory: A longitudinal study. A paper presented in the E.S.E.R.A. Conference, Rome 1997.

Saari, H. (1997). *Learning the structure of matter in secondary school.* Licentiate thesis. University of Joensuu. (In Finnish)

Conceptual Change and Student Diversity: The Case of Volcanism at Primary School

Jean-Louis Chartrain and Michel Caillot
Université René Descartes, Paris, France

Abstract

This paper tries to explain why students from the same class with the same teaching do progress differently in their conceptual change. Here we have studied students' conceptual change regarding volcanoes in a French 5th grade. The different students' preconceptions, before teaching, have been categorized and related to well-known historical and epistemological obstacles. After specific teaching, the objectives, of which were to overcome these obstacles, we have looked at how conceptual change has taken place for each student. In order to explain the observed differences, we firstly considered a possible effect of classical sociological variables (gender, age, parents' jobs and so on). This attempt failed. Secondly, we introduced the concept of individual "relation to knowledge" and determined this relation to knowledge for each student. Taking this variable into consideration, the diversity of conceptual changes is much better explained.

Introduction

Very often, French research in science education does not take student diversity in the classroom into account and considers a student as an epistemic subject without gender, age and social specification. Different hypotheses may be raised: the first would be due to the strong Piagetian influence upon the first generation of French '*didacticiens*' (science educators) who have looked for similarities rather than differences in students' behaviours and representations; the second may be related to the ideology of French republican school, in which all students have the same chance to succeed, regardless of gender and social or ethnic origins. However, several elements should broaden the view of a pure epistemic student to new aspects. First, for a long time, sociological studies have shown that success at school depends on students' social origins. Second, some more recent studies (Broccolichi, 1994) have begun to show how class management and its practices may lead to the production of differentiation among students' learning abilities the so-called "intra-school differentiation". Third, we also have to consider that learning is an interactive process: on the one hand interaction with pieces of knowledge and on the other interaction with individuals (peers and teachers) through social practices, as, for example, the practice of writing at school. Thus, the learning ability of each student is then determined by the relationships that he/she entertains with these social practices. These different reasons can explain student diversity in learning a specific subject matter. In any case, research on student diversity in science learning is practically non-existent in France, with the exception of a recent study made by two psychologists (Rozencwajg & Troseille, 1996).

H. Behrendt et al. (eds.), Research in Science Education – Past, Present, and Future, 265–270.

In an exploratory study, Chartrain (1998) studied how 5th graders change their representations on volcanism and whether these conceptual changes were linked to some students' characteristics as yet undetermined.

Methodology

The class is a 5th grade with 28 students (9 girls and 19 boys). The research includes a teaching section and several data gatherings:

Teaching

The teaching includes several phases:

- the first was the gathering of preconceptions' fell by students on volcanoes through drawings and questions.
- the second was the teaching. The teaching lasted 8 sessions, each of about an hour a week. The students worked on some documents: pictures, videos, a map of ocean floors and a newspaper account of the Piracutin eruption in Mexico. Moreover, tectonic plate motion was introduced at an elementary level and simulated with plastic foam and clay plates. This teaching was aimed explicitly at overcoming the epistemological obstacles students have (see below).
- the third phase – a student assessment - took place three weeks later in order to see how their conceptions on volcanoes had evolved.

Data gatherings

We collected the usual sociological data: gender, age, parents' job and highest degree. Moreover, we gave each student two questionnaires entitled 'learning assessments' (in French '*bilans de savoirs*') in order to determine his/her relation to knowledge. These learning assessments were inspired by Charlot *et al.* (1992) and Montandon and Osiek (1997). The first was given at the beginning of September and the other four months later. Four questions were asked. Only the first differed from the assessment in September: "Since you started school, you have learnt..." and in January: "Since you started your 5th grade, you have learnt..." The three other questions remained the same: (i) "What important 'things' have you learnt so far and what is it important to learn at school?" (ii) "Explain what to learn means for you" and (iii) "What do you feel you can learn this year?"

Conceptual change

Students' preconceptions on volcanism are not far from well known historical conceptions (ASTER, 1995). Therefore there are mainly two ideas: the local one and the central one.

In the local view, students believe that a volcano is always an isolated mountain located in a precise geographical site. The lava is already present inside the volcano chimney. The idea of volcanoes organized in a systematic way on the earth's surface does not exist: a volcano is always seen as a local phenomenon.

In contrast, in the central view, the students believe that there is a huge underground reservoir of hot magma at the Earth's centre, that blows up through active volcanoes spread over the entire earth's surface. In this second belief, volcanism is no longer considered as a set of isolated volcanoes, but rather as a system. The earthquakes are seen as consequences of volcanoes eruption.

Evolution from a local view to a central one requires the student to overcome the epistemological obstacle of considering a volcano as an isolated mountain. Once this obstacle is overcome, the issue of the erupting origin may be raised.

Historically, there is a third conception, which is very recent and has revolutionized Earth sciences in the 20th century. In this view, volcanism is just one of the phenomena related to the motion of rigid plates floating on a viscous underlayer in the mantle. Most of the world's volcanoes are found along the margins of huge plates, into which the Earth's crust is divided. Other phenomena that relate to plate tectonics are ocean opening, continental drift and formation of the world's mountain belts. This global view is outside of the scope of teaching at primary school. However, some elements of this theory can be approached, such as the relation of volcanoes on the earth's globe to plate notions.

Each major view from students' answers showing their conceptions of volcanism before and after teaching were refined into two components for research purposes. For local ideas, we introduced the pure local view (*L*) and a more elaborated local view (*L+*), in which a huge pocket of magma exists under the volcano, in the earth depths, ready to blow out of the volcano in a violent eruption.

In the same way, the central conception is divided into two components: the Central (*C*) and a more elaborated Central (*C+*). In the latter conception, the students know that (i) hot magma is inside the Earth and (ii) the Earth crust plates move but volcanism is not related to this motion.

Finally, after teaching, we introduced a pre-Global conception (called *G*) in which the student is aware of the existence of moving plates on the Earth's surface and links volcanism and earthquakes.

Before teaching, out of 28 students: 21 have the *L* representation, 5 the *L+* and 2 the *C*. After teaching, Table 1 shows how the conceptions on volcanism have changed.

			After teaching			
		n	*L+*	*C*	*C+*	*G*
Before teaching	*L*	21	1	13	5	2
	L+	5	–	1	1	3
	C	2	–	–	–	2

Table 1: Evolution of students' ideas on volcanoes

This table clearly shows that students' ideas on volcanoes changed and there was general progress among the different ideas. The more elaborate the initial idea, the greater the conceptual change. However, of all students obtained the *pre-Global* view, nor the advanced *Central* +. In the second part of this research we try to explain why students' conceptual changes differ.

About sociological variables

The sociology of education has shown for a long time that students' achievement may depend on their gender, ages in class and social and ethnic origin. Thus the first idea was to analyze the results in Table 1 in the light of sociological variables to see if the quality of conceptual change depends on these variables. The chosen variables were (i) gender, (ii) age (the normal age for this class is 10 years old) and (iii) the father's job.

The analysis leads to the following results:

- All students progressed. The progress is greater among the boys: 12 boys out of 19 reach *Central* + and *pre-Global* ideas of volcanoes, whereas only 1 girl out of 9 reaches the *Central* + view. Although the boys' initial views were more elaborate than the girls' views there were proportionally more boys than girls who had achieved a significant conceptual change. Here we find a classical result on the differences between boys and girls in science education.
- When we consider the students' ages and the father's job category, we do not find a real effect.

Other factors must be saught to explain the observed results.

Relation to knowledge

A promising theory and concept enable us to go further in our attempt to understand the students' conceptual changes. Namely, the concept of *Relation to knowledge* (*'rapport au savoir'*) that comes from Charlot's theory (Charlot, Bautier & Rochex, 1992, Charlot, 1997). It is a concept that interface micro-sociology and the subject's identity. Charlot defined it as "*a relation to learning, ...and not only the relation to a knowledge-object, that is one of the components of learning* (1997, 89)". Even though this concept is beginning to spread among French educational researchers, it is still difficult to get a precise and operational definition.

In this paper we have tried to use the notion of 'relation to knowledge' in taking into account different indicators from the two 'learning assessments'. These indicators are:

1 – the use of the first person by a student *("I do...", "I learn...")* or general statements that indicate how the student is involved in school;

2 – the statements used to describe the school entity: either statements showing a utilitarian view of school ("*succeeding", "getting into the next grade", ...*) or statements showing how school can participate in personal development (*"understanding", "progressing", ...);*

3 – the way the students see themselves in the future. This view of themselves can be (i) unconscious or vague, (ii) in short term (at the end of the term, or into the next grade), (iii) long term (at high school or even at university);

4 – the object of learning. What does a student learn? (i) things of everyday life such as playing new video games, (ii) subject matters such as mathematics or French spelling, or (iii) specific objects of knowledge such as fractions or digestion.

5 – the students' conceptions of learning. Does a student consider learning as a 'job' (student's job)? as a series of unrelated and external activities with no meaning? or as a meaningful activity in which he/she is really involved?

The answers to the 'learning assessments' have us to categorize the students into five categories according to the importance they give to knowledge and to their ability to learn something:

- rejection (n=2). These students reject school and what is taught in class. Real life is outside school!

- tourist-type (n=3). These students consider school as a place where they can develop social relationships, make friends and play. Playing is more important than learning!
- intermediary (n=5). In this category, the relation to knowledge is not clear, it varies from the tourist aspect to the utilitarian.
- utilitarian (n=6). These students take school seriously. They do their student job well: they do their homework and they learn the lessons. School is considered as a way to succeed in the future, to get a good job. Learning is primarily seen as instrumental.
- pleasure (n=12). These students are the most involved in their learning. They take pleasure in learning new topics and they have long term views.

Before teaching		*Local* (n=21)				*Local +* (n=5)			*C* (n=2)
After teaching		*L+*	*C*	*C+*	*G*	*C*	*C+*	*G*	*G*
Relation to knowledge	Rejection		2						
	Tourist		3						
	Intermediate	1	2		1			1	
	Utilitarian		4	1		1			
	Pleasure		1	5	1		1	2	2

Table 2: How the students' ideas on volcanoes have changed according to their relation to knowledge

Table 2 shows how the students' ideas about volcanoes have changed according to their relation to knowledge. We can easily see that the students who have the most positive relation to knowledge (the utilitarian and those who find pleasure in learning) make the biggest steps to the most elaborate ideas on volcanoes: the central and the pre-global ideas.

Conclusion

This preliminary case study shows that the diversity of ways in which students change their conceptions does not depend only on the quality of teaching. It depends on other factors. If the traditional sociological factors can explain the observed differences in part, the 'sociological determinism' does not explain all the differences observed in the conceptual changes on volcanism. Here we have explored the student's relation to knowledge as a possible factor that may explain the different conceptual changes observed in a primary school science class. Other studies in progress are necessary to confirm these preliminary results.

References

ASTER (1995). *Représentations et obstacles en géologie* [Representations and obstacles in geology]. N°20. Paris: INRP.

Broccolichi, S. (1994). *Organisation de l'école, pratiques usuelles et production d'inégalités* [School organization, normal practice and production of inequalities]. Paris: EHESS (unpublished doctoral dissertation).

Charlot, B. (1997). *Du rapport au savoir. Eléments pour une théorie* [Relation to knowledge. Elements for a theory]. Paris: Anthropos.

Charlot, B., Bautier, E. & Rochex, J.-Y. (1992). *École et savoir dans les banlieues et ailleurs* [School and knowledge in the suburbs and elsewhere]Paris: Armand Colin.

Chartrain, J.-L. (1998). *Différenciation scolaire et conceptions des élèves* [School differentiation and students' ideas]. DEA dissertation presented at the University René Descartes (unpublished).

Montandon, C. & Osiek, F. (1997). La socialisation à l'école du point de vue des enfants [Socialization in school as seen by children]. *Revue Française de Pédagogie*, 118, 43-51.

Rozencwajg, P. & Troseille, B. (1996). Approches cognitive, didactique et différentielle de la représentation des concepts scientifiques [Cognitive, educational and differential approaches to represent scientific concepts]. *L'orientation Scolaire et Professionnelle*, 25, 2, 285-306.

The Development of Prospective Teachers' Concerns about Teaching Chemistry Topics at a Macro-Micro-Symbolic Interface

Onno de Jong
Centre for Science and Mathematics Education, The Netherlands
Jan van Driel
ICLON-Graduate School of Education, The Netherlands

Abstract

This paper describes an exploratory study of prospective teachers' concerns about teaching a major topic in science education, i.e., linking macroscopic phenomena with microscopic particles and symbolic representations such as formulas and equations. Teaching this topic is often associated with conceptual difficulties for students and, for that reason, may evoke pedagogical content concerns (PCC) among prospective teachers.

The study was designed as a naturalistic case-study. Eight prospective teachers (all had a MSc in chemistry) were interviewed before and after the first two lessons about topics focussing on a macro-micro-symbolic interface. The semi-structured interviews and the lessons were audio-taped and analysed. The results revealed a number of characteristics of the nature of the development of prospective science teachers' PCC.

Introduction

In connection with the knowledge base of teachers, Shulman has introduced the concept of pedagogical content knowledge (PCK) to emphasize the importance of the transformation of subject matter knowledge *per se* into "subject matter knowledge *for teaching*" (Shulman, 1986, p. 9). The subject matter knowledge prospective science teachers acquired during their science studies constitutes one of the main bases from which their PCK may be derived. However, the study of an academic discipline does not provide prospective teachers with the kind of understanding they need to effectively transform their academic knowledge into instructional activities in the classroom (Sanford, 1988; Lederman et al., 1994).

The present study focused on prospective science teachers' concerns in the context of this transformation problem. As an analogue to PCK, the term *pedagogical content concerns (PCC)* is introduced to refer to these concerns. The theoretical framework of this study draws upon Fuller and Bown's (1975) model of concerns of prospective and beginning teachers. In this model, changes in the nature of teachers' concerns are described as stages in the professional development of teachers. According to the model, the first stage contains concerns about issues such as self-image and ways to survive in the classroom: *self-concerns*. The second stage consists of concerns about issues such as teaching performance: *task concerns*. In the third stage, concerns relate to issues such as understanding students and their learning processes: *student concerns*. Fuller and Bown (1975) have also indicated that the first stage of self-concern is preceded by a stage of pre-teaching concerns. In that stage, prospective teachers are concerned about students because they still

H. Behrendt et al. (eds.), Research in Science Education – Past, Present, and Future, 271–276.

identifiy with students. However, little is known about concerns dealing with teaching specific science curriculum topics, that is, pedagogical content concerns (PCC).

In this study, PCC were explored among a group of prospective chemistry teachers. The purpose of this study was twofold. From a theoretical point of view, the aim was to understand the nature of prospective science teachers' PCC in relation to Fuller and Bown's model of concerns. From a practical point of view, the results were expected to have implications for the design of science teachers' preparation courses.

Topic and research question

The present study focuses on prospective teachers' PCC about a central issue in science education, viz., teaching and learning the relationship between the macroscopic domain and the microscopic domain. The macro-domain deals mainly with substances and their properties as well as phenomena. The micro-domain deals mainly with corpuscular models such as molecules, atoms and ions. At the macro-micro-interface, representations play an important role. Common types of representations feature icons, diagrams and symbols. In the present study, the macro-micro-interface was elaborated mainly for *symbolic representations*, such as formulas and equations, in the context of chemistry education only.

When teaching, secondary chemistry teachers are mentally switching between macro- and micro-aspects of science curriculum topics in a routine and often implicit manner (Johnstone, 1993). However, secondary school students often experience difficulties in understanding chemistry topics at a macro-micro-symbolic interface. For them, the conceptual demands of shifting between 'macro' and 'micro' can be overwhelming. Their difficulties have been reported in several studies (e.g., Lee et al., 1993). Students also appear to experience difficulties in understanding symbolic representations (Friedel & Maloney, 1992).

As learning to link macroscopic phenomena with microscopic particles through symbolic representations constitutes an important objective of chemistry education, prospective chemistry teachers need to develop PCK in this domain. It was assumed that these teachers, having been educated as chemists, had developed a habit of making these links in a flexible and implicit manner, thus possibly creating confusion among their students. In this context, the teachers may develop specific PCC as a step towards the development of their PCK. In this study, we wanted to explore the nature and the development of the pedagogical content concerns in this domain, using Fuller and Bown's model of concerns as an interpretative framework. In particular, the following research question was addressed:

What are the major pedagogical content concerns in terms of self-PCC, task-PCC and student-PCC, of prospective teachers before and after teaching chemistry topics at a macro-micro-symbolic interface?

Design

The participants and their context

During the autumn of 1997, eight prospective chemistry teachers (five females and three males; average age: 23) were involved in the research project. All the participants had a master's degree in chemistry and were participating in a teacher preparation course. This post-graduate course (one year) consists of an integrated mixture of university lectures, workshops and field-based activities at secondary schools. The participants entered the project at the beginning of the third month of the course. During the preceding months, no specific attention had been given to the meaning or the use of the macro-micro-symbolic interface. In the third month, the prospective teachers were asked individually to choose a forthcoming topic from the chemistry curriculum with an emphasis on the relationship between macroscopic phenomena, microscopic particles, and symbolic representations. Three participants chose the topic of balancing simple reaction equations, taught to students in grade 9 (age 14-15). The aim was to relate the conversion of substances to the reordering of atoms and reaction equations. Five participants chose precipitation reactions as their topic, taught to students in grade 10 (age 15-16). The aim was to relate precipitation phenomena to the dynamics of ions in solutions and reaction equations.

Procedure

In the context of the project, the prospective teachers were interviewed individually by one of the authors before and after the first two lessons they taught about the chosen topic. During the pre-lesson interview, they were invited to explain their lesson plans. They were also asked to express their expectations regarding the students' conceptual difficulties as well as their own difficulties in teaching the topic. During the post-lesson interview, they were invited to report and comment on their teaching experiences. All the interviews were audio-taped, and the lessons under consideration were also audio-taped. All the interviews and classroom discussions were transcribed into protocols.

Data analysis

First, each author analyzed the pre- and post-lesson interview protocols of the individual prospective teachers in an iterative way. The following main categories were used for this analysis: self-PCC, task-PCC, and student-PCC. Subsequently, within each category, statements were classified into subcategories. In the next phase, analysis results of the individuals were compared to identify common (sub)concerns. In this phase, researcher triangulation (Janesick, 1994) was applied by comparing and discussing interpretations by the authors. The validation of these interpretations was promoted by applying the constant comparative method (Denzin, 1994). This involved the comparison of the analysis of the interview protocols with other sources, in particular, classroom discussions between students and the prospective teachers during the recorded lessons.

Findings

The pre-lesson interviews indicated that the prospective teachers had only a vague idea of the difficulties they might encounter. Many expressed their concerns in rather short statements. A summary of their pre-teaching PCC which emerged from the analysis of the pre-lesson interviews is given in Table 1. After the lessons, the prospective teachers reported that they had experienced more difficulties than they had expected. In addition to the pre-teaching concerns, a number of new concerns were reported after teaching. Moreover, post-teaching concerns were usually expressed in a much more detailed way. A summary of post-teaching PCC emerging from the analysis of the post-lesson interview is also given in Table 1.

Regarding *self-PCC*, the results indicated that this type of concern was lacking before teaching. After teaching, however, all the prospective teachers expressed self-PCC, especially concerns about too fast zig-zag reasoning. This development reflects an emerging awareness of the teachers' own ways of switching between the macroscopic and the microscopic domain, that can hinder the learning processes of students ('too fast'). The absence of this awareness before teaching can be explained by taking into account the prospective teachers' subject matter expertise. To them, switching between a macro-domain, a micro-domain and a symbolic level has become second nature. Their knowledge and skills have accumulated during a long period of learning chemistry. To be conscious of novices' conceptions is not something that comes easily to experts (De Jong et al., 1995).

Table 1: Summary of prospective teachers' pedagogical content concerns (PCC)

Main categories and subcategories of PCC:	**Prospective teachers* expressing concerns:**	
	Before teaching	*After teaching*
Self-PCC		
Too fast zig-zag reasoning between the domains	-	A, B, C, D, E, F, G, H
Dominant orientation towards the micro-domain	-	A, B, C, D, E, F
Mixing up the two domains	-	A, B, C, E, G, H
Task-PCC		
How to teach heuristics for building up symbolic representations	D, H	B, D, F, H
How to handle inappropriate symbolic representations in textbooks	-	A, E, F, G
Student-PCC		
Difficulties in understanding the macro- and micro-meaning of formulas	F, H	A, C, E, F, H
Difficulties in understanding the macro- and micro-meaning of reaction equations	C, F, H	A, C, E, F, H

* Maximum number is 8. Teachers are represented by letters.

As to *task-PCC*, the number of prospective teachers expressing this type of concern after teaching was larger than before. The reported development reflected an increase of concerns about teaching heuristics for building up symbolic representations and handling inappropriate symbolic representations in textbooks. The latter subconcern is interesting because, as Yager (1983) has pointed out, textbooks are usually seen by teachers as very important sources of information, that have a great influence on their knowledge base and teaching decisions. The results of the present study showed that, before teaching, the prospective teachers were not really concerned about the quality of their textbook. However, after teaching, four of them became aware of specific shortcomings in their textbooks, for instance, writing salt formulas as $Na^{+}Cl^{-}(s)$. This kind of writing was considered confusing because the symbol 's' refers to the macro-domain, whereas the ion symbols refer to the micro-domain. A typical statement about such formulas was:

"Confusing, because, in that case, you have charged things in a solid substance and I actually try to teach them that a solid substance consists of charged things but the whole is neutral, otherwise it would not be a solid but an ion."

With respect to *student-PCC*, the results showed that some teachers had already expressed this type of concern before teaching, possibly because of previous teaching experiences during the preceding two months. In any case, the number of teachers expressing student-PCC increased after teaching. For instance, some had noticed that students were inclined to interpret reaction equations in a microscopic context only. An illustrative statement is:

"They do not see that a reaction equation refers not only to molecules, but also to larger quantities..."

Formulas used to refer to non-decomposable substances consisting of multi-atomic molecules were mentioned as an example of students' problems in understanding formulas. For instance, students would often prefer the formula O to represent oxygen as a substance, rather than the prescribed formula O_2.

Conclusions and implications for science teacher education

The present study revealed the major concerns of prospective chemistry teachers as reported by them before and after teaching chemistry topics at a macro-micro-symbolic interface. Their concerns were classified as pedagogical, content-related, self-concern, task concern and student concern. The pre-teaching PCC appeared to consist of student-PCC and some task-PCC, whereas the post-teaching PCC consisted of these concerns to a larger and more elaborate extent. In addition, self-PCC was included in post-teaching PCC. These outcomes are in accordance with Fuller and Bown's model of concerns, which predicts that, before teaching, there is mainly student-PCC. After teaching, however, self-PCC is the major concern; student-PCC is the minor concern and task-PCC is in between.

Science teacher education should pay attention to the concerns of prospective teachers as well as to their PCC. Based on the data presented here, it can be stated that the volume of PCC that prospective science teachers express can increase rapidly after a couple of lessons on a new curriculum topic or issue. This has important implications for the design of training courses for science teachers. These courses should be structured in such a way that prospective science teachers' PCC is taken into account as much as possible. The design could include the following five stages:

1) *Clarifying existing PCC.* During this stage, prospective teachers describe and discuss their own difficulties in understanding a specific curriculum topic or issue. Subsequently, they analyze articles from the research literature or some classroom protocols with respect to students' difficulties in understanding this topic. Next, they discuss the results of their analyses and compare the students' conceptual difficulties with their own conceptions.
2) *From these concerns to teaching intentions.* During this stage, prospective teachers analyze relevant parts of current science textbooks and discuss the quality and usefulness of their content. Subsequently, they formulate intentions for the teaching of a specific topic in their own classes.

3) *From these intentions to teaching experience.* At this stage, prospective teachers prepare lessons and teach these lessons at school.
4) *From these experiences to new PCC.* Prospective teachers reflect on their teaching experiences and express the PCC emerging from this.
5) *[Back to step 2].* The results of reflections on the new PCC are used as a starting-point for new intentions to teach.

This cyclical model will be used by us as a basis for the construction of new courses that take prospective chemistry teachers' PCC into account. In future research, we shall test the effect of these courses on the development of prospective chemistry teachers' PCC and, subsequently, on the formation of their PCK.

References

Denzin, N.K. (1994). The art and politics of interpretation. In N.K. Denzin & Y.S. Lincoln, Eds., *Handbook of Qualitative Research Design* (pp. 500-515). Thousand Oaks, CA: Sage.

De Jong, O., Acampo, J. & Verdonk, A. (1995). Problems in teaching the topic of redox reactions: Actions and conceptions of chemistry teachers. *Journal of Research in Science Teaching* 32, 1097-1110.

Friedel, A.W. & Maloney, D.P. (1992). An exploratory, classroom-based investigation of students' difficulties with subscripts in chemical formulas. *Science Education* 76, 65-78.

Fuller, F.F. & Bown, O.H. (1975). Becoming a teacher. In K. Ryan, Ed., *Teacher Education: the 47th Yearbook of the NSSE, part II* (pp. 25-52). Chicago: Rand McNally.

Janesick, V.J. (1994). The dance of qualitative research design. In N.K. Denzin & Y.S. Lincoln, Eds., *Handbook of Qualitative Research Design* (pp. 209-219). Thousand Oaks, CA: Sage.

Johnstone, A.H. (1993). The development of chemistry teaching: A changing response to changing demand. *Journal of Chemical Education* 70, 701-705.

Lederman, N.G., Gess-Newsome, J., & Latz, M.S. (1994). The nature and development of preservice science teachers' conceptions of subject matter and pedagogy. *Journal of Research in Science Teaching* 31, 129-146.

Lee, O., Eichinger, D.C., Anderson, C.W., Berkheimer, G.D. & Blakeslee, T.D. (1993). Changing middle school students' conceptions of matter and molecules. *Journal of Research in Science Teaching* 30, 249-270.

Sanford, J.P. (1988). Learning on the job: Conditions for professional development of beginning science teachers. *Science Education* 72, 615-624.

Shulman, L.S. (1986). Those who understand: Knowledge growth in teaching. *Educational Researcher* 15, 4-14.

Yager, R.E. (1983). The importance of terminology in teaching K-12 science. *Journal of Research in Science Teaching* 20, 577-588.

How to Enhance Students' Motivation and Ability to Communicate in Science Class-Discourse

Claus Bolte
University of Hamburg, Department of Science Education, Germany

Abstract

To motivate students and to enhance their ability to communicate about scientific topics is a fundamental guideline established by science educators to promote successful learning and scientific literacy (MNU, 1989; BLK, 1997). - The TIMSS-Video-Study and other publications in science education show that (not only) German science teachers are not very successful in this field (Baumert et al., 1997; Bolte, 1996). On the basis of a theoretically based and empirically sound motivation-model (study 1), classroom-observations (study 2) and student-assessments of motivational patterns of their chemistry classes (study 3) this publication will point out possible reasons for unsuccessful science teaching and show what teachers can do to improve their lessons.

Objectives

The objectives of this report are to point out possible reasons for unsuccessful science teaching and to show how teachers can promote their students' motivation and performance.

Design

In order to follow these objectives, this report focuses on three studies:

1st Study No. 1 is the empirical confirmation of a theoretically based motivation-model. It looks into the interrelations between six important motivation variables ('indicators') and the students' performance. Study No. 1 involved 589 students between Grades 7 and 10 from both secondary and grammar schools (German: "Realschule" and "Gymnasium").

2nd Study No. 2 focuses on the students' assessments of six motivation-indicators and on the way in which they would wish them to be implemented in their science lessons. Study No. 2 involved 45 students (22 male and 23 female) from two grade 9 classes at a grammar school.

3rd Study No. 3 is based on classroom-observations and is focuses on the characteristics of teacher-student-communication. 25 chemistry lessons of two grade 9 classes (the same as in Study No. 2) were video taped and analysed.

Theoretical approach

Design of the Kiel Motivational Learning Climate Questionnaire

The "*Kiel Motivational Learning Climate Questionnaire*" (KMLCQ; Bolte, 1996) was designed specifically for a systematical analysis of students' perceptions of their classroom-climate and their preferred learning environment. The

H. Behrendt et al. (eds.), Research in Science Education – Past, Present, and Future, 277–282.

questionnaire is based on the pedagogical interest theory (Prenzel et al., 1988; Deci & Ryan, 1991), on theories of achievement motivation (Heckhausen, 1989) and on reflections from the field of classroom and learning environment research (Fraser, 1989). The instrument focuses on the following six motivation-indicators: comprehensibility/requirements, opportunities to participate, subject relevance, class co-operation, individual student willingness to participate and satisfaction.

This questionnaire contains three or four items per motivation-indicator. Each item is aimed at a certain instructional feature, that is to be assessed between two poles and from two points of view (see example). There is a seven-point-rating-scale to assess the items. The statements that correspond to our ideas about a "good" chemistry lesson are coded with high numerical values ("7" to "5"). Negative statements receive low numerical values (between "1" and "3"); the scale value "4" corresponds to a "neither - nor assessment".

The teaching subjects in chemistry lessons are...

very difficult for me to understand	[ ] [ ] [ ] [ ] [ ] [ ] [ ]	very easy for me to understand

I wish the teaching subjects in chemistry lessons were...

very difficult for me to understand	[ ] [ ] [ ] [ ] [ ] [ ] [ ]	very easy for me to understand

Design of the Kiel Observation Instrument

For the analysis of teacher-student-communication, a computer-assisted and categorically structured observation instrument (the "*Kiel Observation Instrument*"; Bolte, 1996) was designed. The observation instrument is based on the concept of learning and teaching from pedagogical and psychological theory.

The observation instrument has five analytical dimensions. Each dimension corresponds to a question:

- Who is the main agent (analytical dimension No. 1: agent)?
- What type of action does he/she carry out (analytical dimension No. 2: type of action)?
- In what manner and tone does he/she carry out this action (analytical dimension No. 3: manner and tone of acting)?
- To which context is this information related (analytical dimension No. 4: context of information)?
- To whom is this information addressed (analytical dimension No. 5: addresses)?

Each analytical dimension holds various options to choose from, i.e. several observation categories or sub-categories to code the 'behaviour of an agent'. The sub-categories of the organizing/moderating-category (analytical dimension No. 3: manner and tone of acting) will serve as an example:

- calling a person
- giving an order
- transmitting
- making certain
- asking "what" (examining)
- asking "why"
- asking for ideas
- asking for evaluation (feedback)
- focussing on a problem (giving a hint to let a person solve a problem on his/her own)

The observation data can be analysed by quantitative and qualitative criteria. Frequency and duration, as well as the average duration of a registered category or category-combination (t_i*), its share related to all registrations (n_i%) or to the total observed time (t_i%) provide information about the types and effects of classroom-interactions (see figure 3). - Objective- and reliability-analysis prove that both instruments are theoretically and empirically sound (Bolte, 1996).

Design of the motivation-model

Various motivation and interest concepts, such as those by Gräber (1993), Heckhausen (1989), Deci and Ryan (1991) and Prenzel (1988) as well as results from learning environment research (Fraser, 1989) served as a basis for the KMLCQ and the motivation-model (see picture 1). The model has seven variables: three independent variables are characteristical of a teacher's behaviour, whereas the four remaining dependent variables are characteristical of the class in general and of an individual student.

The variable "requirements" describes the intellectual level of the instruction and the standards of the topics taught. It is well known from theories of achievement motivation (Heckhausen, 1989), that a certain standard increases students' efforts and performance. The variable "opportunities to participate" indicates to what degree the teacher is susceptible to students' ideas and opinions. "Subject relevance" indicates whether the students find the topics chosen by the teacher relevant with respect to their every-day life.

These three variables are likely to influence students' behaviour during lessons. A suitable standard, a high level of relevance and the opportunity to participate should increase the variables "class co-operation" and "individual student's willingness to participate". Furthermore, a suitable standard should directly increase student performance (indicated by grades). Relevance of topics should entail a higher satisfaction with the instruction. Finally, active class co-operation should lead to an increase in individual student's willingness to participate and his/her willingness to participate should at least increase his/her performance and satisfaction.

Empirical confirmation of the motivation-model

Path analysis (LISREL 7) applied to the data of four different student-samples (Study No. 1: secondary and grammar school students, male and female students; n=589) shows two things: On the one hand the hypothesis of only one 'general' motivation-model could not be confirmed (correlation between some indicators differ in the analysis differentiated by school-type or gender). On the other hand the analysis confirms that other motivation-indicators in all student-populations may be identified, that influence a student's willingness, satisfaction and performance significantly (see fig. 1). The two most important motivation-indicators which were identified in all analyses are:

- subject relevance
- requirements/comprehension.

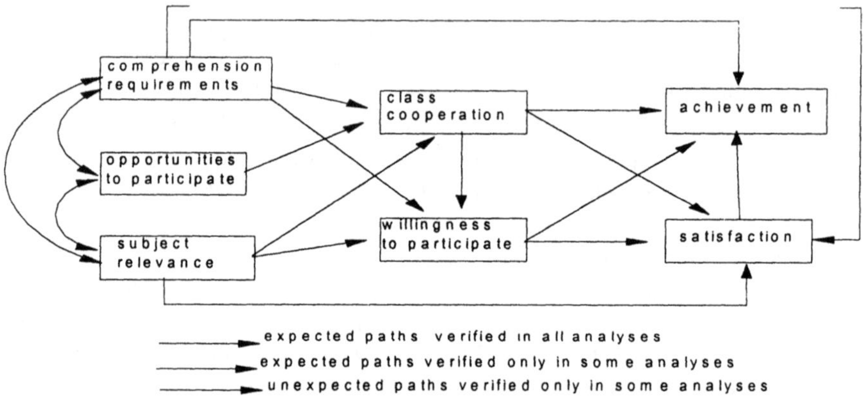

Fig. 1: Model of the interrelations between the motivation-indicators and students' performance

Analysis of students' assessments

Taking into consideration students' assessments (Study No. 2: n=45), some reasons for unsuccessful science teaching become obvious. When comparing the students' inclinations with their perceptions of the motivational indicators it becomes clear that desire and reality do not coincide. Ordinary science lessons do not live up to students' wishes. Of course, it may be doubted that this will ever be possible, but comparison of students' ideas with their perception of everyday instruction offers teachers substantial help. For where the greatest differences occur, change is badly needed. The analysis reveals the most important differences in the field of subject relevance and standards/comprehensibility (see fig. 2).

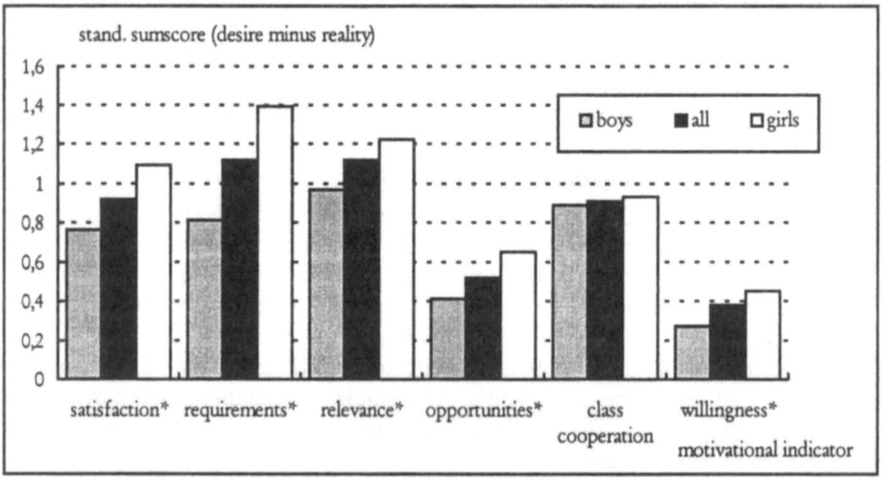

Fig. 2: Students' assessments of the motivational indicators

Furthermore, the analysis indicates significant differences in the assessments of boys and girls. In this study the girls assessed their instruction more unfavourably than the boys and once more: the biggest differences in the assessments of boys and girls occur in the field of subject relevance and standards/comprehensibility (see fig. 2).

Observation of the teacher-student-communication

The 25 video taped lessons were analysed with the help of the Kiel Observation Instrument. A sample of the data is provided in picture 3:

Category resp. Category-combination:	n_i%	T_i %	t_i* [sec]	**Category resp. Category-combination:**	n_i%	T_i %	t_i* [sec]
Teacher	**63,3**	**55,0**	**6,2**	**Male & Female student**	**21,8**	**11,2**	**3,7**
Informing (initiatively)	17,8	24,1	9,7	**Male student**	**15,6**	**7,9**	**3,6**
Giving feedback	10,9	1,8	1,2	Informing (initiatively)	2,1	0,9	3,0
Answering (reactively)	0,9	0,8	6,1	Organizing/moderating	1,6	0,7	3,2
Organizing/moderating	25,9	12,7	3,5	Giving feedback	0,5	0,0	1,2
Calling a person	*8,3*	*0,9*	*0,7*	Answering (reactively)	9,3	4,9	3,8
Giving an order	*4,2*	*1,8*	*3,1*	**Female student**	**6,2**	**3,3**	**3,8**
Asking "what" (examining)	*9,9*	*6,5*	*4,7*	Informing (initiatively)	0,1	0,0	2,6
Transmitting	*0,5*	*0,0*	*1,0*	Organizing/moderating	0,2	0,1	3,9
Asking for evaluation	*0,2*	*0,1*	*2,2*	Giving feedback	0,0	0,0	0,6
Asking "why"	*1,7*	*1,6*	*7,0*	Answering (reactively)	5,0	2,7	3,9
Asking for ideas	*0,7*	*0,4*	*4,5*				
Transmitting	*0,1*	*0,1*	*3,5*				
Making certain	*0,3*	*0,2*	*3,1*	*Time for reflecting*	*7,4*	*4,9*	*4,7*

Fig. 3: Sample of observation data

The data of the classroom observations point out:

- There are (too) few opportunities for vivid student-dominated class-discourses (compare teacher- and male- and/or female-category-combinations).
- The teacher had difficulties in moderating and organizing the class-discourses (see the teacher's organizing/moderating-subcategories): The teacher's discourse-moderation and his context orientation during the classroom-discussion mostly focussed on only one accepted answer (see the teacher category-combinations 'calling a person', 'giving an order' and 'asking "what" (examining)' compared to the other teacher-organizing-category-combinations). Last but not least:
- The sequence of acting demonstrates the (very) high dynamic of the teacher-student-interactions during the lessons (see category-combination 'time-for-reflecting' and the short average of duration of all the other category-combinations).

Conclusions

The teacher's moderating style, his manner and tone of asking questions and his concentration on only one accepted answer - may lead to a mixture of learning and performance-situations, because from the students' point of view, their answers might not be the scientifically 'right' ones. In their opinion, giving 'right or wrong answers' has consequences for their grades and school report. This explains their perception of (too) high requirements and discomfort (see picture 1 and 2). From the students point of view, poor participation during the lesson is an appropriate method to handle this situation (see table 1).

When interpreting the classroom observation data and the results of the students KMLQ-assessment on the basis of the motivation-model, some causes of poor student performance in science (especially chemical) instruction described in TIMSS become obvious: On the one hand, too high requirements and poor relevance of subjects (as the students assessments reveal) together with an unprofessional classroom management (as the classroom-observation-data show) lead to unmotivated and dissatisfied students and at the very least to poor performance (see students' assessments and interrelation between the motivational indicators in the motivation-model). On the other hand appropriate standards and topics in the lessons that fit students' interests together with a classroom organisation that concentrates more on students' conceptions and their vivid participation during the lesson will increase students' motivation and their ability to communicate about science related topics.

Recommendations

When attempting to put down recommendations for science teachers, they might be put as follows:

- Take your students' interests, ideas and conceptions seriously!
- Try to concentrate more on your students' ideas and conceptions during class-discourses! And last but not least:
- Let them speak - don't talk too much yourself!

References

Baumert, J. et al. (1997). *TIMSS - Mathematisch-naturwissenschaftlicher Unterricht im internationalen Vergleich.* Deskriptive Befunde. Opladen: Leske und Budrich.

BLK (1997). *Expertise: Steigerung der Effizienz des mathematisch-naturwissenschaftlichen Unterrichts verfaßt für die BLK (Bund-Länder-Kommission)-Projektgruppe "Innovation im Bildungswesen".* Bonn: BLK.

Bolte, C. (1995). Relationships between different learning climate variables, students' satisfaction and achievement in chemistry instruction - differentiated by school type and gender. Paper presented at AERA-annual meeting, April 1995 in San Francisco, California, USA. (Polyscript).

Bolte, C. (1996). *Entwicklung und Einsatz von Erhebungsinstrumenten zur Analyse der Schüler-Lehrer-Interaktion im Chemieunterricht - Ergebnisse aus empirischen Studien zum Interaktionsgeschehen und Lernklima im Chemieunterricht.* Kiel, Germany: IPN. (Dissertation).

Deci, E. L. & Ryan, R. M. (1991). A motivational approach to self: Integration in personality. In R. Dienstbier, Ed., *Nebraska symposium on motivation* (pp. 237-288). Lincoln, The Netherlands, 38 volumes.

Fraser, B. J. (1989). Twenty years of classroom climate work: progress and prospects. *Journal of Curriculum Studies* 21, 307-327.

Gräber, W. (1993). Untersuchungen zum Schülerinteresse an Chemie und Chemieunterricht & Interesse am Unterrichtsfach Chemie, an Inhalten und Tätigkeiten. *Chemie in der Schule* 39, 270 – 273, 354 - 358.

Heckhausen, H. (1980, reprinted 1989). *Motivation und Handeln.* Berlin: Springer.

MNU (1989). *Empfehlungen zur. Gestaltung von Chemielehrplänen.* MNU-Schriftenreihe, Köln: Dümmler.

Prenzel, M. (1988). *Die Wirkungsweise von Interesse. Ein pädagogisch-psychologisches Erklärungsmodell. Beiträge zur psychologischen Forschung.* Opladen: Westdeutscher Verlag.

How do Boys and Girls use Language in Physics Classes?

Helga Stadler
Institute for Theoretical Physics, University of Vienna, Austria
Gertraud Benke
Institute for Applied Linguistics, University of Vienna, Austria
Reinders Duit
Institute for Science Education at the University of Kiel, Germany

Abstract

Recent literature clearly shows that girls in middle and high schools are not very interested in physics classes. There are a number of proposals now detailing what physics classes should be like in order to take gender issues into consideration (Hoffmann, Häußler, & Peters-Haft, 1997). These suggestions are basically drawn from results of questionnaires investigating the interests and attitudes of boys and girls. Some studies seem to indicate that particularly constructivist approaches lead to a higher acceptance among girls (Stadler, 1998). However, boys and girls are not only different with regard to their interests but also concerning their social behaviour, their working methods and their language (Gilligan, 1982). On the basis of these results, the findings of a study on students' attempts to make sense of a pendulum showing chaotic behaviour (Stadler & Duit, 1997) led us to the hypothesis that boys and girls come to physics classes with different notions of understanding and hence differ in their social and linguistic behaviour.

Data material and method of the study

An instructional unit on basic ideas of chaos theory was taught to some 25 students (typical age 16 years) in a Viennese Grammar school. Throughout the five lessons, (45 minutes each) group work - mostly arranged as open inquiry sessions - played a significant role. The first author of the present paper taught the instructional unit. Whole class activities and the group work of two groups throughout the five lessons were documented with a video camera. Full transcripts of the video document comprise the data of the present qualitative case study. Starting from a preliminary hypothesis about boys' and girls' notions of understanding, as well as being grounded in the Vygotskian tradition of believing that everyone's thinking practices emerge from and are reflected in their discourse practices, we set out to analyse the students' discourse, in detail, using the following methods and theories: (a) question-answer analysis in a conversation analytical framework (van Lier, 1988; Hatch, 1992), (b) an analysis of the use of anthropomorphism in reasoning (Resnick et al., 1993; Keil, 1994) and (c) an analysis of group interaction in the ethnographic tradition (Tannen, 1994). Our analyses of the case study reveal how differences between girls and boys are shaped and perpetuated by the ongoing interactions. The findings presented here are preliminary, as only major trends are summarized in assertions.

H. Behrendt et al. (eds.), Research in Science Education – Past, Present, and Future, 283–288.

Results

In the following, we provide evidences, from our data that support the above hypothesis. The three sections are summarized in assertions that need further investigation in subsequent studies. The topic of the lessons, under inspection here, is the limited predictability of chaotic systems, in particular the strange behaviour of a chaotic pendulum.[1] A pendulum bob (iron) swings over three symmetrically arranged magnets. It is impossible to predict over which magnet the bob will come to rest. In order to support students' understanding certain analogies are employed, such as a ball rolling down a ridge or approaching a wall (for more information on the instruction unit used see Duit et al., 1997).

The different behaviour of boys and girls when answering the teachers' questions.

In whole class situations, the boys dominate the conversation between teacher and students, including situations which are new for both girls and boys. The following scene illustrates this:

Scene 1: The teacher begins the lesson by demonstrating the behaviour of the chaotic pendulum. Students are asked for predictions for what will happen. Then she asks the students to predict what will happen if the pendulum bob is started from the same position again. The number of girls and boys is about the same; usually in all subjects the girls of this class have better marks than the boys. Nevertheless, it is almost exclusively the boys spontaneously offering their opinions. Only two girls begin to speak, one of them is Julia:

Teacher: Different magnet. Chance. Are there other opinions? What would the movement of the pendulum look like? // Julia: Similar. // Teacher: What do you mean by that? Julia: Yes, it will also be a zig-zag-motion. // Teacher: A zig-zag-motion. // Boy: There are too many parameters that have to be taken into account.

The boy's remark stops the discussion between Julia and the teacher. It is only the following group work, that opens a new chance for the girls. Girls and boys work separately, discuss open ended questions and write down their results. Every group has to prepare a final statement. When discussing these statements in front of the whole class the girls are now more involved than the boys.

An analysis of the complete set of verbal interactions under consideration here, reveals, that closed questions offered by the teacher are more frequently answered by the boys, open questions result in a more engaged participation of the girls. The boys tend to use a clipped telegram style more frequently than the girls; i.e., they use only half sentences or merely nominal phrases and they start to speak in technical science terms very early on, whereas girls answer in complete sentences, drawing on vocabulary from everyday language. This, apparently, nails them down to a particular message, that implies certain risks in the context of school life. The hints and half answers of the boys usually including technical terms are frequently interpreted as correct answers by the teacher.

Briefly summarized, it appears that, in general, the boys' use of technical terms, familiar to the teacher, leads her to think that they have something in mind that is correct from the science point of view. Understanding the girls' ideas takes more of the teacher's time and effort, which, in combination with the self-assured dominating behaviour of the boys, leads to the result that their answers are more seriously, taken into consideration.

---

[1] see Figure 3 in the contribution of Wilbers and Duit in the present volume.

Gender specific behaviour in group work

The following results are taken from an analysis of two groups, each consisting of two girls and one boy. In both groups, the boys use language that, at least, potentially leads to domination, e.g. they use instructions and avoid questions which might show them as inexpert. Their questions are rather procedural, therefore they also have a controlling function, at least of the pace of the working process. They provide answers and explanations. Girls on the other hand raise questions about the content, they show uncertainties concerning the understanding of phenomena and they try to overcome these uncertainties in discussions. The following scene demonstrates the differences between the interactions of boys and girls.

Scene 2: The group consists of two girls and one boy. The three students have known each other for a long time and are friends. They are asked by the teacher to design chaotic systems and to explain why the systems they created are chaotic.

Girl 1: Look at a waterdrop flying against a window and running down. One cannot say, wether it will go right or left. // Girl 2: It will always be different. // Boy: Well, the waterdrops combine, or something like that, and a somewhat bigger drop of water is flowing downwards. That's it. // Girl 2: Everytime it combines in a different way. // Boy: Forget it. // Boy: It's a waste of time. // Girl 2: If one allows water to flow down somewhere, for instance if one pours water into a bottle, the water will have no distinct form. // Girl 1: I do not know, where, in this case are the instable states of equilibrium?

Meanwhile the boy is looking at what is going on somewhere else.

In this scene and in others of this group, the boy predominantly uses imperatives or instructions, like "Forget it!" or "Think it over!" He shows no uncertainties, does not raise any content questions, if there are questions, they are procedural. But the girls are self-confident, they do not give any orders, but they reject the orders of the boy. They take part in the conversation as equal partners.

The following example shows the boy looking for concrete solutions to the problem, while the girls look for possible "fields", where they might find a solution. They do not present a solution, but think aloud about possible ways to solve the problem and invite the other members of the group to participate in their way of thinking.

Scene 3: Same group as in scene 2.

Boy (taking his text-book out of his pocket): Let's think it over, chaotic systems. Let's think about balls. // Girl 2: In the universe, chaos dominates. Thermically ...The beams of the sun. // Girl 1: The sun is of no importance within the universe. // Girl 2: I do not know. The planetary system perhaps.

The above findings may be summarized in the following. In the extent to which the boys actually dominate the interaction, (i.e. their style and their intentions are successful), the communication seems to be more "concrete", i.e. they talk about facts rather than about their own uncertainties, guesses and "half understandings". This linguistic behaviour reinforces the disparity between boys and girls in the classroom or in group work: Girls ask, boys answer. It would hardly be surprising if this linguistic pattern would lead both students and teachers to the impression that on the one hand the girls know less, are less competent and on the other hand the boys *are* competent.

Language and argumentation issues

(a) The use of everyday language

Studies (Hoffmann et al., 1997) demonstrate that girls keep using everyday language for a longer time than boys. There is also evidence from research that understanding may be hampered if physics terminology is introduced too early. Our transcripts show that it is particularly important for the girls to have the chance to formulate their initial ideas and thoughts freely in small groups and to develop their thoughts by writing, which gives them the opportunity to formulate their ideas in the technical terminology of physics.

(b) The use of concrete situations in examples and arguments

Already, Gilligan (1982) noticed that girls think about concrete situations in a more personal way, rather than in concise, abstract, idealised rule systems. This becomes linguistically apparent in the use of deictic elements (here, now, I, you). There are a number of examples of this linguistic behaviour in our transcript. The following example demonstrates this form of thinking:

Scene 4: Within a whole class discussion, a boy and a girl look for examples of chaotic behaviour.

Boy: *... when a star explodes, then the gravitation is changing and this influences the curves of the planets. Another example: if one is skiing downhill on a path full of humps, if one falls down then, one does not know in which way one will fall ...*

Girl: *If you fall down a staircase, you cannot predict, where you will fall.*

In a more explicit way, we find this kind of argumentation when students - in order to find explanations of a phenomenon - refer to what they themselves or humans are doing (Scene 5). We found examples of this kind exclusively in statements from girls.

Scene 5:

Teacher: *Which circumstances influence the movement of the pendulum? We discussed magnetic forces ... //* Boy: *The tension of the thread. Friction. //* Teacher: *Andrea? //* Andrea: *It is the human hand, there are small differences ... because the position of my hand is different. //* Teacher: *How can we write that on to the blackboard? //* Andrea: *The starting-position is different.*

The example appears to demonstrates different thinking patterns of the boys and the girls thinking: The boy starts with abstract terms, the girl's starting point are her own feelings and the movements of her body. Later she generalizes by using the correct technical terms. Being asked to write down the results of her reflecions is of importance for her attempt to express her thoughts in correct terms and to find a better understanding.

(c) Anthropomorphisms

It is well known from studies on conceptions of science phenomena, that students quite often use explanations in which the behaviour of certain sets of objects is interpreted in terms for human behaviour. This kind of analogy appears to be typical for human thinking. The science educator Wagenschein (1976) is of the opinion that anthropomorphic formulations do not only appear in the first phases of science lessons but whenever the process of understanding is set into motion (for an account on the heuristic value of anthropomorphic thinking in science learning see

Zohar & Ginossar, 1998). Gender differences are not discussed in the literature on anthropomorphic speech to the best of our knowledge. However, our transcripts indicate that anthropomorphic formulations are almost exclusively used by girls and that girls use this form of analogy again and again.[2]

In the situation of learning about the chaotic behaviour of a pendulum, for instance, the pendulum is seen in analogy to a person and the chaotic behaviour of the pendulum in analogy to the actions of a person. The girls talk about the pendulum, that "*cannot decide*" where it should move, and try to explain, for instance, the strange behaviour by saying "*it seems to like the colour yellow*". This kind of thinking might be seen as an intermediate state, that leads them to correct solutions.

Scene 6: Three girls are discussing the physical entities, that influence the movements of the pendulum.

Girl 1: That it overcomes what the pendulum actually does, that means it influences the movement of the pendulum only because the force is bigger than the will of the pendulum, but we cannot write it like that.

In the following, the girls translate the substantive "will" with "force" and step by step they approach to the correct answer.

Summary: Boys' and girls' notions of understanding in physics

The above preliminary findings of our studies may be summarized in the following table:

Girls	Boys
More frequently ask questions regarding the content in question.	Ask questions on how to further proceed in making sense of a phenomenon.
Are process oriented.	Are result oriented.
Relate physics to their everyday knowledge (use everyday language).	Move into the framework of science (use scientific terminology).
Search for an external relevance on physics and technology.	Accepting physics and technology for their own sake.

To explain these differences we claim that girls and boys have different (tacit) notions of what it means to "understand" in physics. Briefly outlined, girls do not think they understand a concept until they can put it in a broader (non scientific) content. They particularly try to understand the relations of the system of physics to the world seen as a whole – this system can then, possibly, be understood by the coherence of the world. Boys, in contrast, tend to regard technology and physics as valuable in themselves. They are more interested in the internal coherence of the system of physics (and technology) itself. They usually do not have a tendency to relate the formulas and terms to the world (or their understanding of the world) in order to get the feeling that they have understood something. Rather, they "operate" with the objects within physics.

[2] We are surprised that anthropomorphic formulations are used nearly exclusively by the girls in our study. Clearly, further research is necessary to investigate whether this holds for other groups of students, dealing with other topics.

Interpretation of our data is continuing and, hence, the idea of different notions of understanding of girls and boys is developing. In short, we provide preliminary findings from an explorative study - however, we think that the findings gained so far are promising. If further studies support our hypotheses, important consequences for science instruction follow.

(a) Teachers' questions: Teachers need to be aware that open ended and closed questions address either girls *or* boys. They need to use both types of questions in a balanced way.

(b) Group work: If boys and girls participate in a group there is a tendency that boys dominate the discussion. If this happens, group work is not efficient for girls. However, if the girls are self-confident enough to reject the boys' dominance, the boys benefit from the girls' participation as they foster open discussions.

(c) Language and anthropomorphisms: Girls and boys should be given the opportunity to formulate their ideas in everyday language and to use (personal) analogies and anthropomorphisms - not only in the beginning of the learning process. Girls tend to use (personal) analogies and anthropomorphisms more often than boys. They should be viewed as valuable starting points in the learning process.

(d) Writing down ideas: Girls tend to take the task of writing down ideas very seriously. It appears that this activity significantly supports their understanding of science. Hence, enough time should be provided for that.

References

Duit, R., Komorek, M., Wilbers, J., Roth, W.M. & Stadler, H. (1997). Eine Unterrichtseinheit zur eingeschränkten Vorhersagbarkeit chaotischer Systeme für das 10. Schuljahr [A teaching unit on limited predictability of chaotic systems]. In J. Willer, Ed., *Didaktik der Physik. Vorträge Frühjahrstagung Berlin* (pp. 276-281). Bad Honneff, Germany: Deutsche Physikalische Gesellschaft.

Gilligan, C. (1982). *In a different voice. Psychological theory and women's development.* Cambridge, MA: Harvard University Press.

Hatch, E. (1992). *Discourse and language education.* Cambridge, MA: Cambridge University Press.

Hoffmann, L, Häußler, P., & Peters-Haft, S. (1997). *An den Interessen von Jungen und Mädchen orientierter Physikunterricht* [Physics instruction oriented towards boys' and girls' interests]. Kiel, Germany: IPN - Institute for Science Education.

Keil, F. C. (1992). The origins of an autonomous biology. In G. R. Megan & M. Maratsos, Eds., *Modularity and constraints in language and cognition.* The Minnesota Symposia on Child Psychology, Vol. 25 (pp. 103-137). Hillsdale, NJ: Erlbaum.

Lier, L. van (1988). *The classroom and the language learner.* London: Longman.

Resnick, L. B., Salmon, M., Zeitz, C. M., Wathen, S. H., & Holowchak, M. (1993). Reasoning in conversation. *Cognition and Instruction* 11, 347-364.

Stadler, H. & Duit, R. (1997, September). Teaching and learning chaos theory - Case studies on students' learning pathways. Paper presented within the Poster Symposium "Studies on Educational Reconstruction of Chaos Theory" at the ESERA Conference in Rome.

Stadler, H. (1998). Die Bewegung der Erde. Ein Einführungsunterricht in die Mechanik [The movement of the earth-an introduction into mechanics]. *Unterricht Physik* 9, 46, 24-34.

Tannen, D. (1994). *Gender and discourse.* New York, Oxford: Oxford University Press.

Wagenschein, M. (1976). *Die pädagogische Dimension der Physik* [The pedagogical dimension of physics]. Braunschweig, Germany: Georg Westermann Verlag.

Zohar, A. & Ginossar, S. (1998). Lifting the Taboo regarding teleology and anthropomorphism in biology education – heretical suggestions. *Science Education* 82, 679-697.

Part 6: Instructional Media and Lab Work

Improving the Use of Instructional Illustrations in Learning Chemistry

Elke Sumfleth and Lucyna Telgenbüscher
Chemistry Department, Essen University, Germany

Abstract

The investigation focuses on improving the intensity of illustration processing. Cognitive models of learning with text and illustrations (Mayer, 1993) as well as the model of understanding illustrations developed by Weidenmann (1988) provide the theoretical framework for this research. A constructivist learning environment where learners become designers of the instructional material, where they have to design a steps-and-parts illustration, combined with a text from a rich offer of external representations, fosters the intensive processing of learning material. This is supposed to lead to an improvement in retention and problem solving performance.

Introduction

Investigations concerning understanding of chemical terms show that students regard pictorial meaning of content as a first step to understanding (Sumfleth & Körner, 1993; Körner, 1994). A student said: *If you develop an image, you gain insight, otherwise it's only knowledge without understanding.* This might be one reason that illustrations enjoy growing popularity in popular-science literature, school textbooks and science journals. These illustrations often combine different possibilities of visualisation in one figure or reflect different abstraction levels.

Theoretical background

Texts and illustrations help to develop mental models. Illustrations support the understanding of a text and language helps to understand an informative illustration. Schnotz, Picard and Hron (1996) found that successful students use illustrations more intensively than the less successful ones. Illustrations mediate the appearance of objects, the composition of scenes, the flow of events, spatial constellations and the synchronous change of several system components better than verbal descriptions.

According to Mayer (1993), the main cognitive processes that can be supported by explanatory illustrations are selecting, organizing, integrating and encoding of information. Explanatory illustrations direct the learners' attention to relevant information, they help towards understanding relations between different system elements and they support the integration of new information into already existing knowledge. Most successful are "steps-and-parts"- illustrations. They consist of several, single illustrations, each presenting the state of the system at various points

H. Behrendt et al. (eds.), Research in Science Education – Past, Present, and Future, 289–294.

in a process (Mayer & Gallini, 1990). Its cognitive function is to help to envision a mental model of the system, that is consistent with the learners' pre-knowledge. Above all, illustrations help especially students with poor knowledge.

Following Schnotz (1996) text and illustration are regarded as complementary sources of information, because they represent an object not only in a symbolic way but also in an analogous way. They support the construction of a mental model in different ways. Schnotz explains learning with texts and illustrations as an interaction of propositional representations and mental models. Understanding verbal information, means to construct a mental model based on propositional representations. That affords a bigger cognitive effort than starting with illustrations. Then the mental model can be used to understand new information and to elaborate propositional representations. This information processing model explains not only an improvement of retention but also a deeper understanding of concepts if illustrations are used.

Thereby other characteristics of the learners play important roles, like individual competence and subjective assessment of the relevance of the learning material. In addition, one has to consider the social context, learning habits and personal aims. The results of previous investigations show that illustrations tend to be recalled incorrectly by novices, that pictorial information without a complementary textual description leads to misunderstandings and may induce non scientific conceptions and that a distance between text and illustration causes separate usage of both sources of information. In addition, there are several variables influencing effective learning with illustrations apart from the illustration itself. This is for example learners prior knowledge. For instance, experts and novices judge the same illustration differently as shown in the following by two quotations from interviews. The expert says: *I think it is quite clear what happens in the mass spectrometer. What could be added, because it isn't clear here, is that the particles should have a certain velocity before they enter the magnetic field (...). Although it would become more complicated. The illustration is clear for basic understanding.* On the contrary the novice comments: *It isn't clear to me when I read here: 'Ionaccelerator' and the arrow points to an empty space. It looks as if the sample is evaporated on the glowing electrode. I can't imagine that the arrows depict the electron beam.* Besides learners' prior knowledge, his or her intensity of illustration processing, his or her attitude towards the illustration, his or her visual literacy, like the ability to decode the symbols and of course the quality of the text presented with the illustration, influence the effective use of illustrations.

Following Weidenmann (1994) there are two kinds of illustration processing, an ecological one and an indicatorical one. Ecological processing describes an automatic realising of the contents of the illustration. The viewer uses the same procedures as for perceiving natural surroundings. When the illustration seems to correspond to the viewer's own pre-knowledge, the viewer stops the illustration processing. If the illustration does not correspond to the pre-knowledge it will be processed more intensively. The indicatorical processing goes beyond pure recognition and is directed to the reconstruction of the arguments visualised in the illustration. In daily-life, situations learners look at illustrations according to a personal task orientation. This task orientation is influenced by the supposed relevance and by elements of surprise. Therefore the processing intensity will increase when the learner recognises an illustration as provocative and stimulating.

Usually, the viewer tends towards a minimal mental effort and a minimal processing intensity according to an economical principle (Hatfield & Epstein, 1985).

Regarding chemistry, the main reasons for superficial processing of illustrations seem to be first that teaching is directed to successful manipulation of formulas, second that tests normally assess knowledge of formulas, third that realistic illustrations do not stimulate reasoning and fourth that school-book authors and perhaps teachers too believe that illustrations are self-explanatory. Often illustrations are scarcely described in the text. Learners with poor pre-knowledge, however, are not able to fill the gap. To improve indicatorical illustration processing constructivist learning environments may be effective because they support active construction and reconstruction of knowledge.

Hypotheses and design of the investigation

A learning environment in which low prior-knowledge learners design themselves a 'steps–and-parts' illustration combined with a text from a rich offer of external representations fosters the intensive processing of learning material. This is supposed to lead to an improvement in retention and problem solving performance. The idea is that learners use the presented text and illustration as tools to create instructional material. In doing so learners have to use many types of thinking skills. First there are skills like determining the nature of the task and organising the investigation, analysing and interpreting the information collected. Second organisation and representation skills are needed, like deciding how to sequence information and how to represent the information. Finally, reflection skills are important like, evaluating the process used to create it.

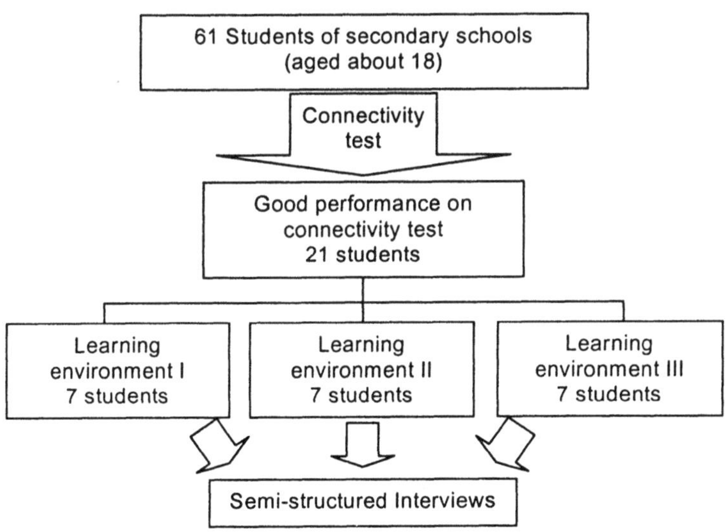

Fig. 1: Design of the investigation

61 students are given a written test (Fig. 1) which refers to the special chemical knowledge needed to understand the following instructional material (connectivity

test, see Sumfleth 1988). A relatively homogenous group of 21 students who show good performance on this test, is chosen and divided into three groups of seven students. Each group works in a different learning environment (I,II,III). Learning environment I (LE I) consists of a text describing electrophile addition on a symbolic level. Learning environment II (LE II) consists of text and symbols as well as illustrations visualising the reaction on a molecular level (space-filling models) and on a macroscopic level (realistic illustration of apparatus for bromination of ethene). LE III consists of a text divided into parts, illustrations of three different models on molecular and symbolic levels to choose from as well as seven different illustrations of the bromination experiment. The students of LE I and II receive the instruction to read the text and to look at the illustrations. Students of the LE III are to design instructional material by combining text passages with illustrations of their choice.

Procedure

After exploring the learning environments all students have to work on the same tasks during a semi-structural interview. This interview consists of four parts, of a prediction-observation-explanation-task, a recall task, a problem-solving phase and a reflection phase. The prediction-observation-explanation-task intends to test the conceptual understanding on the macroscopic level. The recall and the explanation of the reaction mechanism shall document the conceptual understanding on the molecular level. Two new problems concerning the electrophile addition to olefines shall be solved by the students to show that they are able to transfer their knowledge to new situations. And finally they shall judge the quality and usage of text and illustrations. The interviews are recorded on audio and video cassettes and fully transcribed including gestures.

Findings

Students of the third learning environment generate more connections between terms which go beyond the text content while those of the LE I merely tend to memorise and recall symbols. The latter group displays lack of understanding on the molecular level. The majority of all students recalls the stereospecific step of addition, but only the majority of students of the third environment is able to explain it. This leads to the conclusion that individuals with different internal representations create similar external representations. The majority of "wrong" statements produced by students of the learning environments I and II results from memorising formulas without conceptual understanding. Students tend to handle words without taking the meaning into consideration.

The results concerning problem-solving do not show large differences regarding the solutions achieved by students from different learning environments (Fig. 2). The second problem is more often solved by students of the third learning environment. However, there are differences regarding the way they achieve their solutions: In most cases students of the learning environments I and II use trial and error strategies to find the right formula. Students of the learning environment III, however, more often activate the knowledge necessary for solving the problem (2a,b), they take into account the submicroscopic level (3) and recognise analogies (5) There is for example the analogy between bromide and chloride ions, as they are both negatively charged and are from the same main group, showing similar chemical properties. This strategy of analogical reasoning is more successful than trial and error processes. Achieving the correct solution (7) is not the decisive criterion. In addition, even students with low pre-knowledge are able to solve the problems successfully. Learners, who design their own learning material, come to a deeper information processing beyond surface characteristics of text and illustration. They achieve a deeper understanding.

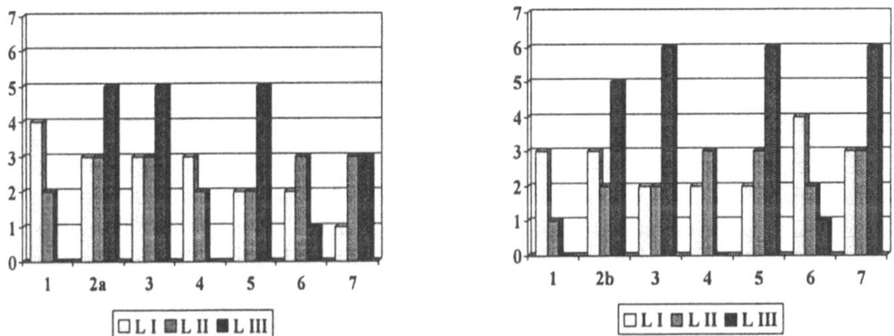

a) Problem 1 b) Problem 2

Fig. 2: ***Categories of problem solving strategies***:
1-Solution is given immediately, 2a-The water-molecule contains polar bondings (problem 1), 2b-Similarity of the chemical properties of chloride- and bromide-ions (problem 2), 3-Reflection on the submicroscopic level, 4-Rules to set up Lewis formulae, 5-Remembering the solution of an analogous problem, 6-trial & error, 7-Achieving the correct solution

Students from learning environments I and II memorise symbols without conceptual understanding and manipulate formulas while solving a problem in order to get to a plausible solution. For example one student of learning environment I argues, concerning the problem 'bromination in the presence of chloride ions': *So, I think that we obtain bromine chloride....(...) Is it possible that we get HCl? (....) Theoretically we've got this ring* [bromonium ion] *and then this chloride could be added here and we 'll obtain chloride instead of br.... And the product could have chloride and bromine or choride-chloride* (draws formula, see below). *The ions are exchangeable: there is either bromine-chloride, or bromine-bromine or chloride-chloride. I can't imagine other possibilities here.*

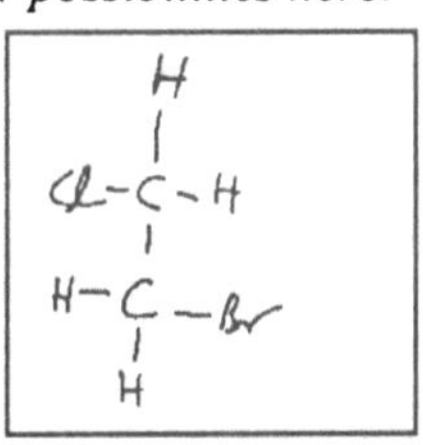

And another student of learning environment II explains the problem solution of the 'bromination in water': I can imagine that an alcan could be obtained from hexene and bromine oxid or something like this:

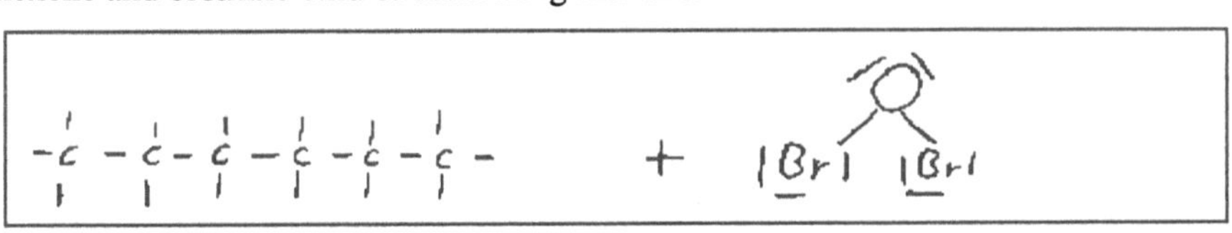

But I don't know if it exists and if it is logical. ... Or this one [bromonium ion] could be bound to water molecule ... and bromine here and OH here ... And one bromine with water (draws formulae). And this one [HBr] could be negative and that one [bromine alcohol] positive, but I'm not sure because oxygen took the electrons.

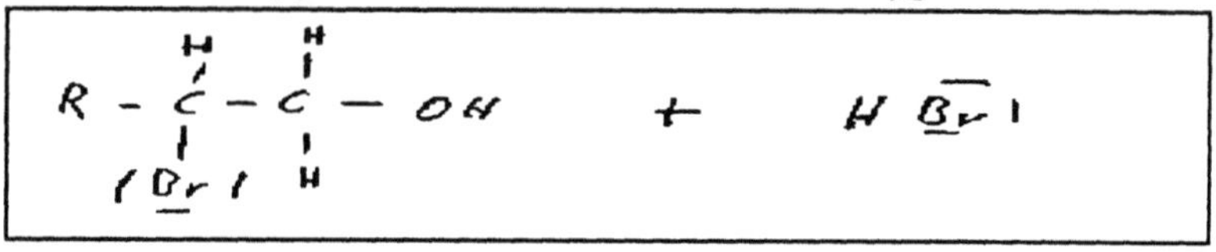

Summary

Students who get involved in designing instructional material and are provided with suitable external representations of reactions on the molecular level are more likely to generate mental representations on this level. In addition, these students have the chance to work with different modes of representations. Therefore, they are able to recognise limits of every representation. Using symbols only or models, which are very simplified, so that they are prone to be processed superficially leads to focusing attention on surface characteristics of the external representations and creating superficial internal representations. Consequently, learners have to be exposed to situations where they are forced to take well-founded decisions.

References

Hatfield, G. & Epstein, W. (1985). The status of minimum principle in the theoretical analysis of visual perception, *Psychological Bulletin* 97, 155-186.

Körner, H. D. (1994). *Vorstellen und Verstehen. [Imagination and Understanding].* Frankfurt/Main, Germany: Lang.

Mayer, R. E. & Gallini, J. K. (1990). When is an illustration worth more than ten thousand words? *Journal of Educational Psychology* 82, 715-726.

Mayer, R.E. (1993). Illustrations that instruct. In R. Glaser, Ed., *Advances in Instructional Psychology, 4* (pp. 253-284). Hillsdale, N.J.: Lawrence Erlbaum Associates.

Schnotz W. (1996). Psychologische Ansätze des Wissenserwerbs und der Wissensveränderung. [Psychological approaches of knowledge acquisition and development]. In R. Duit & C. von Rhöneck, Eds, *Lernen in den Naturwissenschaften* (pp. 15-36). Kiel, Germany: Institut für die Pädagogik der Naturwissenschaften.

Schnotz, W., Picard, E. & Hron, A. (1993). How do successful and unsuccessful learners use texts and graphics? (pp. 1-22) Forschungsbericht Nr. 60, Deutsches Institut für Fernstudien an der Universität Tübingen, Arbeitsbereich Lernforschung,.

Sumfleth, E. (1988). *Lehr- und Lernprozesse im Chemieunterricht. [Teaching and learning processes in chemistry instruction].* Bern-Frankfurt-New York: Lang.

Sumfleth, E. & Körner, H. D. (1993). Wodurch wird ein Atom zu einem konkreten Begriff? [How does an atom become a concrete concept?] In H. Kramers-Pals & G. Niehaus, Eds., *Chemiedidaktische Forschung – Lopend Onderzoek in de Chemiedidaktiek* (pp. 89-92). Essen, Germany: Westarp-Wissenschaften.

Weidenmann, B. (1988). *Psychische Prozesse beim Verstehen von Bildern. [Psychological processes and understanding of illustrations].* Bern, Switzerland: Huber.

Weidenmann, B. (1994). Informierende Bilder. *[Illustrations that inform].* In B. Weidenmann, B., Ed., *Wissenserwerb mit Bildern* (pp. 9-58). Bern, Switzerland: Huber.

Acknowledgements

Financial support of the German Science Foundation (DFG, SU 187/2-1) is gratefully acknowledged.

Computing in Stereochemistry – 2D or 3D Representations?

Slavica Pavlinic, Paul Buckley, Janet Davies and Tony Wright
Massey University, Palmerston North, New Zealand

Abstract

This paper reports part of a preliminary phase of a study examining students' responses to a computer-based (CB) stereochemistry task and to different representations of organic molecules. Six themes in the way students used the CB tutorial have been identified: manipulation of models, multiple representations, use of colour, simplicity, familiarity and types of isomerism. It has been found that 3D animations and the colourful molecular images encouraged students to practise the use of different representations. Students' preferences for either two-dimensional (2D) skeletal formulae or three-dimensional (3D) ball and stick representations depended on the focus of the exercise and also the type of isomerism they were investigating. The study has demonstrated that a suitably-designed, computer based task encourages students to make use of the molecular representation that, for them, best suits the problem.

The problem

Understanding chemistry means being able to combine the macroscopic, symbolic and microscopic dimensions of chemistry (Johnstone, 1991; Gabel, 1999). As suggested in Figure 1, the threefold nature of chemistry needs to be integrated with the use of scientific language, a feature crucial to gaining understanding (Laws, 1996).

Stereochemistry, that concerns the nature of molecular shape and its significance for physical and chemical properties, has been identified as difficult to understand (Barta & Stille, 1994; Black, 1990). It has been suggested that this difficulty arises because understanding stereochemistry requires the use of specific stereochemical terms, switching between different ways within the threefold nature of chemistry and making use of 3D visualisation ability (Tuckey et al., 1991; Tuckley & Selvaratnam, 1993). According to Kleinman et al. (1987), students may be unable to learn chemistry the classroom communication may fail, because they cannot relate to or form the image appropriate to a particular concept. The most frequent ways of representing molecules involve chemical formulae or ball and stick models aimed at to stimulating 3D visualisation (Tuckey et al., 1991). Barta and Stille (1994) declared that the translation of 2D drawings of a molecule into three dimensions, or the reverse, when solving stereochemical problems, tends to be the most problematic step for students. Success depends on the student's background, experience, intellectual and 3D visualisation skills and is frequently overwhelming for novices (Barta & Stille, 1994; Tuckey and Selvaratnam, 1993).

Computers are a particularly attractive platform for teaching stereochemistry because of the possibility of generating interactive experiences, that model the microscopic world of the molecule, as well as taking advantage of the Hawthorne

H. Behrendt et al. (eds.), Research in Science Education – Past, Present, and Future, 295–300.

effect in which a new teaching tool can bring positive learning outcomes (Pankuch, 1998). It is well documented that understanding and achievement in chemistry and particularly stereochemistry, are closely related to 3D visualisation ability (Barta & Stille, 1994, Tuckey et al, 1991, Tuckey & Selvaratnam, 1993, Bodner et al., 1986). It is assumed that students enter university having both visualisation and manipulative skills and these are not specifically taught. In this study a CB tutorial with a non-linear task design was written, that gave the students the opportunity to switch between the 2D symbolic representation of a molecule and 3D models that could be rotated on screen.

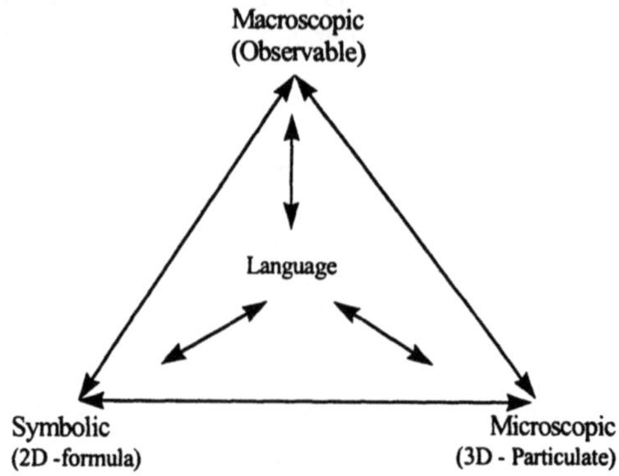

Fig. 1: The threefold nature of chemistry integrated by the use of language

Design and procedure

A non-linear task design has been described as the "one most prominent element that makes the computer a valuable tool for individualising instruction" (Schaefermeyer, 1990) or as a more student-determined learning approach (Goforth, 1994). The tasks were composed using Java Script with ChemDraw for 2D drawings and Chem3D computer models for 3D representations. It consists of i) a glossary, covering the major stereochemistry concepts and ii) exercises, that give students the chance to test and refine their conceptions about stereochemistry. Adaptive feedback, involving links between the exercises and the glossary, are available within every screen. The feedback involves branching, that directs the student down a pathway that matches individual needs (Goforth, 1994). Unlike much commercially available software (Pavlinic et al., 2000), links provide easy navigation forward, back and across different concepts.

Volunteers for testing the software study were drawn from a class of 300 students working on a first year unit on stereochemistry. Eight females and three males volunteered, four working in pairs, and the others individually. Final course grades showed that the participants ranged across the spectrum of performance.

The stimulated recall technique (O'Brien, 1993) was used in the data collecting process. The students were video taped while performing the task and the picture was combined with an image of the computer screen. Both were recorded on one tape. An observation sheet was used to record significant events as the task took place. Immediately after the task, students were shown their task performance tape and simultaneously interviewed. These interviews, in which the students talked reflectively about their task experience, were also video taped.

Verbal transcripts of student talk (while interacting with the computer based tutorial and during the reflective interviews), field notes and completed observation sheets were analysed simultaneously to identify themes in the way students used the CB tutorial (Merriam, 1998). The contribution of each theme was assessed from students' discourse in which they described the way they went about a particular aspect of a task.

Findings and discussion

Analysis of the students' talk, action and choices, lead to the identification of six themes in the way students used the CB tutorial which are summarised in Table 1.

Manipulation of models. The use of molecular models in grasping abstract chemical (Laws, 1996; Ingham & Gilbert, 1991) and particularly stereochemical (Baker et al., 1998) concepts has been found helpful as a hands-on approach and encouraged students to think of molecules and construct models on their own. The software allowed students to *rotate* 3D models and *observe* them from different views. This interactivity gave students the chance to investigate structures themselves, rather than just watching as in lectures: ".. it's good when you have to work out things like shape of molecules because you can turn on and have a look at what shape it is ..".

Table 1: Themes in the student use of the stereochemistry tutorial

Theme	Dimensions of theme	Evidence in student talk
Manipulation of models	* Rotation, inversion and reflection of 3D models and enlarging. * Observation along different axes.	* ".. it's good when you have to work out things like shape of molecules.." * ".. see how they look like from different views."
Multiple representation	* Ability to move between different representations. * Change of model display * Navigation options from model to meaning	* ".. you can move the ball and stick one, .. which is good, like put them next to each other, but once you don't really need to move, it's good the skeletal one, it's pretty easy to see .." * from ball and stick to space filling model * from Exercise to Glossary and back
Use of colour	* Different coloured atoms make the molecules easier to recognise.	*".. and it's quite handy how you got the oxygen in a different colour, that makes it a lot easier to recognise it." or ".. like the carbons and the hydrogens are pretty much the same colour and sort of, get a bit confused."
Simplicity	* For complex molecules, 2D skeletal formulae simplify. * 2D skeletal and structural formulae show the type of bond explicitly (3D models implicitly).	* " ... there are so many different ways to turn them on and sometimes you don't really know what are you looking at ..."(3D models) * ".. easier to see the positions of double bonds .." or * " .. clear to see the actual bonding .."
Familiarity	* For complex molecules, 2D skeletal formulae are familiar.	* " ... we got practised because that's what has been used in exams ..."
Type of isomerism	* Preferred representation depends on stereochemical concept.	* skeletal formulae for structural isomers (".. easier to count atoms" as ".. the ball and sticks hide each other..") or diastereomers ("..clear to see double bonds..") and 3D models for enantiomers

Three different dimensions of the theme ***Multiple representations*** have been identified. Students appreciated the ability to *move between different representations* leaving them to investigate a molecule using either one (2D or 3D) or both of them, as necessary: " ... you can move the ball and stick one, ... , but once you don't really need to move it, it's good the skeletal one". Switching between 2D and 3D representations as appropriate is an important feature of the professional chemists approach to solving problems associated with molecular structure. Having both kinds of representation students described as ".. good to see the relationship between them" and ".. because if you weren't sure about hydrogens or something you just have to look at the other molecule" or ".. because you can, like, count the carbons on the skeletal one and then go to the ball and stick and count hydrogens". If we want students to develop 3D visualisation, it is important to offer students more than one way of representing molecules so they can select the representation that better suits their own view. Ability to *change the model display* gave students the chance to see and explore the same model using ball and stick or space filling representation. After experiencing interactive 3D models (often for the first time) students expressed their appreciation of recognising a "sort of true representation". In addition, the *navigation* options in the tutorial, as shown by Figure 2, enabled students to associate the representation with its meaning, for instance.

Use of colour. The influence of colour on student learning is well known and described. Shubbar (1990) referred to the earlier study (Shubbar, 1984) in which the use of multicoloured diagrams improved the student ability to visualise rotation. In this study, students described having differently coloured atoms (e.g. red for oxygen) as "quite handy" making the molecule easier to recognise, unlike the monochrome gray-white carbon-hydrogen frames of the more complex molecules: ".. like the carbon and the hydrogens, they are pretty much the same colour and sort of get a bit confused".

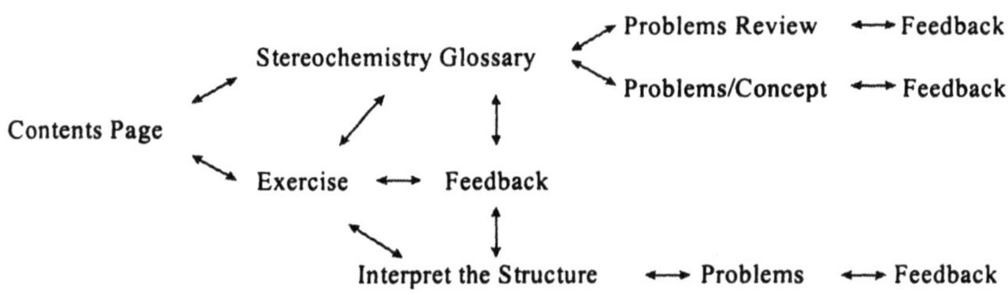

Fig. 2: Navigational links available in the stereochemistry tutorial

The theme ***Simplicity*** has been characterised by the use of 2D representations in two ways. For the more *complex molecules*, it is important to simplify the representation using symbols. In some situations the power of being able to rotate the molecules confused students because of the multitude of images they generated ".. there are so many different ways to turn them and sometimes you don't really know what are you looking at". Furthermore, 2D representations show the *type of bonds* explicitly. Students described the use of skeletal formulae as "easier to see" or "more clear" when counting atoms or looking for double bonds.

Familiarity. The symbolic level of representation is extensively used at high school, but skeletal formulae are not introduced (in New Zealand) until university. Skeletal formulae appeared more familiar to students: " .. we got practised because that's what has been used in exams .." Students who had mastered skeletal formulae found it ".. a lot easier to write them, rather than hypothesise on them" and much easier to use them when analysing structural isomers and diastereoisomers. The former were more easily distinguished, because students found skeletal formulae as an ".. easier way to count the atoms" due to the fact that ".. you get mixed up with the ball and stick model, you have to rotate it because you can't see properly and then you forget the numbers and all that". The latter is recognised by the clear presentation of double bonds. However, being able to use or interpret skeletal formulae, requires an understanding of this type of 2D representation. One student explained ".. the lines are confusing and sometimes the double bonds just confuse me more .." or another "I am sometimes confused, so I can never remember when the hydroxo group is on the end, I can never remember whether carbon is on the end, on that stick as well."

Finally, the *preferred representation depended on the stereochemical concept* being analysed describing the theme ***Type of isomerism***. The data gave clear evidence that experience with 3D representations helps when discerning the shape of the molecule, for instance: ".. because you can see where the hydrogens are", but not if the counting of atoms was required: ".. the balls and sticks hide each other" or ".. was a bit confusing because all the hydrogens are getting in the way a little bit, they sort of obscure the picture". Table 2. shows the common preferences for 2D and/or 3D representations in understanding particular type of isomerism.

Table 2: Preferences for the use of 2D/3D representations in understanding types of isomerism

Aspect being analysed	**3D interactive models**	**2D skeletal fomulae**
Structural Isomers	✓ for simple molecules	✓ for complex molecules; easier to count atoms
Enantiomers	✓ rotation possible, highlighting different atoms with different colours	
Deastereomers		✓ the positions of double and single bonds are clearly represented
Conformational Isomers		✓ the positions of double and single bonds are clearly repesented

Conclusions

The findings suggest that multiple, 2D and 3D representations, with the possibility to move between them, depending on the aspect or type of isomerism being analysed, are useful and desirable for helping students learn stereochemical topics. The wider the range of structures representing the phenomena, the more likely the student will develop an understanding of these abstract concepts.

The combination of software features like colour, 3D shape and interactivity are powerful when refining ideas about the more complex concepts in stereochemistry (such as recognising enantiomers). The 3D animations and colourful molecular images available on computer encouraged the students to practise the use of different representations. Suitably designed computer based tasks enabled the students to integrate 2D symbolic and 3D microscopic levels of understanding and enabled them to select the representation that, for them, best suited the problem. On a given problem students, of course, bring together a number of themes to get a solution.

Very positive responses were expressed for the task design (with an adaptive feedback and hints available). Most of the participants started with the exercises, visiting the glossary as needed. However, some of them chose the glossary first. The navigation options and particularly the possibility of checking terms (which have to be learned together with the concepts) if students ".. mixed them up" were very well received and frequently used.

References

Baker, R.W., George, A.V. & Harding, M.M. (1998). Models and Molecules - a workshop on stereoisomers. *Journal of Chemical Education* 75(7), 853-855.

Barta, N.S. & Stille, J.R. (1994). Grasping the concepts of stereochemistry. *Journal of Chemical Education* 71(1), 853-855.

Black, K.A. (1990). Flow chart determination of isomeric relationship. *Journal of Chemical Education* 67(2), 141-142.

Bodner, G.M. & McMillen, T.L.B. (1986). Cognitive restructuring as an early stage in problem solving. *Journal of Research in Science Teaching* 23(8), 727-737.

Gabel, D. (1999). Improving teaching and learning through chemistry educational research: a look to the future. *Journal of Chemical Education* 76(4), 548-554.

Goforth, D. (1994). Learner control = decision making + information: a model and meta-analysis. *Journal of Educational Computing Research* 11(1), 1-26.

Ingham, A.M. & Gilbert, J.K. (1991). The use of analogue models by students of chemistry at higher education level. *International Journal of Science Education* 13(2), 193-202.

Johnstone, A.H. (1991). Why is science difficult to learn? things are seldom what they seem. *Journal of Computer-assisted. Learning* 7, 701-703.

Kleinman, R.W., Griffin, H.C. & Kerner, N.K. (1987). Images in Chemistry. *Journal of Chemical Education* 64 (9), 766-770.

Laws, P.M. (1996). Undergraduate science education: a review of research. *Studies in Science Education* 28, 1-85.

Merriam, S.B. (1998). *Qualitative research and case study applications in education.* San Francisco: Jossey-Bass Publishers.

O'Brien, J. (1993). Action research through stimulated recall. *Research in Science Education* 23, 214-221.

Pankuch, B. (1998). Multimedia in lectures and on the World Wide Web. Part I. *Computers in Chemical Educational Newsletter*, Spring, 4-7.

Pavlinic, S., Wright, T. & Buckley, B. (2000). Students using chemistry courseware - insights from a qualitative study. *Journal of Chemical Education* 77(2), 231-234.

Schaefermeyer, S. (1990). Standards for instructional computing software design and development. *Educational Technology*, June, 9-15.

Shubbar, K.E. (1990). Learning the visualisation of rotations in diagrams of three dimensional structures. *Research in Science & Technological Education* 8(2), 145-154.

Tuckey, H., Selvaratnam, M. & Bradley, J. (1991). Identification and rectification of student difficulties concerning three-dimensional structures, rotation and reflection. *Journal of Chemical Education* 68(6), 460-464.

Tuckey, H. & Selvaratnam, M. (1993). Studies involving three-dimensional visualisation skills in chemistry: a review. *Studies in Science Education* 21, 99-121.

Learning Physics with Multimedia- and Experimental-Supported Workshop Instruction

Dieter Heuer
Institut für Didaktik der Physik, Universität Würzburg, Germany
Kurt Blaschke,
Gymnasium Bad Kissingen, Germany

Abstract

A concept of Multimedia- and Experimental-Supported Workshop Instruction is introduced and evaluated. Dynamic Iconic Representations are used as a new methodical element. In order to support students in building an appropriate image of how physical quantities determine an experimental process, we measure all relevant quantities with sensors or calculate them from measured values. Together with a schematic representation of the experimental process, we visualise these quantities and their connections as vectors, lines, areas, etc. on the computer screen. Two test classes were instructed according to the Bavarian Syllabus with the help of the new concept. Both classes show significantly higher performance (65% to 85% gain in the Hake-Plot) than control classes (10%-40%) even though these classes worked with interpretation of graphs or additionally with computers. We used teacher-constructed Multiple-Choice-Tests and the FCI-Test as assessment tools. When using more complex problems, similar differences in performance resulted. Students themselves reported Dynamic Iconic Representations to be very helpful for their understanding.

Introduction

Since the beginning of the eighties, science education researchers have intensively explored learning difficulties—especially in Kinematics and Dynamics—and discuss and test learning strategies to overcome them (Schecker, 1985; Thornton, 1996; Redish et al., 1997; Hake, 1998). Frequently, pre-instructional images from everyday life interfere with the concepts to be learned in Physics instruction so that only limited success results. Our evaluation with 800 11th grade high school students in Bavaria (Heuer & Wilhelm, 1997; Blaschke, 1999) shows that German students have difficulties similar to the students assessed by R. Thornton and D. Sokoloff (1990).

How can student achievement be improved?

One of the students' learning difficulties is the ability to gain information from physical experiments. Therefore it is very important to represent this information in a way that minimises the learner's cognitive efforts to explore experiments. Nowadays, this information is easy to represent when using the computer as an aid for kinematics and dynamics instruction. The common form of representation is through diagrams that, unfortunately, have some substantial disadvantages. The mental connection of the observed experiment with the diagrams has to be reconstructed first, which is normally hard, for novices. It should be more helpful to,

H. Behrendt et al. (eds.), Research in Science Education – Past, Present, and Future, 301–306.

first, manipulate the results in real-time in a way that they are visually understandable through the chosen encoding system. This is possible without lengthy logical implications. Iconic visual elements are especially suitable for this purpose as they are able to dynamically represent experimental situations and physical quantities. Compared to static figures in textbooks those visualisations explore a new learning potential: Such representations not only show the concrete experimental situation, reduced to the most important elements, but, additionally, they show structural quantities and structural connections in a form students can easily work with (Heuer, 1996).

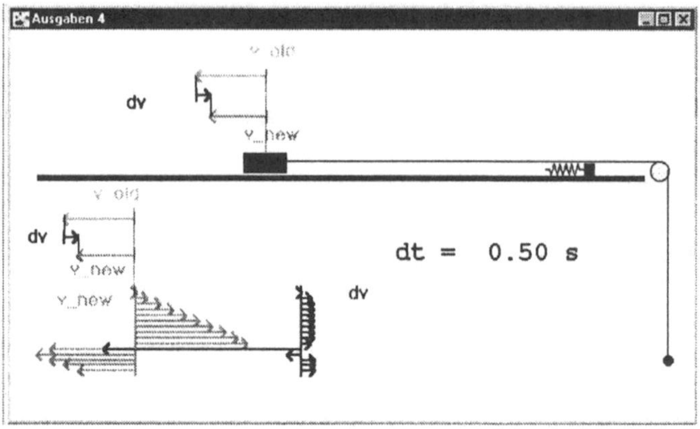

Fig.1: A glider moves on an air track. The following vectors are displayed at any time: the old velocity v_old, the momentaneous velocity v and their difference dv. Additionally, after every fixed time-interval the v-vectors are marked on the screen.

One example for this method of representation is the accelerated motion of an air track glider pulled by a weight in fig. 1, upper part. Of course, this is only a static picture and the Dynamic Iconic Representation of new and old velocity and its change dv are "frozen". When seeing how the quantities change during the whole experiment it is easy to make statements about the velocity of the glider before and after reflection at the end of the track. Using this method, it is possible to clearly represent surface features characterising the experiment as well as in-depth structure showing the underlying physical concepts at the same time.

Methodical elements

An instructional concept that enables students to use new ways of learning through visualisations also allows new methodical elements.

- Statements about physical quantities and their connections can be processed more directly through Dynamic Iconic Representations than through graphs because they are shown together with the experimental process.
- Simultaneous use of several encoding systems gives the learner opportunities to change from an abstract code to a more pictorial one that he or she is more used to. With this additional information, it is possible to eliminate difficulties in understanding the new content, s. fig. 1. Additionally, simultaneous representation of all surface properties of the experiment helps to build connections between real-world situations and structural content.

Easy-to-realise iconic representations that visualise important structural content during an experiment or its reproduction invite the learner to use mental "meta-levels" of reduced complexity. Fig. 1 shows dv the change of v as such a mental "metalevel" to the way of understanding acceleration.

- With the help of iconic representations, visualising relevant structural content, new subject-specific methodical procedures can be created.

Fig.2 shows another example: A glider moves on an inclined air track with a slope that can be varied during the experiment. Force F_H , velocity v and acceleration a are measured continuously. The experiment and its reproduction - eventually including single-step mode - make the proportionality of force F_H and acceleration a (independent of v) obvious through iconic dynamic representations.

Of course, the representations, produced with help of the computer, are primarily an opportunity for the learner to test his or her ideas of physical content and to work on it. Representations should help the student to actively think about his or her imaginations, i.e., to express and discuss them and to match them with reality. Suitable instructional modes are discussion with teacher, demonstration experiments, group work with student lab experiments and group work with corresponding simulations. These instructional modes can be used alternately.

Integrated use of these methodical elements leads to a new methodical concept characterised as Multimedia-Experimental supported Workshop (MEW)-Instruction. We call this instructional method multimedia-experimental because a variety of different media like computer-supported experiments and simulations, with their different codes are used in parallel. Workshop Instruction symbolises the open nature of the learning processes and the active work of the student on a basis of already existing ideas and mental models. This reinforces the constructive component of learning.

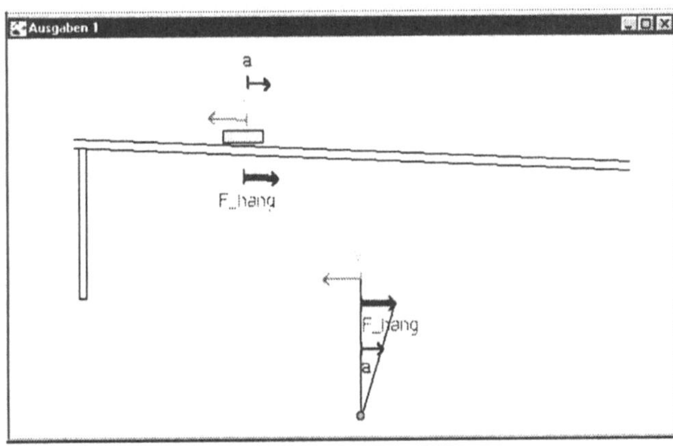

Fig.2: Changing the inclination of the air track permanently causes a variable net force F_H on the glider. The experiment directly shows the proportionality of F_H and acceleration a independent of velocity v.

Multimedia- and experimental-supported workshop instruction

Blaschke realised the MEW-concept in two pilot studies - each in two 11th grade Science classes including 43 and 37 students. As a tool for the computer-supported experiments with Dynamic Iconic Representations as well as for the simulations, we used the program PAKMA[1]. In the first study in 1996/97, we connected MEW-instruction with student lab experiments (SLE). In the second study in 1997/98 we used teacher demonstration experiments (TDE) only as instructional mode, but let students work on selected PAKMA simulation projects in the computer lab in

[1] PAKMA can be loaded from the web: http://didaktik.physik.uni-wuerzburg.de. PAKMA Projects and material for the course can be ordered from the authors (heuer@physik.uni-wuerzburg.de).

groups of two after the intermediate test (Blaschke & Heuer, 1999). Including all experiments, the sequence conducted in accordance with the syllabus takes about 23 lessons.

In the following, we will present our realisation and report our experiences concerning the MEW-TDE concept and the results of the studies. Results of the test after MEW-SLE instruction were similar to the ones after MEW-TDE (1997/98).
The instructional method is characterised by the following learning steps: make predictions; discuss in groups; observe experiments without and with the aid of the computer; analyse statements from Dynamic Iconic Representations (DIR), later including diagrams; work out a solution in class discussion and analyse variations of the experiments. Comparing with the work of Thornton and Sokoloff (1990) there is one important difference: in many learning steps we use DIRs. The essential quantities: change of distance, velocity, change of velocity, acceleration, and force are visualised as vectors and integrated into the schematic animation of the experimental process. At the same time, structural connections are shown through the arrangement of the elements, connecting lines, etc., Blaschke (1999).

Topics concerning kinematics and dynamics of 1-dim movements

The essential kinematical quantities: position, change of position, velocity and change of velocity, were introduced in situations like starting, rolling and braking a bicycle and going up and down in front of a sonar ranger.

Students got to know the concept of acceleration through observing a motion on an incline with variable slope (reflection at the end of the incline, rhythmic tipping). This concept is usually much more difficult to understand for the students than the concept of velocity. In order to deal with these difficulties, we examined motions, where velocity and acceleration have different directions at some points and where negative acceleration occurs. In dynamics, we performed traditional experiments on force and acceleration (Newton's Second Law) as well as experiments with velocity-dependent friction and constant friction. Students reported them to be very important for their physical understanding and for clarification of still existing questions.

At the end of the instructional sequence we carried out the experiment „pulling a bicycle with constant pulling force and braking", to analyse all the acting forces. Together with the friction-experiments, this related to the introductory experiments as well as to motions from everyday situations - an important and motivating experience for the students.

Hypotheses, tools of the survey and assignment of groups

In order to determine how far images about the concepts of velocity, acceleration and force have changed more after MEW-Instruction than after traditional instruction, we used Multiple-Choice-Tests (KD-Test), mostly including diagrams, as pre-test intermediate test and six months later as post-test. We additionally assigned the Force Concept Inventory (FCI)-Test (Hestenes et al., 1992) as pre- and post-test. 27 other classes with traditional instruction were included in the study to compare achievement. We divided them into three control groups (560 students total) depending on intensity and form of computer use: MK1 (traditional instruction, interpretation of graphs), MK2 (occasional computer use, interpretation of graphs), and MK3 (frequent computer use, frequent interpretation of graphs). These groups took the KD-Tests at the end of 11th grade. Students from the test group MK4 additionally filled out questionnaires about the instructional process and the usefulness of the animation elements (especially the vectors) and graphs.

Evaluation

Evaluation of the questionnaires shows: nearly all students think Dynamic Iconic Representations are important for understanding the ideas of velocity and acceleration. When given more complex problems, students with higher performance (on the tests) seem to have had more use of the representations. Students judge the presentation of the experimental process, together with the developing graphs, to be very helpful.

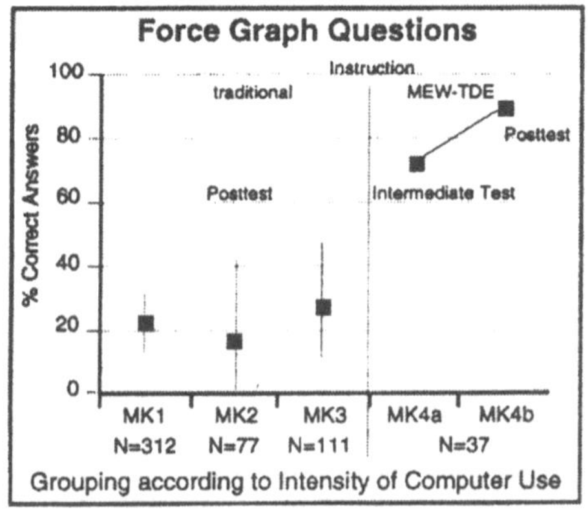

Fig. 3: Percentage of right answers on the post-test and the intermediate test, referring to the item-cluster "Forces" (control groups MK1-MK3 and test group MK4 (MEW-TDE).

Evaluation of the test shows that the control groups performed similarly to the ones in Heuer and, Wilhelm (1997). If we group the classes according to the category „Integration of Computer Use“, the results differ substantially. Only the students in the test group performed remarkably well (see fig. 3, similar results for the other item clusters). In order to take into consideration the different previous knowledge, we used a Hake-Plot (Fig. 4). It shows the gain of performance as percentage of the possible gain of performance depending on the percentage of right answers in the pretest. The Hake-Plot displays the results of the test group (MK4) on our KD-Test as well as on the FCI-Test and as examples for the control groups the results of the best group (MK3) on the KD-Test.

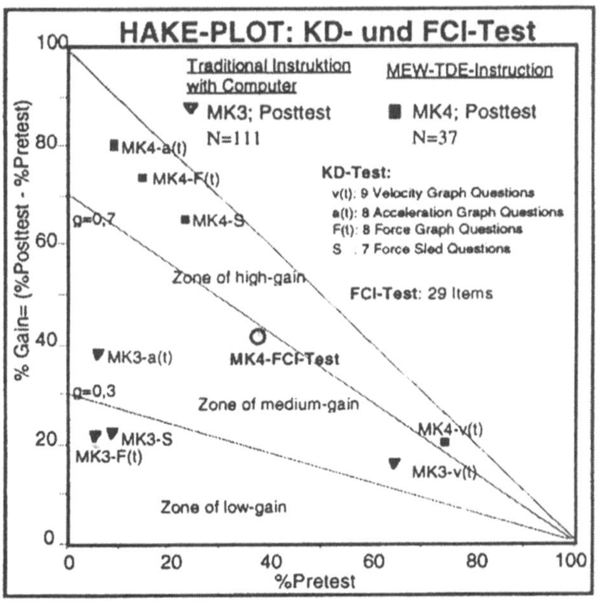

Fig. 4: Hake-Plot, displaying the gain of performance for the teacher-made KD-Test and the FCI-Test as function of the pre-test results.

In order to assess students' knowledge not only in standard situations, we included a more complex problem: Two connected gliders move due to the influence of two constant forces, after some time the connection breaks (Redish et al., 1997). Again, only the test group performed very well. Analysis of the control groups shows that no (MK1) or only occasional use of the computer (MK2) does not provide sufficient help (see fig. 5).

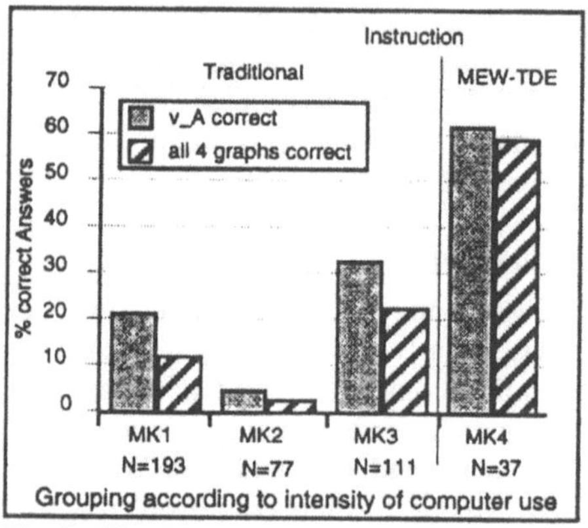

Fig. 5: Like fig. 3 percentage of right answers for the more complex problem described in text.

Conclusion: significance of MEW-instruction

The two basic elements of our instructional concept are the workshop-style and the use of Dynamic Iconic Representations of experiments and simulations. They make high achievement possible only through their interaction. From the viewpoint of Educational Psychology, it is immediately obvious that the workshop-style is very important. Image-oriented thinking facilitates building-up of internalised structures and mental models (Weidenmann, 1994) that are easier to access when predicting, comparing and analysing new statements.

References

Blaschke, K.(1999). Dynamik-Lernen mit multimedial experimentell unterstütztem Werkstatt (MEW)-Unterricht - Konzepte, Umsetzung und Evaluierung, Diss. Uni Würzburg.

Blaschke, K. & Heuer, D. (1999). Physik-Simulationen im Computerraum - Ein Lernpotential, das genutzt werden sollte. *Praxis der Naturwissenschaften-Physik* 48 (8), 39-44.

Hake, R.R. (1998). Interactive-engagement versus traditional methods: A six-thousand-student survey of mechanics test for introductory physics courses. *American Journal of Physics* 66 (1), 64-74.

Hestenes, D., Wells, M. & Swackhammer, G. (1992). Force concept inventory. *The Physics Teacher* 30, 141-158.

Heuer, D. (1996). Dynamic physical representation of real experiments. In R. Tinker, Ed., *Microcomputer-based labs: educational research and standards* (pp. 259-270) Berlin: NATO ASI Series F.

Heuer, D. & Wilhelm, Th. (1997). Aristoteles siegt immer noch über Newton. *MNU* 50, Heft 5, 280-285.

Redish, E., Saul, J. & Steinberg, R. (1997). On the effectiveness of active-engagement microcomputer-based laboratories. *American Journal of Physics* 65 (1), 45-54.

Schecker, H. (1985). Das Schülervorverständnis zur Mechanik. Eine Untersuchung in der Sekundarstufe II unter Einbeziehung historischer und wissenschaftlicher Aspekte, Diss. an der Universität Bremen.

Thornton, R. K. & Sokoloff, D. R. (1990). Learning motion concepts using realtime microcomputer - based laboratory tools. *American Journal of Physics* 58 (9), 858-867.

Thornton, R. K. (1996). Using large-scale classroom research to study student conceptual learning in mechanics and to develop new approaches to learning. In R. Tinker, Ed., *Microcomputer-based labs: educational research and standards* (pp. 89-114) Berlin: NATO ASI Series F.

Weidenmann, B.(1994). Informierende Bilder. In B. Weidenmann (Hrsg.), *Wissenserwerb mit Bildern* (pp. 9-58). Bern: Hans Huber-Verlag.

Generating Hypotheses in Scientific Enquiry

Isobel J. Robertson
Faculty of Education, University of Strathclyde in Glasgow, UK

Abstract

Ten years ago, assessment of investigations was introduced into Scotland's Standard Grade (SG) science courses, for students aged 14-16. Recent 'harmonisation' of assessment procedures for national certification included retention of the researched model of investigation (introduced in Biology and Physics in 1991) for use in all SG sciences. Comparison of assessment demands with those of the General Certificate of Secondary Education in England, has revealed greater emphasis given, in Scotland, to hypothesis generation. Analysis of open-ended investigations of 150 biology students resulted in classification of 'working hypotheses'. Data from 66 students in one school are used to describe the characteristics of these and the use of working hypotheses within a formative assessment model. It is postulated that failure to establish conceptual clarity at the generation stage can result in difficulties in later processes of investigation. Some comparisons are made with students' performance on *Problem solving* questions in the national examination.

Introduction

In defining scientific enquiry, Rachelson (1977) drew attention to the lack of attention given, in school science, to hypothesis generation in comparison to hypothesis testing. With current emphasis on enquiry in science education (as evidenced by the current OECD Programme for monitoring student knowledge and skills in the new millennium), has this imbalance been addressed?

Internal assessment of investigations, as part of the *Practical Abilities* element of some Standard Grade science courses, was introduced into Scottish schools in 1988. The first model suggested by the Scottish Examination Board (SEB) was found to be flawed mainly because, in placing an initial demand on satisfactory planning, many students were prevented from carrying out an investigation. Between 1991 and 1998, in Physics, then Biology and later, Science and Chemistry, the model for assessment of investigations developed by Bryce, McCall, MacGregor, Robertson and Weston (1991) was introduced. However, each subject specified slightly different requirements for weighting objectives, aggregating marks and collating evidence for moderation.

'Harmonised' arrangements for all sciences (taking effect in 1999) retained this model, with standardisation across subjects. Some regretted the decision to reduce the weighting for *Practical abilities* from having equal standing with *Knowledge and Understanding* and *Problem Solving* course elements to 1:2:2 made by the Scottish Qualifications Agency (1997) - the successor of SEB. Robertson (1996) had argued that concerns about inflated grades could be addressed by raising cut-off scores – initially set, by SEB, relatively low – to better differentiate performance and raise standards.

H. Behrendt et al. (eds.), Research in Science Education – Past, Present, and Future, 307–312.

Emphasis is placed on open-ended investigation, in which each student suggests an idea to investigate (stimulated by a 'starter question') and generates a 'working hypothesis' to test and evaluate. The model was described as possessing "clear construct validity"; investigative skill objectives (on which it is based) were recognised as "transparent", assessment of students' performance, "reasonably straight-forward" and "inter-marker reliability generally high" (SEB, 1997).

An altogether different history of practice in assessment of scientific enquiry for the General Certificate of Secondary Education (GCSE) unfolded in England and Wales, with several examining groups and politically driven changes of policy and curriculum (Donnelly, Buchan, Jenkins, Laws & Welford, 1996). The new commonality in structure has allowed comparison of assessment demands in science in two distinct education systems. Each assessment framework (exemplified in Northern Examinations and Assessment Board 1997) describes four 'skill areas' which show similarities but obvious differences. More emphasis appears to be placed on the generation of hypotheses in SG and more emphasis on using scientific knowledge in GCSE. Both systems are concerned with evaluation – from 'predictions' (GCSE) and 'hypotheses' (SG). Directions to use detailed scientific knowledge appear to provide a focus, for GCSE students, on evaluation of a correct (textbook) answer rather than the relationship of experimental results to students' predictions.

Design, procedures and data analysis

Research into investigations in 20 different biological contexts involved the participation of 150 SG biology students in three schools. Data reported here were drawn from content analysis of written responses in semi-structured investigation booklets completed by 112 students in one school. The researcher had observed students' actions, clarified the meaning of written statements and made and recorded assessment decisions during investigation sessions.

The SG model for investigation currently defines 13 investigative skill objectives. Nine of these were considered relevant in this research context to obtain 'scores' in four categories of investigation and overall. For 81 students, final examination scripts were inspected to allow comparisons of performance on investigations with that on elements of written *Problem Solving*. Analysis, by SEB, of final grades in all three course elements, showed performance of student samples as very representative of the national examination cohort.

Generation of a working hypotheses

Assessment of the *Generative* component of investigations was devised to help students produce an approximation to scientific hypotheses, open to testing and falsification. In generating an 'idea', students were effectively being asked to identify an independent variable. By stating the 'aim' of the investigation, they had to identify the dependent variable (stated in the starter question) and link it to the independent variable. In answering, *What do you expect to happen?* students had to indicate the direction of any predicted effect, in order to facilitate later evaluation of experimental findings. Taken together these statements were considered a 'working hypothesis'.

Analysis of written responses of 150 students suggested a classification of aim statements as A: well expressed, B: insufficiently specific or clearly expressed, C: seeking the 'best answer' to a particular problem or D: indicating linguistic and/or conceptual difficulties (which were found impossible to separate). Table 1 shows examples of these for *Catalase* investigations together with the directional statements made to complete working hypotheses. The starter question referred to the action of hydrogen peroxide on samples of plant tissue and asked, *What might affect the rate at which oxygen is produced?*

Aim statement *I will try to find out:*	**Completion of working hypothesis** *What do you expect to happen?*	**Direction clear?**
A: well expressed		
R1... if the surface area... will affect the rate at which oxygen is produced	The larger the surface area the quicker oxygen will be produced.	Yes; use of quicker.
R2... if temperature will affect the rate at which oxygen is produced.	The oxygen rate increases as temp increases until can no longer.	Yes, after clarifying
B: insufficiently specific or clear		
R3 ... how the surface area of a substance affects how it will react.	... the potato which has been cut up and has a larger surface area will produce more oxygen bubbles.	Assumes same times.
R4... if the same plant will produce more oxygen at different temperatures.	... more bubbles to be produced at higher temperatures less at lower temperatures.	
C: best answer to particular problem		
R5... what temperature the cucumber produces the most oxygen.	... the room temperature would make the cucumber produce more oxygen.	Yes, after clarifying.
R6... how much and how little hydrogen peroxide affects the rate at which oxygen is produced.	... by adding a little drop of hydrogen peroxide a few bubbles will come from it and adding a lot of hydrogen peroxide a lot of bubbles will come from it.	Pre-scientific view?
D: linguistic/conceptual difficulties		
R7(different vegetables) ... produce of oxygen at different rates as the breakdown takes longer levels.	... bubbles will come off the food at different times I will try to measure this.	'times' meaning rates?

Tab. 1: Examples of students' responses (R1-R7) in Catalase investigations with classification of 'aim statement' and evaluation of 'working hypothesis'.

The formative assessment model demanded that each student was assessed on having achieved the objectives relevant to the generation of a working hypothesis prior to carrying out the investigation. The assessor could ask for verbal or written clarification of a written response before making an assessment decision. The assessor then provided sufficient remediation to allow students not achieving a particular objective to proceed.

The majority of students with a well expressed aim produced a clear expectation (R1), some after clarification (R2). Aim statements R3 and R4 were assessed as sufficiently acceptable to allow experimentation to begin. Student R5 appeared to be seeking a 'best answer'. R6 shows an unsatisfactory working hypothesis which required revision. Most students with aim statements indicating conceptual or linguistic difficulties produced working hypotheses which reinforced this classification and some remediation was provided. Analysis strongly suggested that such responses contributed to confused experimentation and/or evaluation and that an appropriately rewritten hypothesis should be a necessary requirement to allow the investigation to proceed.

Aim statement category	**Ind. Var.**	*Catalase* n=17	*Froth* n=12	*Yeast* n=18	*Milk bacteria* n=19	% success n=66	Combined % n= 66
A	cont.	3 of 5	2 of 4	4 of 6	8 of 11	26	**31**
	discont.	1 of 1	0 of 0	0 of 2	2 of 2	5	
B	cont.	0 of 3	0 of 0	2 of 3	0 of 1	3	**8**
	discont.	1 of 1	1 of 2	0 of 0	1 of 1	5	
C	cont.	1 of 3	2 of 2	1 of 2	0 of 0	6	**8**
	discont.	1 of 1	0 of 0	0 of 0	1 of 1	2	
D	cont.	0 of 1	1 of 3	0 of 5	0 of 0	2	**2**
	discont.	0 of 2	0 of 1	0 of 1	0 of 3	0	
% success	cont.	24	42	56	42	36	***49***
	discont.	18	8	0	21	12	
	total	**42**	**50**	**56**	**63**	***48***	

Tab. 2:* *Numbers of appropriate 'working hypothesis' for four categories of Aim statement, showing nature of Independent Variable (Ind. Var.) for four sets of investigations ($n_{students}$=66).

Table 2 presents the results from one school, from four investigations in which the majority of students' identified a continuous independent variable. It shows that, for five such *Catalase* investigations, three 'well expressed' aim statements were combined with appropriately directional expectations. Performance varied across contexts and with the type of variables selected. Some *Catalase*, *Froth* and *Yeast* investigations involved measurement of two continuous variables while *Milk bacteria* required the use of a chemical indicator to show changes in the dependent variable. Overall, for the generation of an appropriate working hypothesis:

- just under 50% of the sample were successful;
- performance over four contexts was broadly similar;
- over two-thirds of those with a well-expressed aim were successful (31% of sample);
- half of those with aim statements lacking specificity or clarity had success (8% of sample);
- two-thirds of those seeking a 'best answer' were successful (8% of sample);
- only one student with a very flawed aim statement was successful (2% of sample).

Some areas of conceptual difficulty were identified across contexts. One was interpretation of starter questions including 'rate of a reaction'. There was confusion between the rate of reaction and the time taken to react and between speed of bubbling and number of bubbles released. Some students became confused when a test procedure such as a chemical indicator was introduced. Some conceptualised investigations as a comparison of discrete conditions determined by the presence or absence of a particular factor e.g. light, sugar, bacteria or an enzyme.

Comparisons of different aspects of scientific enquiry

Table 3 shows the Pearson product moment coefficients computed to measure relationships between 'generating hypotheses', 'working with variables', 'drawing a table and graph' and 'evaluation' (in investigations), a measure of successful performance given by the total of Inv. objectives achieved and scores on aspects of written *Problem solving (PS)*. High correlations between practical work and performance on written tasks were not expected (Bryce & Robertson 1985). The highest correlation found was 0.52, between total scores on *PS* questions and total Inv. scores. Correlations of 0.40 and 0.43 were found for *Generating ideas* and *Making hypotheses/predictions* (in *PS* tasks) and total Inv. scores. Some relationship was also shown between these *PS* categories and *Working with variables* and between *Making hypotheses/ predictions (*in *PS*) and *Evaluation* (in investigation). The higher correlations between the first three PS categories and *Working with variables* than for those with *Generating hypotheses* might suggest that the generation process is difficult to replicate in a written context than recognising and controlling variables.

	Categories of practical investigation skills				Total
Aspects of Problem Solving questions in final examination	Generating hypotheses	Working with variables	Drawing table graph	Evaluation	score on Inv. objectives
Generating ideas	0.20	**0.34**	0.24	0.29	**0.40**
Making hypotheses / predictions	0.22	**0.32**	0.27	**0.37**	**0.43**
Identifying variables	0.03	**0.30**	0.26	0.18	0.27
Drawing conclusions	0.15	0.16	0.04	0.30	0.26
Drawing graphs	0.21	0.27	**0.43**	0.29	**0.41**
Reading scales	0.21	0.22	0.02	0.19	0.25
Interpreting graphs	0.25	**0.34**	0.13	0.22	0.36
Mathematics concepts /processes	0.29	0.15	0.18	0.14	0.26
Total Score	-	-	-	-	**0.52**

Tab. 3: Relationships between total scores on aspects of Problem 1 solving questions with total scores on Investigative objectives ($n_{students}$ =81).

Conclusions and looking forward

Analysis of the contribution of the experimental stage to evaluative processes are discussed in Robertson (1999). It is argued here that the generative phase is important for teaching and learning processes of scientific investigation. Failure to establish conceptual clarity at an early stage was found to contribute to problems in operationalising variables and/or failure to retain their identification during experimentation. This led some students to evaluate a different hypothesis from the one articulated! Wenham (1993) concluded that it was necessary to recognise different kinds of hypotheses, loosely defined as tentative solutions to problems. Although some students conceptualising the investigation as a problem with a particular solution were able to generate an appropriate working hypothesis, it was found that such a conceptualisation could prove counter-productive in evaluation. Helping students to consider a hypothesis as a statement of the relationship between two variables which is testable and falsifiable may provide a better basis for developing scientific literacy.

The framework for assessment of practical investigations described is a formative one, used to inform summative decisions on final grades. The success of the model demands that teachers allow sufficient freedom to their students to learn by making mistakes. Sensitive formative assessment has been shown to allow the teacher to intervene at critical points, enabling students to do this. Difficulties in promoting such a system include the long standing tension between those who see assessment as essentially formative and those more concerned with grading. Another issue relates to the place of scientific knowledge and understanding in practical investigations. Should this be a prerequisite for investigations or should processes of scientific enquiry be the main focus? It is hoped that this paper will provide a stimulus for discussion in European countries, that are now attempting to introduce assessment of investigations into their curricula.

References

Bryce, T.G.K., McCall, J., MacGregor, J., Robertson, I.J. & Weston, R.A.J. (1991). *TAPS 3: How to assess open-ended practical investigations in biology, chemistry and physics.* Oxford: Heinemann Educational.

Bryce, T.G.K. & Robertson, I.J. (1985). What can they do? A review of practical assessment in science. *Studies in Science Education* 12, 1-24.

Donnelly, J., Buchan, A., Jenkins, E., Laws, P. & Welford, G. (1996). *Investigations by order.* Leeds: Centre for Studies in Science and Mathematics Education.

Northern Examinations and Assessment Board (1997). *GCSE Science Framework syllabuses for 1998 onwards: Science.* Manchester: NEAB.

Rachelson, S. (1977). A question of balance: A wholistic view of scientific inquiry. *Science Education* 61, 109 - 117.

Robertson, I.J. (1999). Key evidence in testing hypotheses. In M. Bandiera, S. Caravita, E. Torracca & M. Vicenti (Eds.), *Research in Science Education in Europe* (pp. 193-200). Dordrecht, The Netherlands: Kluwer Academic Publishers.

Robertson, I.J. (1996). Are practical investigations in the Standard Grade Sciences achieving their intended purposes? In *SERA Conference Proceedings: Education 1995* (pp. 50-54). Glasgow: University of Strathclyde.

Scottish Examination Board (1997). Assessment of practical abilities in the sciences at standard grade. Internal consultation document 970926.

Scottish Qualifications Agency (1997). *Standard Grade Arrangements in and after 1999.* Dalkeith: SQA.

Wenham, M. (1993). The nature and role of hypotheses in school science investigations. *International Journal of Science Education* 15, 231-240.

Using Laboratory Work for Purposeful Learning about the Practice of Science

Amanda Berry, Richard Gunstone, John Loughran, Pamela Mulhall
Monash University, Australia

Abstract

Much laboratory work in school science has been criticised for its highly ritualised nature, in which students are led "cook-book" style through a set of instructions to a predetermined end point. Along the way students collect data and answer questions that are presented in a formal report. It has been argued that such an approach is not only an ineffective means of developing students' understanding of science concepts, but also presents a misleading picture of the way in which scientific knowledge develops. This paper describes two different approaches to laboratory work. In each the teacher's purpose was to provide her students with some insight into ways in which knowledge in science is generated, shared and validated.

Subject and problem

Most laboratory work follows a familiar rubric. Students are presented with an aim, a suggested hypothesis, steps for carrying out an experiment (method or procedure), observations and/or measurements that should be recorded and questions that lead to the conclusion, ostensibly drawn from the experiment. Generally, the teacher's main purpose rests on the belief that, in following this rubric, students will learn (and believe) a particular science 'fact' because they 'see' it through the experiment, i.e. the experiment reinforces and consolidates learning of 'theory'. An additional teacher purpose may be that students will be following a method similar to that which led to the original discovery of the relevant fact, i.e. students may learn something about the way that scientists perhaps just understand the world, and therefore understand how scientific knowledge is developed.

Research raises many questions about laboratory work as an effective means for students to learn science content (Hodson, 1993; Millar, 1991) or to obtain a valid picture of how scientific knowledge is produced (Hodson, 1993; Woolnough, 1991). Recent studies (Driver, Leach, Miller & Scott, 1996) have advocated the inclusion of practical investigations in science that help students understand more about the nature and status of scientific knowledge. It is argued that unless students are able to develop and articulate an understanding of the nature of scientific knowledge, they will be ill equipped to interpret the validity of knowledge claims made in the name of science.

Here we report on two different approaches to laboratory work introduced by two teachers working together at the same school. For each teacher, an important purpose was to develop her students' understanding about science beyond the dominant picture of propositional knowledge and routine procedures.

Data used in this study have been drawn from a larger project that explored student learning in laboratory work. In the larger study we identified several factors

H. Behrendt et al. (eds.), Research in Science Education – Past, Present, and Future, 313–318.

as important in promoting student learning from laboratory work - the extent of students' prior content knowledge related to the laboratory task, their perceptions of the purposes of the laboratory task and their knowledge of experimental procedures. In addition, a separate but related issue emerged; the impact of students' perceptions of the nature of science on their approach to, and learning from, laboratory tasks. Many of the students we observed appeared to operate from a formula driven view of science. Following the procedure and completing the task became an end in itself, and there was little thinking about the procedures used (for example making careful measurements or repeating measurements) or the importance of these in what students were trying to do. This paper explores what we learned from the work of the two very different teachers about the ways in which laboratory work, designed for specific purposes, impacted on students' understanding. In this case of the practice of science. We use these two situations as specific cases that purposely challenged the traditional approach to laboratory work.

Our approaches to the exploration of laboratory work in our research have been focussed by information processing and constructivist views of learning, as elaborated by White (1988) and Fensham, Gunstone and White (1994). Within this broad framework, ideas of the importance to conceptual learning of informed intellectual engagement (metacognition) that are advanced in the Project for Enhancement of Effective Learning (e.g., Baird & Northfield, 1992) have been of particular influence on our thinking. The aims of the study being described here were focussed by the broad question "How can laboratory work be used more effectively in the teaching of science?" Specific research questions were

when teachers have clear purposes for laboratory work and design sequences that reflect these purposes, what student learning occurs?

how is this student learning influenced by students knowing the purposes of the teacher?

It was of obvious value to the research that the two teachers shared our concerns and questions. Such teachers have been rare in our work on learning from laboratories.

While the two teachers shared the same broad purposes for the laboratory work described in this study, each had different specific goals. The first, Amanda (a co-author of this paper and at the time a teacher at the school), wanted her year 9 students to become more aware of the socially constructed nature of science knowledge. She designed a chemistry unit in which students were involved in generating, sharing and validating scientific information. She also intended that through using a chemistry context for the laboratory work her students would learn some important chemistry concepts. The second teacher, Susan, disillusioned with the use of laboratory work as a vehicle for improving students' understanding of science concepts, designed a chemistry unit for her year 10 students which had, as its goal, the development of students' understanding of the role of experimental work in establishing scientific fact. In her case, the chemistry context was simply the content vehicle for this role, she did not expect specific chemistry to be learnt. Amanda's expectation that her students would develop an understanding of chemical concepts as a result of their experiences is an important difference between the goals of the two teachers.

Outline of the units

Amanda's year 9 unit was set in the context of a forensic mystery and required students to solve chemical puzzles using specific content knowledge acquired in the unit. Initial activities were designed to stimulate students' thinking about the ways in which information is collected, organised and interpreted in science and centred on the identification of some common household white powders (e.g., table salt, bicarbonate of soda). In small groups, students were required to devise tests to identify the different powders, design a table for representing their results, carry out tests, then communicate their procedures and results in a class forum following the testing.

The initial activities framed the purposes of the unit as illustrating the parallels between the ways laboratory work contributes to the construction of science knowledge (e.g., through careful and multiple observations, repetition by other scientists, shared agreement about results and their interpretation) and students' own subsequent investigations to determine the nature of some mystery substances.

In Susan's unit, students selected one experiment from a large number, conducted this experiment, recorded their results and then documented their approach in such a way that another group of students could replicate their work. This 'verifying' group then provided feedback to the original group about the efficacy of the procedure with which they had been supplied. This approach aimed to help students understand that communicating results and having one's ideas re-tested by others is an integral part of developing and accepting science knowledge (see Hart et al., 1999 for a full description). By coincidence a student teacher working in Susan's class was a recent Doctoral graduate in physics. He presented his work to the students and explained the role of conferences/journal articles and the science research community in developing and validating knowledge claims in science.

Through their curriculum organisation and teaching practice both Susan and Amanda demonstrated an approach to laboratory work which challenged the stereotypic 'recipe' approach and emphasised the importance of students' developing some insight into where knowledge in science comes from and how it is established. Common to both teachers' approaches was an emphasis on students' understanding the need for reproducibility of experimental results by others as one characteristic aspect of science.

Design and procedure

Data were collected from:

- Regular observations of lessons during the units, and detailed field notes
- Informal interviews with pupils during laboratory sessions
- Post lesson debriefing discussions with each teacher, with detailed notes taken
- Audiotaped interviews with small groups of students at the end of each unit (Amanda's class: 3 groups, total of 6 of the 28 students; Susan's class: 4 groups, total of 11 of the 22 students)
- A written survey conducted with Susan's class halfway through her unit

Interview questions were quite direct, e.g. "Why, do you think, you were asked to write a report for another group to use?", and approaches to analysis were straightforward. Taped interviews were transcribed, analysed and coded according to the teacher's purposes for each unit, as was the written survey. The transcripts and other data sources were inspected in terms of our questions, and considered in terms of triangulation.

Data analysis and discussion

Data analysis suggests that the experiences that comprised each unit had an impact on students' thinking about the practice of science in ways congruent with each teacher's aims. Our analysis is given in terms of the teachers' purposes for the units of work. Where student quotes are used these are representative of the statements of many students.

Testing/validating findings

An important purpose for both teachers in the units of work was to help their students learn about the role of testing in validating scientific findings. In every interviewed group, in both classes at least one student commented on the role of testing (replication) in response to a question about what students had learnt from the unit. Students from both classes referred to the importance of accurately recording the results of a scientific experiment in order that others might repeat it.

A student from Amanda's class highlighted several different factors related to the role of testing and verification that she had learnt from her work.

> Lisa: "...when you've finished testing, it's just test results. You need to compare your test results to somebody else's or do the test over and over...and then you'd be able to call it (a table of) facts."

Lisa demonstrates her understanding of the:

- need for multiple observations. (Replication by the group performing the task and by other groups was a key component of students' work.)
- tentative nature of initial findings. (Data collected by groups were considered to be 'on trial'; they could not be called definite until verified.)
- difference between 'test results' and 'facts'. (Following replication of results through multiple testing and peer scrutiny, collective consensus meant that findings could be considered as scientific 'fact'.)

A student from Susan's class noted an additional factor related to the role of testing:

- reproducibility of results. (Following the experimental procedure should lead to achieving the same results.)

> Annie: *"That was the whole point [of the experiment]– to make sure we got the same results."*

It is interesting to compare the statements from these two students for the assumptions implicit in each regarding the outcome of repeating others' experimental procedures. Lisa implies that through multiple testing and comparison of results the weight of supporting data leads to the correct result, or a scientific "fact". Annie, on the other hand, refers only to the importance of testing for the "*same* result(s)", rather than the 'correct' one. The difference between obtaining the 'same result' and the 'correct result' may well arise from the different foci taken by Amanda and Susan in teaching about the role of testing in validating scientific knowledge. One of Amanda's central activities involved students gathering information that was collated into a "table of facts" then used as the basis for further experimental investigations. Use of the word "fact" in this context, together with class discussions which emphasised that facts recorded in the table were those that the class had agreed upon as true, may well have been the basis for Lisa's beliefs.

Links with the work of scientists

Both teachers anticipated that their students would learn something of the work of scientists and the processes associated with the establishment of scientific knowledge. When asked how their work in the unit related to what scientists do, students in Amanda's class found it difficult to describe an application beyond the specifics of the forensic context of the unit. Nevertheless, in two of the three groups interviewed at least one student referred to the importance of scientists collecting and verifying information with peers.

In contrast, students from Susan's class saw the work they had been doing as a means of achieving an understanding of the way that scientists work.

Tita: *"I think the aim was different this time, it wasn't just to...see the result of the prac* [common Australian term for laboratory], *it was to experience something greater,* [to experience] *the role of the scientist."*

For Susan's students this sense of purpose emerged as the unit progressed. It appears that the student-teacher/physicist's talk played an important role in helping students to realise that purpose. For example, one student described the task of repeating someone else's results as "*a bit weird at first*" but after the student-teacher physicist's talk "*it made sense*" because she saw that what they were doing was "*what scientists do all the time to make sure their results are accurate.*"

Communication in Science

An important component of both units was to help students learn something of the social processes of science through the activities of communicating procedures and findings. For Amanda, this was established through frequent, whole class discussions, particularly in the initial stages of the unit. These focused on encouraging students to clarify meanings for words they used and developing a shared language that facilitated meaningful communication amongst class members. This is supported by the researcher/observer who noted that the teacher frequently asked her students questions such as: "What do you mean by...?", "How sure are you of these results?" "Do we need more evidence?" This encouraged students to publicly discuss, argue and justify their observations and beliefs. This in turn led to the refinement of the language they used in communicating their experimental procedures and results to others.

For Susan, communication of experimental findings was established through each student group writing a procedure which was passed to a separate 'verifying' group who provided both verification of findings and written feedback about the clarity of the experimental design. Some students had not previously considered this as an element of their own work, or the work of scientists.

Betty: *"It shows you how clear everything has to be and how sometimes tedious it is...if you've discovered something, how thorough you have to be when rewriting it."*

Through acting out this role, Betty has achieved a new understanding of the way in which scientific knowledge is constructed, particularly the importance of clarity in communication.

Collaboration in laboratory work

The opportunity to collaborate with peers and to work independently was valued by students in both classes. This was evident from comments made during the interviews and observations of students made by each teacher and by the researchers. While co-operation in group work is an espoused value of much school laboratory work, this is not always achieved since the tasks carried out by students are often routine, or unproblematic. Here, students were motivated to help each other either through explaining work or through contributing expertise to solve problems that arose in carrying out experiments.

Conclusion

Laboratory work occupies a significant amount of time in many secondary science curricula. While seen as important and enjoyable by both teachers and their students, it appears that the kind of learning perpetuated by most school laboratory work neither promotes students' learning of science concepts, nor helps students to learn about the practice or purpose of science itself.

The significance of the work reported in this paper lies in its investigation of a challenge to the usual "script" (White, 1988) for laboratory work. The two teachers in this study were concerned to actively engage their students' thinking about their science learning and to provide students with insights into the way that scientific knowledge is established, through involving them in the construction, communication and validation of scientific information.

The experiences provided by the teachers had an impact on the thinking of at least some of the students in their classrooms. Students from Susan's class developed a stronger sense of the ways in which their learning related to the broader context of scientists' work. This was less evident in the responses from students in Amanda's class where the purposes of the unit were more diverse, and included learning of chemistry content. By comparison, Susan's purposes were more singular – learning of content was not important, only the increased understanding of the role of communicating and verifying in the establishment of science knowledge. It is likely that this more singular focus resulted in greater achievement of this purpose than was the case for Amanda's class. This suggests that limiting the purposes for laboratory work can help students to learn more from it by enabling a stronger focus on these more limited purposes.

(This research was supported by an Australian Research Council Large Grant.)

References

Baird, J. & Northfield, J., Eds. (1992). *Learning from the PEEL experience*. Melbourne: Monash University.

Driver, R., Leach, J., Millar, R. & Scott, P. (1996). *Young people's images of science*. Milton Keynes, UK: Open University Press.

Fensham, P., Gunstone, R. & White, R., Eds. (1994). *The content of science*. London: Falmer.

Hart, C., Mulhall, P., Berry, A., Loughran, J. & Gunstone, R (1999). *What is the purpose of this experiment? OR can students learn something from doing experiments?* (Manuscript Under Review).

Hodson, D.K. (1993). Rethinking old ways: Towards a more critical approach to practical work in science. *Studies in Science Education* 22, 85-142.

Millar, R. (1991). A means to an end: The role of processes in science education. In B. Woolnough , Ed., *Practical Science (pp. 43-52)*. Milton Keynes, UK: Open University Press.

White, R.T. (1988). *Learning Science*. Oxford: Blackwell.

Woolnough, B., Ed. (1991). *Practical Science*. Milton Keynes, UK: Open University Press.

University Students During Practical Work: Can We Make the Learning Process Intelligible?

Per-Olof Wickman and Leif Östman
Department of Educational Processes and Learning, Stockholm Institute of Education and Department of Education, Uppsala University, Sweden

Abstract

In this paper we suggest and apply a mechanism for the learning process based on Wittgenstein's later work, pragmatism and sociocultural perspectives. The theory is used to analyse recordings of student discourse during a practical an insect morphology. These are the first results of a research project on student discourse during practical work in the laboratory and in the field. The project is intended as a long-term study of biology and chemistry students during their first three years at Uppsala University.

Introduction

Here we present the theoretical background and results from a research project on student discourse during practical work in the laboratory and in the field. The project began in 1998 by developing theory and methods. The project is intended as a long-term study of biology and chemistry students during their first three years at Uppsala University. It is a collaborative endeavour financed jointly by the teacher education department and the science departments. One aim is to develop science education research for university level biology and chemistry, another to develop forms for feedback between science education researchers and science teachers (cf. Roberts & Östman, 1998).

With these aims in focus, a theoretical framework has been developed for analysing the learning process as it is expressed through talk and action during practical work. The following results are a summary of a more elaborate manuscript that will be published elsewhere (Wickman and Östman unpublished).

Our approach can be used to illuminate two issues, viz. 1) how discourses change and 2) how we enter new practices. By understanding the mechanisms behind such changes, it is possible to study how the content and form of the curriculum interact with students to produce learning *in situ* in the classroom.

Learning as situated in practices

Learning is generally viewed as a process where earlier knowledge influences the meaning of new experiences. This view of learning has resulted in numerous constructivist studies, describing young people's ideas about different scientific phenomena and concepts (Driver et al., 1994; Pfundt & Duit, 1994). In most of these studies, the learning process has been identified with conceptual change and often, concepts are seen as parts of conceptual schemes (cf. Davidson, 1985), reminiscent of scientific theories or paradigms (Posner et al., 1982; Vosniadou, 1994). Recently, however, an increasing interest has been paid to situational aspects and the significance of contextual factors on learning (Hennessy, 1993; Aikenhead, 1996;

H. Behrendt et al. (eds.), Research in Science Education – Past, Present, and Future, 319–324.

Säljö & Bergqvist, 1997; Schoultz, 1998). Studies like these have been accused of not explaining how learning is possible if knowledge is situational (Halldén, 1999). However, the observation that knowledge is situated in practices cannot be ignored by saying that learning cannot be explained by observation. Instead new theoretical approaches are needed, that can accommodate the situational and continual aspects of learning. Dewey (1997) viewed the tension between these aspects to be at the heart of learning.

Analyses of learning, although they involve what people say and do in specific situations, are often put forward as analyses of how individuals think. Their concepts or their way of giving things, events etc. a meaning is seen as mental structures, independent of the practice in which talk and action were recorded. In order to create a link between the object of study — i.e. language usage and other actions — and the result formulated in terms like conceptual change, students' conceptions etc., one has to introduce a metaphysical relation between the world, language and thinking. Quite often this metaphysical relation is seen as dualistic. Thinking and the world are seen as primary and essential and language as secondary. Or to put it in other words: first comes the interaction between thoughts and essential elements of the world and then we use language to express the concepts or meanings that are formed in the interaction. Language is, in this perspective, perceived as a medium with which we express our thoughts: we dress our thoughts in language. This view implies that meaning or our concepts can be separated from language and the practice where talk and action occur. By perceiving language like this, it can be argued that a study of action and language can reveal how people think and how their concepts change.

This view of the relation between the world, language and meaning is one of many possible views. For example, it can be argued that in our daily practice we do not separate language usage from our meaning making process. Meaning making is situated in practices with language and action as inseparable parts. In the actual process of talking and acting, we do not separate language and action from meaning. It would be impossible to read a text, listen to somebody or talk to someone if we, at the same time, separated the process of understanding — i.e. making meaning out of the text or the speakers or the conversation — and the process of reading, listening or talking.

A logical consequence from this line of reasoning is to focus on how different circumstances influence how people make meanings and how these circumstances help people in entering new practices. This means that with an analysis of language usage we do not want to say anything about how people understand the world in general or how understanding is related to language, except when people are actually talking about such things. To focus on the meaning making process is not unique to our research. Bruner (1990), for example, has written a whole book on this issue. And to view the actual practice of using language as inherently connected to making meaning and thereby to acting in the physical and social world has been the core of the work of the later Wittgenstein (1967, 1969) and many pragmatists (e.g. Dewey, 1958).

Our theoretical framework is close to the sociocultural perspective in that we view action as a proper unit for analysis:

> I propose that mental functioning and sociocultural setting be understood as dialectically interacting moments, or aspects of a more inclusive unit of analysis— *human action* (Wertsch, 1995, p. 60).

In this study, we are focusing on one specific form of action, namely language usage. Language usage is connected to practices, that have purposes. For example medical doctors have a certain language usage and this language usage is connected to their profession as doctors, to their practice of curing people. And without their language usage they could not take part in this practice. In this respect language usage is not something, that can be isolated from other actions or practices. So our focus is on language usage in specific activities.

This approach means that we view learning as an acquisition of a language usage and its associated practice that, when successful, gives us the possibility to act differently in the world. This is a pragmatic view on learning (Dewey, 1997).

Learning a language-game

To analyse the learning process, we have used certain aspects of Wittgenstein's discussions of knowledge and meaning in *Philosophical investigations* (Wittgenstein, 1967) and *On certainty* (Wittgenstein, 1969). The presented mechanism is not an interpretation of Wittgenstein but an elaboration for our particular purpose.

The mechanism of learning is what happens when people construe new meaning in encounters by relating to similarities and differences to what is fixed in already established language-games. *Language-games* can be seen as "uses of language in which meanings of words are clearly understood" (Hardwick, 1971). They are practices with purposes involving action and utterances. In language-games some things are *fixed* (Wittgenstein, 1969). What fixed is, is immediately intelligible, "what is and happens when we do" in encounters between people and the physical world. Operationally, it is what students do and say without questions or hesitation. What is fixed is so because of the *relations* (*similarities* and *differences*) that surround it in the language-game. When relations of a language-game do not suffice to make sense of an encounter, we talk about a *gap*. These are encounters where people try to construe new relations to what is already fixed in the language-game. A gap is filled when new relations are construed to what is fixed in the language-game. If this micro-process of filling gaps proceeds, the total change can be described as entering a new language-game and its associated practice.

Methodology

In agreement with our aims and a number of other studies we have chosen to study the learning process as it presents itself *in situ* during students spontaneous talk and actions during an authentic educational setting (e.g. Edwards & Westgate, 1987; Barnes & Todd, 1995; Roth, 1998). We have chosen this procedure (instead of interviews for example) because we are interested in understanding the learning process as a situated practice.

We have used our theoretical approach to analyse audio-recordings of talk between two students during a one-hour long practical, where they, without teacher aid, were to study the morphology of five insects. A central aim of the practical was that students should learn by observing the physical insects. The teacher told the students to practice their own powers of observation by looking generally on the insects' antennae, wings, mouthparts, compound eyes, ocelli and hair cover. They were also to look for anything particular to do with the different parts. Hence, in the language usage adopted here the purpose of the practical was that students, in encounters involving physical insects, should construe new relations to the organs mentioned above. We restrict the present analysis to the students' efforts to make the morphology of a bumble bee intelligible.

Those parts of the students' talk that treated the morphology of the bumble bee were divided into 17 meaning exchanges. Each meaning exchange treated a particular insect organ. When analysing student talk, we firstly examined what is fixed in relation to bumble bee morphology. Secondly we examined what relations in form of differences and similarities students construe in their conversation and in their actions. Thirdly we examined how specific encounters contribute to students noticing gaps and to filling them with differences and similarities. What roles do encounters with the physical bumble bee play? Finally we examined what gaps can be filled and cannot be filled with similarities and differences that are fixed. By this methodological approach, based on student discourse, the learning process of students can be analysed. The analysis is made in light of the particular purposes of classroom practice and the particular relations to the physical insects that the curriculum aims at.

Application

A short example illustrates our methodological approach. The two students (M and L) and a student next to them (X) are studying the wings of their bumble bee:

L: But what were those vessels called?
X: Yes, we checked that too.
L: Just so we learn.
M: Wing veins. That is nothing that we should…
L: Wing veins. Okay, let's check. Should we look in this manner, or how should we do it?
M: They become more apparent perhaps?

In this meaning exchange M and L discussed the wing veins of bumble bees. In an earlier meaning exchange they established that they were studying a bumble bee. As mentioned above, the analysis begins by examining what is fixed to the students. In this encounter with the physical bumble bee 'vessels' were fixed. They used this word without hesitation. At the same time a gap occurred between vessels and their proper name. In the encounter between M and L, they construed a similarity between what looks like vessels and wing veins. This encounter also referred to prior encounters involving experiences that obviously helped them to construe relations to the physical bumble bee. Hence, in an encounter between the physical bumble bee and L and M – that involved prior encounters – they construed a similarity between bumble bees and wing veins. Hence, learning can be analysed as a sorting of prior and present encounters for the purposes of present and future practices.

By relating to similarities and differences to what was already fixed, students gradually, in encounters that involved the physical bumble-bee, construed similarities between bumble bees and large compound eyes, transparent wings with veins, antennae, brightly coloured and rich hair cover, mouth parts with jaws, a sucking mouth and small appendages and three pairs of legs. They also construed a difference between bumble bees and powerful claws. The relation between 'bumble-bees' and 'large compound eyes' was established in two steps. 'Compound eyes' and 'large' was already fixed from former language games. They used these words without hesitation or questions. In an interaction that involved the physical bumble bee and a real beetle they construed a similarity between 'compound eyes of the bumble bee' and 'compound eyes of the beetle'. They also construed – in an encounter that again involved the physical insects – a difference between the eye sizes of the two insects. The result was the similarity between bumble bees and large compound eyes.

Gaps remained for bumble bees and ocelli, number of wings and the scientific terms for different mouth parts. The interactions that occurred were not sufficient for establishing differences and similarities that filled the gaps. The students' talk made it evident that they, in all of these cases, could not fill the gap between their particular bumble bee and the generalised definitions they found when looking for help in text books. Students did not refer to evolutionary theories that could have aided them in filling these gaps with similarities and differences. Instead they used text books as if they contained simple definitions and facts about the appearance and location of all organs in all insects. They remarked that there was nothing to find about the morphology of bumble bees in their book.

Despite the students' problems with the language game of zoologists, there were examples of how students, in small steps, acquired a new language game. One such conquest was how they made 'mouth parts' intelligible by using similarities and differences to that which was fixed in earlier language games. This term at first created a gap. Mouth parts were something that appeared to be situated on the mouth, so how could they be parts of the mouth? Apparently this conflicted with their former relationships with mouths as openings. Finally, they solved this by relating the mouth to the similarity between the insect mouth and 'a sort of outgrowth of it all, of the whole mouth'. Such changes in language usage are important parts of establishing new ways of acting in practices (Östman, 1998).

Summary

This short article has only touched several issues of interest when studying learning in formal and informal settings (see Wickman and Östman in manuscript for more details). We think the present theoretical approach can be used to study some of these issues further. These include: how we generalise from single observations, how esthetical and moral relations are construed as a part of the scientific discourse and how questions of power influence meaning making. How people enter new language-games or discourses needs to be studied in a large variety of discourses and situations. Studying how talk and action vary between students during the same practical or how discourse varies between practicals in different subject domains is one way to make more general conclusions. Of special interest would be to see how student talk during practicals changes over the years. Our approach can also be used to look at socialisation in a wider context than science.

References

Aikenhead, G. S. (1996). Science education: border crossing into the subculture of science. *Studies in Science Education* 27, 1-52.

Barnes, D., & Todd, F. (1995). *Communication and learning revisited: making meaning through talk.* Portsmouth: Boynton/Cook Publishers, Inc.

Bruner, J., (1990). *Acts of meaning.* Cambridge: Harvard University Press.

Davidson, D. (1985). On the very idea of a conceptual scheme. In J. Rajchman & C. West Ed., *Post-analytic philosophy* (pp. 129-144). New York: Columbia University Press.

Dewey, J. (1958). *Experience and nature* (2nd ed.). New York: Dover.

Dewey, J. (1997). *Experience and education.* New York: Touchstone. First published 1938.

Driver, R., Squires, A., Rushworth, P., & Wood-Robinson, V. (1994). *Making sense of secondary science: research into children's ideas.* London: Routledge.

Edwards, A. D., & Westgate, D. P. G. (1987). *Investigating classroom talk.* London: Falmer Press.

Halldén, O. (1999). Situating the question of conceptual change. *In Abstracts. 8th European Conference for Research on Learning and Instruction.* Göteborg, Sweden.

Hardwick, C. S. (1971). *Language learning in Wittgenstein's later philosophy.* The Hagne, Netherlands: Mouton.

Hennessy, S. (1993). Situated cognition and cognitive apprenticeship: implications for classroom learning. *Studies in Science Education* 22, 1-41.

Östman, L. (1998). How companion meanings are expressed by science education discourse. In D. A. Roberts, & L. Östman Ed., *Problems of meaning in science curriculum* (pp. 54-70). New York: Teachers College Press.

Pfundt, H. & Duit, R. (1994). *Students' alternative frameworks and science education (4th ed.).* Kiel, Germany: Institute for Science Education.

Posner, G., Strike, K., Hewson, P. & Gertzog, W. (1982). Accommodation of a scientific concept: towards a theory of conceptual change. *Science Education* 66, 211-227.

Roberts, D. A. & Östman, L. (1998). Preface. In D. A. Roberts, & L. Östman Ed., *Problems of meaning in science curriculum* (pp. ix-xii). New York: Teachers College Press.

Roth, W.-M. (1998). Learning process studies: examples from physics. *International Journal of Science Education* 20, 1019-1024.

Säljö, R., & Bergqvist, K. (1997). Seeing the light: discourse and practice in the optics lab. In L. B. Resnick, R. Säljö, C. Pontecorvo, & B. Burge Ed., *Discourse, tools and reasoning: essays on situated cognition* (pp. 385-405). Berlin: Springer.

Schoultz, J. (1998). *Kommunikation, kontext och artefakt - studier av elevers behärskning av naturvetenskapliga diskurser.* Linköping: Linköping University.

Vosniadou, S. (1994). Capturing and modelling the process of conceptual change. *Learning and Instruction* 4, 45-69.

Wertsch, J. V. (1995). The need for action in sociocultural research. In J. V. Wertsch, P. del Río, & A. Alvarez Ed., Sociocultural studies of mind (pp. 56-74). Cambridge: Cambridge University Press.

Wittgenstein, L. (1967). *Philosophical investigations (3rd ed.).* Oxford: Blackwell.

Wittgenstein, L. (1969). *On certainty.* Oxford: Blackwell.

Learning About Investigations - The Teacher's Role

Lisette M.M. van Rens and Peter J.J.M. Dekkers
Vrije Universiteit, Amsterdam, The Netherlands

Abstract

A group of Dutch physics and chemistry teachers have been co-operating since 1997 in the development of pupils' investigative skills. We studied how much attention investigative skills received when these efforts started. Practical tests were designed to develop investigative skills, that previously received insufficient attention. This paper discusses the existing teaching practice and the classroom trials of these practicals. Conclusions are given as a tentative plan, specifying how fostering the development of investigative skills can be incorporated in physics and chemistry teaching.

Research question

The curricula of higher ability streams in Dutch secondary education are under reform. In the new programme pupils have to demonstrate their mastery of 'investigative skills' in science subjects and learn to conduct scientific investigations. This aim is not new (Hodson, 1992; Tamir, 1989), but views still diverge as to what is to be learned and how it is best taught. The exam programme implies that 'being able to conduct an investigation' is equivalent to mastering a set of investigative skills, a view similar to the views expressed in, e.g., Lawson (1994) and Klopfer (1990). However, the claim that 'investigative skills' can be identified and distinguished from each other, from other knowledge (e.g. common sense knowledge) or from other skills (e.g. technical skills) is not a claim we believe to have been adequately tested at this point.

Together with teachers, we designed learning activities in physics and chemistry to teach pupils about investigations. The activities were used in class and evaluated. This paper discusses the first explorative phase of our research, directed at mapping the teachers' teaching prior to the new exam programme. We construct this map by answering the following questions: To what extent was the development of investigative skills an educational aim in physics and chemistry prior to the reform? How is that reflected in actual teaching and learning? What views and plans do teachers have in order to foster 'learning how to carry out investigations'? and what kind of support, if any, do teachers need in that area? The answers led to a plan of action that is a compromise between aims and experiences of teachers and our (preliminary) insights regarding an effective educational strategy for learning about investigations. The next stage of our study involves a classroom test of the plan.

Research Method

The network

In September 1999 all schools have introduced the new curriculum. Our study was carried out in a 'network' where preparations were made for this educational reform. In this network, teachers of chemistry and physics of five schools cooperate

H. Behrendt et al. (eds.), Research in Science Education – Past, Present, and Future, 325–330.

with science education specialists of the 'Vrije Universiteit' in Amsterdam. The rationale for this cooperation is that teachers can develop new teaching practices more efficiently if they can make use of each others' expertise and the support of an educational specialist. Premises guiding the network are:

1. Changes in existing teaching practice may take place if the teacher sees these as desirable, feasible and possibly fruitful in the light of his existing teaching practice.
2. Pupils learn how to investigate by doing investigations.
3. The theory of an educational strategy for 'learning about investigations' will form and grow in the process of: formulation of an educational aim(s) – design of materials – classroom trials – reports - evaluation - adjustment of aims etc.

The first meetings involved an exchange of views regarding the first premise. When agreement was reached about aims and approach, activity shifted towards the third premise. Consensus existed from the start about the second premise, which is why 'investigative' practicals recieve much attention. Network meetings occur bimonthly. Reports, based on letters, transcripts or notes of the meetings, provide information on the development of aims and approach. (The documents of this study were translated into English by us.) Of these reports, a summary was made focusing on:

1. Aims and views of the teachers regarding investigations by pupils;
2. Teachers' approach to designing and implementing investigative practicals;
3. Experiences during trials of investigative practicals;
4. Evaluation of the activities within and of the network itself.

Inventory and interviews

Teachers produced an inventory of the practicals in their teaching (Tamir, 1989) by listing which investigative skills their pupils practised prior to the new programme. To carry out an investigation, pupils must have the necessary theoretical and procedural knowledge and apply it correctly at the right time. Typical for investigations is that actions of a particular kind are executed. Some of these 'investigative' actions are of a physical nature, but most actions, and the most important actions, require a mental performance. These actions can be grouped in the following 'phases', present in many, but not necessarily all investigations:

A. Develop aim of research	D. Collect data	G. Draw conclusions
B. Plan analysis of data	E. Work out results	H. Report
C. Plan collection of data	F. Interpret results	

Each phase consists of several investigative actions. For example, phase F consists of: (1) explain phenomena; (2) interpret properties of graphs; (3) compare results with other sources; (4) discuss accuracy and precision. This list, used in both subjects and based on Feiner-Valkier (1993), is just one of many that can be found in the literature. However, it is constructed with the teachers as a list they find adequate. For each practical and stream, teachers noted which action received attention and specified whether pupils carried it out under guidance or independently. To provide a background for the inventory, the teachers answered questions about the role of investigative actions in their teaching.

Teachers were asked whether the list of investigative actions was adequate and complete. They were asked for their views regarding the educational aim of developing investigative skills. LvR interviewed 5 chemistry teachers, PD 5 physics teachers. Each interview lasted from 1.5 to 2.5 hours. The interviews were transcribed and summarised. Relevant statements were added verbatim. Finally, some of the lessons during which teachers tried out the investigative practicals were attended by us. Illustrations from these lessons are used in this paper.

The teachers

All teachers teach age-groups 16-18 and have at least 5 years of teaching experience. We observed differences between subjects, teachers and schools especially regarding views about education, educational aims, approach and innovation-mindedness. From experiences in in-service courses and workshops we conclude that our findings, plans and approach are useful to at least a certain group of physics and chemistry teachers.

What is 'carrying out an investigation'?

Our view on 'learning how to investigate' formed gradually and then influenced the network activities. Here, we do not discuss the literature that contributed to it. The account given below shows how this view was used in the network.

Pupils have a certain investigative skill if they can execute with success the specified investigative action at an adequate level. 'Success at an adequate level' is determined by the theoretical and procedural knowledge the pupil is expected to possess. Applying knowledge to actions requires that links between knowledge and actions are established and maintained. The pupil can establish and maintain these links by posing, at appropriate times, the following questions regarding his actions:

1.What am I doing? (What have I done? What shall I do next?)
2.Why am I doing this? (Why did I do this? Why should I do this?)
3.How can I do this differently, better?

These questions provide only a pattern, they have to be adapted to the various investigative actions before the answers can guide pupils' further actions. A teaching practice in which pupils develop investigative skills is one, we believe, in which they ask themselves questions derived from the three questions given here and learn to use the answers. Acting in accordance with the answers means guarding and optimising the quality of the investigation, a quality which is reflected in the validity and reliability (Millar, 1994; Smits, 1995) of the outcomes.

Findings

Current teaching practice

The teachers in this study regarded the list of investigative actions to be adequate. In the inventories the difference in numbers of practicals is striking, varying in physics from 3 to 26 over the course of 3 years. The inventories also show that during practicals, pupils are chiefly involved in the phases D, E and F of collecting, representing, manipulating and interpreting data. Phases G and H, drawing conclusions and reporting, receive less attention.

Phases A, B and C, of phrasing an aim and planning the investigation, are hardly ever carried out by pupils on their own. During interviews, teachers gave as reasons: 'I just follow the textbook'; 'the current exam program does not really demand it'; 'I am not familiar with pupils carrying out investigations.' A lack of time, adequate teaching materials, technical assistance and laboratory space are mentioned as additional hindrances. The teachers feel that, to start investigations at age 16, pupils should have skills related to reading, writing, arithmetic, measuring, processing data and writing reports.

Teachers' aims and plans

Teachers consider the development of pupils' investigative skills to be a relevant educational aim. About the new exam programme, they say that 'the demands are clear, but how to teach pupils these skills is not'. The teachers believe that practicals can teach pupils to be conscious of the relevance of their actions, to learn how to think logically, to discuss with and consult each other, to improvise and to experiment. Most teachers do not want to increase the time spent on phases D, E and F and doubt whether there is enough time to address the other phases properly. The teachers plan to follow a systematic approach, gradually increasing the independence and quality of the investigations. As a first step in working towards this aim, the teachers expanded their existing practicals by letting pupils carry out the planning of investigations.

In the first network meetings, plans were made to (re)design investigative practicals in accordance with the teachers' aims and experience in investigative tasks. In these practicals, pupils were to be given more freedom and responsibility in the design and execution of investigations. Teachers wanted the practicals to relate to the theory concurrently discussed in class ('recognisable for pupils'), to have a size of 2 or 3 periods ('small scale changes') and to consist of complete investigations ('otherwise motivation is lost'). Teachers wanted the practicals to relate to the pupils' "lifeworld" knowledge and 'intellectual capacity' ('so the pupils remain motivated'). They wanted to monitor and assess the investigations by using pupils' plans and reports.

This approach resulted in nine (chemistry) and eleven (physics) investigative practicals. Design, trials and adjustments of these practicals are time consuming; the focus remained on year 4 (age 16).

Experiences in class

Dekkers and van Rens (1999) have reported the classroom experiences. Here we only outline the process that teachers went through during trials, evaluation and adjustment of the practicals. Before the trials, teachers expressed concerns; 'Will they find the tasks to be sufficiently challenging?'; 'Will they enjoy it?'; 'We should make certain that they can obtain some results'; 'The experiment should function properly, I will ask the technical assistant to check it first.' In class, these concerns were allayed. The pupils were enthusiastic and engaged, responding creatively to questions like: what will you investigate, how will you do it, what do you need to do so? These matters were evident in the pupils' plans of operation. The teachers then turned to the problem of handling differences in approach and quality. Unsafe plans were dismissed. However if plans are of a poor quality, should pupils be allowed to run into trouble, or should guidance be given? The teachers decided to suggest improvements to the plans and give the pupils time to discuss these. As it turned out, the pupils made hardly any use of these suggestions. As one pupil proudly remarked: 'Our plan is simply the best in our class.'

During the experiments, many pupils learned that experiments can turn out differently from expectations. This usually caused them to focus on technical details, while the teachers mainly helped pupils overcome their uncertainties.

The pupils wrote individual reports of the chemistry practicals. In these reports some pupils attempted to explain what they had observed, but did not relate this to their research question. Others did remember their research question, but adjusted it so that their observations could be used to answer it. The teachers noticed this, but did not go into the matter during discussions of the reports. The teachers indicated that they did not yet know how to deal with this issue, it will require attention in the future.

The teachers agreed that it was important to focus on pupils on describing their investigative actions, on the quality of these actions and on improving this quality. Yet they often failed to do so. For example, when the pupils did not improve their plans, the teachers tried to stimulate external (referring to e.g. the height of marks) rather than internal motivation (by making pupils see the point of a good plan). Pupils were not urged to draw explicit conclusions about investigative actions (e.g. by answering the question: How can I plan better next time?). It was unclear whether pupils learned things which could be carried over to future investigations.

At present, we cannot show that improving teaching requires paying more explicit attention to investigative actions. However, some experiences suggest that we are on the right track. For example, one chemistry teacher noticed that stating the importance of planning the approach and formulating expectations had no effect on the quality of the plans. He then decided to have pupils account for their chosen approach and formulate their expectations as a regular part in their reports of investigations. His pupils, who did three investigative practicals within two years, clearly improved in choosing a reasoned approach and specifying expectations.

When teachers attempt to focus on investigative actions, a balance must be found between the extent to which pupils are left to resolve their problems and the time given to do so. Another question is how much practice pupils need to acquire a skill before mastery of the skill is assessed. To answer questions of this kind, we need more experience with activities in which investigative skills are the *explicit* aim.

We see that once teachers gain experience and insight in one area, a new concern surfaces. This process has not yet ended. Several teachers question the outcomes of learning in this area as compared to the invested time and effort. They put justifiable demands on investigative practicals.

Main conclusions

The teachers in the study are prepared to pay more attention than before to 'learning about investigations'. They all designed and implemented practicals for this purpose. They feel that 'learning how to do investigations' is an educational aim that fits science subjects. They actively adjust their teaching to provide a place for 'learning about investigations' in line with the new exam criteria. In investigations, the teachers pursue a gradual, systematic increase of independence and scientific quality. Teachers think that the best way for pupils to acquire investigative skills is to carry out investigations and to try to resolve their problems by themselves.

To pay more attention to investigative actions other than observing, data handling and representing results, it often suffices to adjust existing practicals. The teachers are developing the skills needed to adjust, design and teach these new practicals. The main worries of the teachers, concern the available time and personal support. How much time do pupils need to solve their problems and how can the solutions be made transferable? The teachers are becoming aware that our knowledge of learning about investigations is still limited. A view of what is required in 'learning about investigations' develops gradually.

We have designed a plan for subsequent research that takes into account several of the above issues. The feasibility and effectiveness of this approach will now be empirically tested. A series of investigations is to be designed for the age-groups 16-17, working systematically towards self-reliant investigations of high quality. Several practicals designed and piloted in the network can be of use in this series.

The series consists of complete investigations. One phase in each investigation is chosen to receive special attention. The skills pertaining to it are specified as aims of learning. This phase is open on the one hand; pupils are given time to experience problems and solve these. On the other hand, it is also guided; pupils are asked to describe their problems and evaluate and compare the solutions.

The remaining phases are either open-unguided or closed-guided. A phase that has received special attention is open-unguided; higher standards are maintained concerning the independent performance and quality. A phase which has not yet received special attention is closed-guided, performed with the class as one team with strong teacher guidance. The most difficult phases of investigations are dealt with last and during the series, a gradual integration of investigative skills is pursued.

This plan of research, which is only roughly outlined here, does not state *how* pupils and teachers can be enabled to make the relevant questions (what am I doing, why and can I do it better?) become the central questions in their activities. We intend to find ways to get teachers and pupils to focus on these questions, on answering them and using the answers. This, however, requires that both the teachers and researchers develop a better understanding of how pupils handle investigations. We will have to determine what pupils learn in practicals, what we want them to learn in investigations, and how we can motivate them to engage in learning this.

References

Dekkers, P. & van Rens, L. (1999). Drogen en rollen, hoe onderzoek je dat? *NVOX.* 4, 171-175. email: rens@ido.vu.nl; p.dekkers@eudoramail.com

Feiner-Valkier, S. (1993). *Een Programma van Experimenten.* Enschede: SLO.

Hodson, D. (1992). Redefining and reorienting practical work in school science. *School Science Review* 73, 65-78.

Klopfer, L.E. (1990). Learning scientific enquiry in the student laboratory. In E. Hegarty-Hazel (Ed.). *The Student Laboratory and the Science Curriculum* (pp. 95-118). London: Routledge.

Lawson, A.E. (1994). Research on the acquisition of science knowledge: epistemological foundation of cognition. In D.L. Gabel (Ed.). *Handbook of Research on Science Teaching and Learning* (pp. 131-176). New York: MacMillan.

Millar, R. , Lubben, F., Gott, R., & Duggan, S. (1994). Investigating in the school science laboratory: conceptual and procedural knowledge and their influence on performance. *Research Papers in Education* 9 (2), 207-249.

Smits, T. (1995). Open onderzoek in actie. NVOX. 8, 387-381.

Tamir, P. (1989). Training teachers to teach effectively in the laboratory. *Science Education* 73 (10),59-69.

Point and Set Paradigms in Students' Handling of Experimental Measurements

Saalih Allie and Andy Buffler
Department of Physics, University of Cape Town, South Africa
Fred Lubben and Bob Campbell
Department of Educational Studies, University of York, UK

Abstract

The procedural understanding of first year university students before and after instruction has been investigated in the context of experimental work in physics. A written instrument was used to probe the students' ideas about data collection, processing and comparison. The responses of the students are analysed in terms of "point" and "set" paradigms which are proposed as a framework for evaluating the effectiveness of laboratory curricula.

Introduction

Experimentation and measurement are fundamental to knowledge production in the natural sciences. Meaningful engagement by students in scientific activities that are experimentally based, requires an understanding of the reasons for the procedures that are followed. However, a growing body of work indicates that students at both school (Coelho & Séré, 1998) and university (Evangelinos *et al.*, 1999) carry out the procedures without such a basic understanding. In a study of undergraduate physics students, for example, a clear disjuncture between virtuosity in applying the formalism of data analysis and the level of understanding were observed (Séré *et al.*, 1993). The current work forms part of a research programme in which procedural understanding of experimental work of entering undergraduate science students at the University of Cape Town is being documented. The ultimate aim of the exercise is to use the findings to inform the development of an introductory physics laboratory curriculum. This paper reports on procedural understanding, in the context of experimental work in physics, of a group of students, both at entry to the university and after a twelve-week laboratory course. The laboratory course in question is part of the physics component of the Science Foundation Programme (SFP) which was set up at the University of Cape Town to assist educationally disadvantaged students. Details of the SFP and the laboratory course are contained in Allie and Buffler (1998).

Since most of the SFP students have little or no first hand practical experience, a prime aim of the laboratory course is to develop the notion of measurement. Aspects of data collection and data processing are addressed by exercises, such as drawing up tables, taking several measurements of a quantity, plotting graphs, fitting straight lines, and calculating the mean and the standard deviation from the statistical formulae, as well as graphically, from a Gaussian curve. The idea of spread in data is introduced by getting the class to measure the time of travel of a sound pulse over a given distance. The readings are processed to form a distribution (a Gaussian curve results) from which the key ideas of mean and uncertainty are introduced. The

H. Behrendt et al. (eds.), Research in Science Education – Past, Present, and Future, 331–336.

laboratory course consists of weekly three-hour sessions over twelve weeks. About half of this time is spent in the laboratory carrying out experiments while the remainder is used for the exercises described.

Methodology

The research instrument comprised a set of nine written questions (probes) based on those developed for a previous study (Allie *et al.*, 1998). All these probes related to the same posited experiment which was presented as follows, together with a detailed diagram (not shown here). *"An experiment is being performed in the Physics laboratory. A wooden slope is clamped near the edge of a table. A ball is released from a height, h, above the table as shown in the diagram. The ball leaves the slope horizontally and lands on the floor a distance, d, from the edge of the table. Special paper is placed on the floor on which the ball makes a small mark when it lands. The students have been asked to investigate how the distance, d, on the floor changes when the height, h, is varied. A metre stick is used to measure d and h*". The situation was also demonstrated using a large-scale model. The probes focused on decisions to be made while collecting data, processing data and comparing two different sets of measurements of the same quantity. Each probe was of the same form as the example shown in fig. 1 below. Thus, a situation was presented where a procedural decision was required and a number of alternative actions (A, B, C) were suggested. Most importantly, a detailed reason for each choice was requested. The probes were answered under formal examination-type conditions, strictly in the sequence presented. As each answer sheet was completed, it was placed into an envelope and never reviewed by the student. The instrument was completed by 70 students before and after they had completed the laboratory course.

The students work in groups on the experiment. Their first task is to determine d when h = 400 mm. One group releases the ball down the slope at a height h = 400 mm and using a metre stick, they measure d to be 436 mm.
The following discussion then takes place between the students.

With whom do you most closely agree? (Circle ONE): **A - B - C**
Explain your choice below.

Fig. 1: Example of a probe: in this case the reasons for repeating measurements

Analysis

The analysis of the probes consisted of categorizing the student responses according to the answer choice (A, B, C), *together* with the different types of reasoning put forward by the students. The coding of the responses was undertaken using an alphanumeric scheme, that was developed and tested previously (Allie *et al.*, 1998). This enabled the underlying reasoning to be identified for each student. Earlier work (Buffler *et al.*, 1998) suggested that the actions and reasoning employed by students could be classified into two groups by defining two constructs, namely a "point" paradigm and a "set" paradigm as discussed below. Using this framework, the pre- and post-test results were analysed by looking for patterns across the three areas that the probes addressed, namely, data collection, data processing and data set comparison.

The *point paradigm* is characterised by the notion that each measurement results in a single, "point-like" value which could, in principle, be the true value. As a consequence, each measurement is independent of the others and the individual measurements are not combined in any way. In its most extreme form, this way of thinking manifests itself in the belief that only one single measurement is required to establish the true value, as indicated in the work of (Séré *et al,*. 1993). Responses were coded as being associated with the point paradigm when, for example: (a) it was stated that measurements are repeated in order to find a recurring value or to perfect the measuring skill in order to finally take one 'perfect' measurement; (b) a specific measurement was selected (e.g. the highest, the recurring, the first or the last) to represent a series of numerical readings; (c) specific points (such as the origin, the extreme points or any three aligned points) were chosen through which to draw a straight line to represent a collection of plotted points; or (d) two sets of data were contrasted either by comparing individual measurements in the set, or by treating the mean values of the data sets as points to be compared.

The *set paradigm* is characterised by the notion that each measurement is only an approximation to the true value and that the deviation from the true value is random. As a consequence, a number of measurements are required to form a distribution that clusters around some particular value. The best information regarding the true value is obtained by combining the measurements, using theoretical constructs in order to describe the data collectively. The operational tools, that are available for this purpose, include the formal mathematical procedures that can be used to characterise the set as a whole, such as calculating the mean and the standard deviation. In turn, these quantities become tools for making comparisons with other data-sets or theories. Responses were coded as being associated with the set paradigm when, for example: (a) it was stated that repeating measurements was aimed at taking a mean; (b) the mean and the spread were calculated to represent the data; (c) a 'line of best fit', that took account of all points, was drawn for plotted data; or (d) different sets of measurements were contrasted by comparing the degree of overlap of the intervals defined by the mean and some measure characterising the spread of the data.

Results

Table 1 summarises the results from the pre- and post-tests with regard to students' understanding about repeating measurements during data collection (3 probes). It shows that before instruction the large majority of students (76%) subscribed to the point paradigm while after instruction there appeared to be a large shift (16% to 71%) towards the set paradigm. However, it is not clear that these students have embraced the set paradigm as a whole. For example, many students indicated that the purpose of repeating measurements is to allow for a mean to be *generated* (rather than a mean being a way of dealing with the inherent scatter in the data). This suggests there is a strong possibility that elements of the set paradigm are being used by rote or on an *ad hoc* basis. The degree to which this is the case requires the combined analysis of the other probes.

Table 1: Students' use of paradigms for data collection		**Paradigm after instruction**			
		Point paradigm	Set paradigm	Not codeable	**Total**
Paradigm before instruction	Point paradigm	9 (13%)	40 (57%)	4 (6%)	**53 (76%)**
	Set paradigm	4 (6%)	6 (9%)	1 (1%)	**11 (16%)**
	Not codeable	2 (3%)	4 (5%)	0 (0%)	**6 (8%)**
	Total	**15 (21%)**	**50 (71%)**	**5 (7%)**	**70 (100%)**

Table 2 summarises the pre- and post test findings for 3 probes dealing with the comparison between two data-sets. The first of the three probes required students to compare two sets of data with the same mean but different scatter. While the second probe provided two sets of data with different means but the same (overlapping) spread. The third probe presented two data-sets with different means and different but overlapping spreads. In Table 2 students are grouped according to whether or not their responses, *across the three probes*, were consistent with the set paradigm. As expected from the background of the students, none were classified as using the set paradigm consistently prior to instruction. After instruction only 26% responded consistently in terms of the set paradigm while more than two thirds (70%) resorted to both paradigms, possibly indicating either rote or ad hoc application of the elements associated with the set paradigm.

Table 2: Students' use of paradigms when comparing data-sets		**Paradigm after instruction**			
		Mixed paradigms	Consistent set paradigm	Not codeable	**Total**
Paradigm before instruction	Mixed paradigms	48 68%)	18 (26%)	3 (4%)	**69 (98%)**
	Consistent set paradigm	0 (0%)	0 (0%)	0 (0%)	**0 (0%)**
	Not codeable	1 (1%)	0 (0%)	0 (0%)	**1 (1%)**
	Total	**49 (70%)**	**18 (26%)**	**3 (4%)**	**70 (100%)**

Tables 3 and 4 compare various aspects of the post-probes. In Table 3 the post-probe results of Table 2 (data-set comparison) are contrasted with the combined results of the post-probes dealing with data collection (one probe) and data processing (two probes). One of the data processing probes required a mean to be calculated from a set of numerical data, while the other required a straight line to be drawn to a set of graphical data. It is interesting to note that there appears to be a strong link between the paradigms used for these two probes. For example, students who joined individual data points (i.e. did not fit a straight line to the data as a whole) often evidenced point paradigm use in the other probe by choosing the recurring value to represent the data rather than calculating a mean. (Space limitations preclude showing the evidence in detail). Table 3 shows that only about a fifth (21%) of the students based their responses on the set paradigm for all of data collection, data processing and data-set comparison and that the largest group (37%) were inconsistent in their use of the paradigms.

Table 3: Students' use of paradigms for data collection / processing and data-set comparison after instruction

		Paradigms used in data collection / processing				
		Consistent point paradigm	Mixed paradigms	Consistent set paradigm	Not codeable	**Total**
Paradigms used in data-set comparison	Mixed paradigms	9 (13%)	25 (37%)	13 (18%)	2 (3%)	**49 (70%)**
	Consistent set paradigm	0 (0%)	3 (4%)	15 (21%)	0 (0%)	**18 (26%)**
	Not codeable	0 (0%)	1 (1%)	2 (3%)	0 (0%)	**3 (4%)**
	Total	**9 (13%)**	**29 (42%)**	**30 (43%)**	**2 (3%)**	**70 (100%)**

Table 4 shows results for students who have been classified on the basis of all the probes discussed thus far, together with the results of the final probe (mean/sd probe), in which students were asked to compare two data-sets described in the formal manner of a mean and a standard deviation of the mean.

Table 4: Students' action and reasoning for data collection, data processing and data-set comparison after instruction

		Point or set action for mean/sd probe			
		Point paradigm action	Set paradigm action	Not codeable	**Total**
Classification of student reasoning from all previous probes	Consistent point paradigm reasoning	3 (4%)	6 (9%)	0 (0%)	**9 (13%)**
	Inconsistent paradigm reasoning	11 (17%)	17 (24%)	1 (1%)	**29 (42%)**
	Consistent set paradigm reasoning	14 (20%)	16 (23%)	0 (0%)	**30 (43%)**
	Not codeable	1 (1%)	1 (1%)	0 (0%)	**2 (3%)**
	Total	**29 (42%)**	**40 (57%)**	**1 (1%)**	**70 (100%)**

From Table 4 it is clear that although more than half the students (57%) carried out an *action* associated with the set paradigm, fewer than half of this group (23% of the sample) provided a *reason* that was also consistent with this paradigm. In other words 33% of the students (57%–23%–1%) appear to have used the correct set paradigm action either by rote or in an *ad hoc* way. In summary, only a quarter of all the students (100%–42%–33%) can be regarded as having completely embraced the set paradigm.

Conclusions

In terms of the constructs of the point and set paradigms, the purpose of laboratory instruction can be regarded as attempting to shift students' actions and reasoning away from the point paradigm to those commensurate with the set paradigm. Both the present study and the previous work (Buffler *et al.*, 1998) strongly suggest that students come from school firmly located within the point paradigm, and that any set paradigm actions (such as calculating a mean) are most often a rote response. Even after a six month laboratory course, only about one quarter of the students seem to have reached the required instructional goals. The present probes and the analysis framework appear to offer useful research tools that can be used to evaluate the effectiveness of any laboratory curriculum that aims to address procedural understanding in the context of experimentation.

References

Allie, S. & Buffler, A. (1998). A course in tools and procedures for Physics 1. *American Journal of Physics* 66 (7), 612-623.

Allie, S., Buffler, A., Kaunda, L., Campbell, B. & Lubben, F. (1998). First year physics students' perceptions of the quality of experimental measurements. *International Journal of Science Education* 20 (4), 447-459.

Buffler, A., Allie, S., Campbell, B. & Lubben, F. (1998). The role of laboratory experience at school on the procedural understanding of pre-first year science students at UCT. In N. A. Ogude & C. Bohlmann, Eds., *Proceedings of the 6th Annual Meeting of the Southern African Association for Research in Mathematics and Science Education (pp.* 495-502). Pretoria, South Africa: University of South Africa Press.

Coelho, S. M. & Séré, M.-G. (1998). Pupils' reasoning and practice during hands-on activities in the measurement phase. *Research in Science and Technological Education* 16 (1), 70-96.

Evangelinos, D., Valassiades, O. & Psillos, D. (1999). Undergraduate students' views about the approximate nature of measurement results. In M. Komorek *et al.*, Eds., *Proceedings of the Second International Conference of the European Science Education Research Association Volume I (pp.* 208-210). Kiel, Germany: IPN.

Lubben, F. & Millar, R. (1996). Children's ideas about the reliability of experimental data. *International Journal of Science Education* 18 (8), 955-968.

Séré, M.-G., Journeaux, R. & Larcher, C. (1993). Learning the statistical analysis of measurement error. *International Journal of Science Education* 15 (4), 427-438.

Acknowledgements

The research reported in this paper was supported by the British Council (DFID), the University of Cape Town and the University of York. We also thank Indresan Govender and Fiona Gibbons for their assistance.

Beyond the Laboratory – Learning Physics Using Real-Life Contexts

Elizabeth Whitelegg and Christopher Edwards
Centre for Science Education, The Open University, U.K.

Abstract

This project is investigating 'use of knowledge' by students who are involved in a course where knowledge is taught, in specific contexts taken from life outside and inside school, but beyond the laboratory. In order to investigate this, the research is examining how successful students are in conventional exams where real world contexts are trivialised and laboratory/school contexts dominate, and compare this with an assessment, where context varies but is not trivialised and is real world. In particular, the research is investigating, how effectively students are able to transfer their learning of physics concepts across the differing, real-life contexts found in the Supported Learning in Physics Programme (SLIPP), and, finally, onto a non-contextual situation such as an examination question. This paper focuses on the initial study, which is part of an on-going research project. It reports on students' attitudes to learning physics at post-16 level, through a real-life context approach as revealed through interviews and observation. It also considers preliminary data on the effect on student learning using context rich and context poor assessments.

Background to the research

The SLIPP curriculum project consists of eight books of text-based student learning material, that provide a supported learning programme in physics for post-16 students to use with the support of their teachers. By using particular learning strategies, the project aims to motivate students to learn, by increasing their interest in physics. The learning material in each SLIPP unit is set within a real-life context. For example in the unit *Physics for Sport*, the concept of equilibrium of forces is taught through consideration of the way rock climbers use hand and foot holds at various angles on a climbing wall; pressure laws are taught through a discussion of SCUBA diving and both circular motion and simple harmonic motion are taught in the context of springboard diving. These contexts are introduced alongside the physics content being taught, they are not introduced, as they all too often are in more conventional physics texts, as disparate applications of physics knowledge that can be separated from the physics content and easily ignored. In the development of the SLIPP materials, the real-life context was chosen *before* the associated physics content. This means that content is not ordered in the traditional way found in most syllabi and textbooks. This has led to the same physics concepts being taught in more than one unit, but in different real-life contexts. This has the advantage of providing more practice for students, so reinforcing learning and enabling students to make links across different domains. It acknowledges that students learn in different ways and may help those who had difficulty with the approach in one context and prefer another. Sometimes one unit may not cover a particular physics concept in depth, but the full treatment can be found in another of the units

H. Behrendt et al. (eds.), Research in Science Education – Past, Present, and Future, 337–342.

(Whitelegg, 1996). The context therefore 'serves as an organiser for the science content' (Aikenhead, 1994, p.55).

The research problem

Research by Lave (1988) suggests that learning is context dependent. Other research on learning science using everyday life contexts (Murphy, 1994; Hennessy 1993) suggests that this way of learning can be a motivating factor for students (particularly girls) helping them to learn more effectively. Further research by Murphy et al (1995) suggests that context-based learning, within the classroom, is only effective if the contexts are *appropriate* for individual students and relate to their own experiences outside the school environment.

In SLIPP programme, the main aim of the approach is to use everyday-life contexts as a motivating factor for students. To encourage students to engage with the physics content so that learning physics does not appear as purely an academic exercise, but has links to life outside school and knowledge which is transferable beyond the laboratory. However, the research aims to examine whether the approach of the SLIPP materials can improve students' learning. There is now a large body of literature that suggests that situational interest has a powerful affect on students' learning (Hidi & Berndorf, 1998). Those who oppose context-based learning, claim that it is context-bound, whereas concepts taught in traditional ways are transferable. However, as Varella has shown using an STS approach, students can become more capable of applying science concepts to new situations (Varella, 1992). In fact, many supposedly abstract concepts are very much bound to the sterile school laboratory contexts, in which they are learned, and students cannot transfer their learning to their real world experience. SLIPP accepts that knowledge is acquired in context, hence, the context of learning matters as it is not to be separated from the knowledge that is acquired through activity. Transfer of knowledge assumes that knowledge is separable from context and can be disembedded for use. The issue of transfer is, therefore, problematic and contentious in debates about how knowledge is constructed particularly from a situated perspective.

This research aims to investigate the effect of learning physics through an everyday life approach and to examine the transfer phenomenon by looking at how effectively students are able to transfer their learning of physics concepts across the differing real-life contexts found in SLIPP and to transfer learning to a new context not featured in SLIPP, then finally onto a non-contextual situation such as an examination question. This paper reports on an initial study which looked at students' attitudes to learning physics through everyday contexts, and the first phase of the assessment data, when students undertook context rich and context poor assessments.

The research design and procedure

The research has started with a 12-month initial study, that commenced in September 1998. This study involved 38 year 12 students (aged 17+) and 6 teachers in three case study schools. The three schools were very different in nature. School 1 was a mixed-sex state school in a suburban environment. There were 13 boys and 4 girls in the class, academic standards were good, though covering a wide range and two male teachers shared the teaching. The Head of Science was enthusiastic about the approach of SLIPP and was keen to emphasise the relevance of physics in everyday life contexts. School 2 was an independent, fee-paying girls-only school. There were only 7 girls in the class, also taught by two teachers, one male, one female. Academic standards here were the highest of all 3 schools. Both teachers were enthusiastic about the approach of the project, particularly as it aimed to develop independent learning skills amongst students. They did not afford as much attention to the everyday context approach as the teachers at school 1. School 3 was a boys-only state school, with 18 boys in the class, taught by two male teachers. This was the most traditional of the three schools, where teachers and students referred to the physics topics they were studying according to a reference to their syllabus section, rather than describing the physics content. This group showed the greatest range of academic standard.

For this initial study, an intensive data collection phase took place over 12 weeks of the 1998 Autumn term, with two of the schools being visited by the researchers for 50% of their physics lesson time each week and the third slightly less so. The data was collected via observation of physics lessons; audio recording of individual students using radio microphones during practical investigations in the laboratory; video taped, classroom-based sessions; pencil and paper assessments using context-rich and context-poor questions and at the end of term, open-ended, audio-taped interviews using a structured protocol with additional probing of responses. (In school 1 we also returned at the end of the year to do further interviews with the students.)

Three contexts and associated physics concepts were examined in detail in this study. In school 1, students used the *Physics, Jazz and Pop* unit to study sound wave transmission, oscillations and simple harmonic motion with one teacher and the *Physics on the Move* unit to study forces, Newton's laws etc. with the other teacher. In school 2, students studied electromagnetism in *the Physics Phones Home* unit and forces, Newton's laws, moments etc. in the *Physics for Sport* unit. In school 3, students also used *Physics, Jazz and Pop* in addition to *Physics for Sport* and covered the same topics as described above, plus some work on materials (i.e. stress and strain).

This wide ranging methodology is enabling the researchers to build up a detailed picture of the students' learning of the physics concepts they were being introduced to, of how the everyday-life context embedded within the SLIPP materials interacted with the physics concepts and the effect of this pedagogy on learning.

Results

The results of the semi structured interviews from all three schools have been analysed once categorized, using QSR's NUDIST software. The students' responses to the interview questions were divided into 22 categories and those comments that referred to the everyday-life approach of their study were sub divided into 28 categories. For the purposes of this paper, we will focus only on those categories of responses that relate to students' attitudes to the everyday context approach, its role in motivating students, gender and school effects that relate to these issues and initial outcomes from the context rich and context poor assessments from the three schools.

In answer to questions about the relationship of physics concepts to everyday contexts, several students felt that approaching the physics concept through these contexts made the physics more understandable and interesting. They were able to see how physics concepts related to their interests outside school, such as playing a musical instrument in a band. It enabled them to talk to 'scientifically-minded' peers out-of-school about the physics of music and students were better able to predict the outcome of experiments, that were related to students' knowledge of music and so do the experiments in more depth.

Several girls, in particular, mentioned how this relationship between the physics concepts and their out-of-school interests made the physics more interesting and understandable for them. They talked about the need to put physics concepts in concrete settings - they were more interested to learn about real, rather than abstract things.

Some boys at the single sex school (school 3), however, were concerned that approaching physics learning in this way might lead them to learn more ideas than they might need to for their syllabus and that some of the context material was not 'relevant' to the course. But not all worried about this as they said that they could always ignore the extraneous material!

Despite some boys worrying about 'extraneous' material, others at this school also had difficulty in recognising that an everyday context was interwoven with the physics content and it took a lot of probing to get students to recognise the existence of a context.

At the end of the 12-week term using the SLIPP materials, students were tested using pencil and paper assessments taken from old A-level (18+) examination papers. Questions were selected from these papers that covered the areas that the students had been studying during that term. The initial results of the pencil and paper tests from all three schools showed that students did slightly better on the context rich questions that they did on the content poor. Their average score on the content rich questions was 43.6% compared to 40.3% on the context poor questions, a trend towards context rich questions of 3.3%.

In the interviews, the majority of students expressed views that were in favour of learning through everyday contexts and students suggested that it enabled them to learn in more depth if the contexts used were appropriate for them and matched their out-of-school interests (e.g. music, rock climbing). This seemed to be particularly true for girls, although some boys in school 1 (the mixed sex school) had similar views.

Our analysis of the context-rich and context-poor assessments is on-going and we will need to do both further analysis and undertake further assessments of this sort with more refined questions, to clarify the outcomes of this dimension of the research. Our results so far show a trend towards improved scores on the content-rich questions, but this is not large enough to be considered significant and a larger sample is needed.

Boys in the most traditional of the three schools, school 3 (one of the single-sex schools), were the only ones who mentioned concerns about learning things outside their syllabus. In addition, some boys at this school also had difficulty in relating to a context – so they chose to ignore it, for fear of wasting time on non-physics content, or else they ignored it to the extent that it didn't exist for them! Their teacher also contributed to this, sometimes dismissing or undermining the context, although he stated that he liked to teach physics via an historical approach and we did observe him introducing some historical developments of physics into his lessons.

Even though SLIPP is a supported learning scheme, that is aiming to develop students' independent learning skills, how it is used is still very much in the hands of the teacher. This effect is also noted by Aikenhead who says that the 'teacher will influence student outcomes far more than specific curriculum, textbooks or teaching strategies. Thus, student outcomes from the same STS course can vary significantly from one teacher to another.'(1994, p.170.) At school 1, where the teacher was in sympathy with the philosophy of the project, students were all aware of the everyday context approach and in favour of the approach, or else were happy to go along with it. In school 3, the most traditional of the three schools, the teacher made it possible for students to ignore the context and in some cases actually undermined the contexts. Despite this, a number of students did find the approach interested them, but there were those who didn't notice it!

The girls in school 2 were mixed in their views of the approach. Several were quite sceptical at the beginning of their involvement in the project – sceptical of the supported learning approach, rather than the context aspect. Hildebrand (1999)talks of teachers who experience resistance from their students, because teachers '(break) the pedagogic contract' when they move from a model based on transmission to one based on constructivism. She feels that girls are particularly vulnerable to this as it threatens their own self-perceptions as learners.

We believe that this is what we experienced at the girl's school when the students expressed resistance to using SLIPP as a new way of learning. This resistance had modified by the end of term.

Conclusion

Despite the materials all being presented to the students in a form that they can use almost independently of a teacher (although this isn't the intention of the project) the research still encountered the problem of traditional, teacher attitude and provided evidence of the ability of teachers to strip away the context in order to lay bare the 'real' physics. Whilst this may be OK for some (we suspect a diminishing minority of students) most will flourish if physics is made more interesting to them and connects with their lives out-of-school.

This initial study has provided the basis for future work, using similar methods where the focus of the research will further examine the transfer of physics learning across the contexts used within the learning programme and onto examination-style context-rich and context-poor assessments.

References

Aikenhead, G. (1994). What is STS Science Teaching? In J. Solomon & G. Aikenhead (Eds.), *STS education: International perspectives on reform* (pp. 47-59). New York: Teachers College Press.

Aikenhead, G. (1994). Consequences to learning science through STS: A research perspective. In J. Solomon & G. Aikenhead (Eds.). *STS education: International perspectives on reform*(pp. 169-186). New York: Teachers College Press.

Hennessy, S. (1993). Situated cognition and cognitive apprenticeship: Implications for classroom learning. *Studies in Science Education* 22, 1-41.

Hidi S. & Berndorff D. (1998). Situational interest and learning. In L. Hoffmann, A. Krapp & K. A. Renninger (Eds.), *Interest and Learning*. Proceedings of the Seeon Conference on Interest and Gender (pp. 74-90). Kiel: Institute of Science Education University of Kiel (IPN).

Hildebrand, G. (1999, March). Breaking the pedagogic contract: Teachers and student voices. Paper presented at NARST Boston.

Lave, J. (1988). *Cognition in practice.* Cambridge University Press.

Murphy, P., Issroff, K., Scanlon, E. with Hodgson, B. & Whitelegg, E. (1995). Group work in primary science. In A.M. Andersen, K. Schnack, & H. Sørensen (Eds.), *Science-Natur/Teknik, Assessment and Learning.* Copenhagen: Royal Danish School of Educational Studies.

Murphy, P. (1994). Gender differences in pupils' reactions to practical Work. R. Levinson (Ed.), *Teaching Science.* London: Routledge.

Varella, G.F. (1992). Greater ability to apply concepts using an STS approach to teaching science. In R.E. Yager (Ed.), *The statue of STS: Reform efforts around the world* (pp. 87-92). ICASE 1992 Yearbook. KnappHill, South Harting, Petersfield, U.K

Whitelegg, E. (1996). The supported learning in physics project. *Physics Education,* 291-295.

Name Index

N

O

P

Q

R

Subject Index

MIX
Papier aus verantwortungsvollen Quellen
Paper from responsible sources
FSC® C105338

If you have any concerns about our products,
you can contact us on
ProductSafety@springernature.com

In case Publisher is established outside the EU,
the EU authorized representative is:
Springer Nature Customer Service Center GmbH
Europaplatz 3, 69115 Heidelberg, Germany

Printed by Libri Plureos GmbH
in Hamburg, Germany